D0214361

DEVELOPMENTAL AND CELL BIOLOGY SERIES

EDITORS
D. R. NEWTH   J. G. TORREY

NEOPLASTIC AND NORMAL CELLS IN CULTURE

NEOPLASTIC
AND NORMAL CELLS
IN CULTURE

J. M. VASILIEV

Cancer Research Centre of the USSR, Moscow

I. M. GELFAND

Laboratory Korpus A, Moscow State University, USSR

CAMBRIDGE UNIVERSITY PRESS

CAMBRIDGE

LONDON   NEW YORK   NEW ROCHELLE

MELBOURNE   SYDNEY

Published by the Press Syndicate of the University of Cambridge
The Pitt Building, Trumpington Street, Cambridge CB2 1RP
32 East 57th Street, New York, NY 10022, USA
296 Beaconsfield Parade, Middle Park, Melbourne 3206, Australia

© Cambridge University Press 1981

First published 1981

Printed in Great Britain at the
University Press, Cambridge

British Library Cataloguing in Publication Data
Vasiliev, J M
Neoplastic and normal cells in culture. –
(Developmental and cell biology; 8).
1. Tumours   2. Cell culture
I. Title   II. Gelfand, Izrail' Moiseevich
III. Series
616.9′92′0028   RC254.5   80–40075
ISBN 0-521-23149-3

Dedicated to the memory of
M. Abercrombie

Contents

Preface

After many decades of studies we are still unable to distinguish reliably between single normal and neoplastic cells on the basis of their structural or molecular markers. The main reason for this lack of established diagnostic methods at the cellular level is that phenotypic differences between normal and neoplastic cells are differences of the most complex forms of cell behaviour: those of cell interactions with their environment. These are the cell–environment interactions leading to the formation and maintenance of organized multicellular systems. More specifically, they include processes regulating cell proliferation, differentiation and morphogenesis. At present, cell biologists are only just beginning to gain some understanding of the general features of these processes in normal cells and of their alterations in neoplastic cells. We are only now beginning to develop an adequate 'language' for the description of these processes. As yet none of the main problems in the field has been solved in molecular terms; in fact, we are only learning how to formulate these problems.

The aim of this book is to describe and discuss comparative characteristics of the interactions of normal and neoplastic cells with their environment in cell cultures. Cell culture is the only method that permits the study of behaviour of individual cells in more or less controlled environmental conditions. It would be hardly an exaggeration to say that the main achievements of cancer cell biology during the last two decades have been associated with the use of cell cultures.

Most studies in this field have been performed with different varieties of fibroblastic cells, because they are relatively easy to cultivate. Our group has also experimented mostly with cells of this type; therefore, cultured fibroblastic cells have become the principal characters in this book and we have chosen epithelial cells as the second type for discussion. We hoped that parallel descriptions of the cells of two tissue types might help to distinguish general and tissue-specific features in their behaviour. Two aspects of cell–environment interactions are described at length in the two central parts of the book: regulation of cell shape and locomotion (morphogenetic processes), and regulation of cell proliferation. Unfortunately, there is not enough material for discussion of the regulation of differentiation in fibroblastic and epithelial cultures. Therefore, we included selected references to cultures of other cell types such as haematopoietic cells and teratoblastoma cells. Biochemical, virological and genetic data are described only as far as they are related to the main subject of the book, that is, to the analysis of complex forms of cell reactions to the environment in normal and neoplastic cultures. Differences between an artificial, man-made environment of cultured cells and their natural environment *in vivo* are obvious. Nevertheless, the basic rules of cell behaviour remain the same both in culture and in the organism. Therefore, the results obtained *in vitro* are relevant and important for an understanding of neoplastic processes *in vivo*. To substantiate this important point we included chapters 1 and 13 describing general properties of neoplastic cells *in vivo*, comparing them with the properties of transformed cells in culture.

Analysis of complex forms of cell–environment interactions in culture has lead us to suggest that they can all be regarded as combinations of a few types of cellular reactions designated as 'basic morphogenetic reactions' and 'growth activation reactions'. These reactions can be induced by diverse factors present in the environment. The same cell may have either a normal or transformed phenotype, depending on its environment and reaction norm. Genetically stable neoplastic transformations involve alterations of mechanisms controlling the cellular reaction norm to external factors. Needless to say, these and related considerations advanced in the book should be regarded not as final conclusions but as attempts to define principal unsolved problems.

Acknowledgements

All the illustrations in this book are original photographs made in the laboratories where the authors work. Many of these photographs were kindly given to us by our collaborators and we wish to express our gratitude to A. D. Bershadsky, E. E. Bragina, A. P. Cherny, L. V. Domnina, V. I. Guelstein, O. Y. Ivanova, A. V. Lyubimov, T. G. Moizess, O. I. Pletyushkina, J. A. Rovensky, T. M. Svitkina and I. S. Tint.

Stimulating discussions with many of our friends and colleagues during various phases of preparation of this book were invaluable to us; we wish to thank especially G. I. Abelev, A. D. Bershadsky, I. L. Chertkov, V. I. Gelfand, V. I. Guelstein, L. B. Margolis, E. L. Prigogina and A. A. Stavrovskaya. The laborious preparation of the manuscript could have been performed only with the help of L. A. Lyass, A. S. Serpinskaya, A. A. Stavrovskaya, T. M. Svitkina, and E. N. Vasilieva, to whom we are deeply grateful. We are also indebted to E. N. Vasilieva for preparing the author index. We gratefully acknowledge the editorial work of Drs G. Dunn and A. J. Colborne.

Part I

Introduction

1

General properties of neoplastic cells

Introduction

The term 'neoplasms' includes all true tumours and leukaemias as well as related lesions associated with selective growth of abnormal cell lines in multicellular organisms. Neoplasms may develop in many, possibly in all, groups of multicellular organisms. There is no standard definition of neoplasms but typical neoplasms are characterized by at least three general features: (*a*) growth of each lesion by multiplication of one or several cell lines (neoplastic cell lines); (*b*) ability of cell lines to grow selectively in certain conditions compared with surrounding non-neoplastic cells; (*c*) formation of deficient multicellular structures by neoplastic cells. The first part of this chapter is devoted to a more detailed discussion of each of these features; then we will discuss which normal cells may be progenitors of neoplastic cells, how properties of neoplastic cells are changed in the course of their growth and the nature of genetic changes leading to their formation.

Growth of neoplasms by multiplication of one or a few cell lines

This property of neoplasms may be formulated as follows: each neoplastic lesion contains one or a few lines of self-maintaining cells with relatively stable properties. The growth of each neoplasm is accompanied by the multiplication of these cell lines; therefore each neoplastic lesion, at any moment of its existence, contains the descendants of those cells that were present in the lesion at the beginning of its growth. Besides self-maintaining neoplastic cells, a neoplasm may contain the descendants of these cells which have lost the ability for self-maintenance. These are irreversibly differentiated and/or dying neoplastic cells and they may be designated as 'neoplastic non-stem cells'. A neoplastic lesion may also contain non-neoplastic cells: stromal cells or surviving cells of the normal tissues invaded by neoplasm. Stromal cells, by definition, never descend from a neoplastic cell line.

Data confirming the presence of self-maintaining cell lines in different neoplasms are manifold. First of all, human pathology has shown that the cells of metastatic nodules of all types of malignant tumors have similar morphological and biological properties to the cells of the parent primary

3

lesion. For instance, the cells of metastatic hepatic tumours may continue to form bile and liver-specific proteins; myeloma cells produce specific immuno-globulins; neoplasms of endocrine organs may continue to secrete hormones, etc. On the basis of innumerable data of this type, pathologists long ago reached the conclusion that a true tumour grows by the multiplication of its own cells (Ribbert, 1914). Experiments with tumour transplantation in animals also confirm that neoplastic nodules formed in the new hosts may retain, in many passages, all the main properties of the original neoplasms including morphological, biochemical, antigenic and cytogenetic markers. Recent investigations have demonstrated the clonal character of many neoplasms; that is, they have shown that all the neoplastic cells of each primary neoplastic lesion are the descendants of one cell. This can be shown by different methods. One of them is the demonstration of identical chromosome markers in various cells of one neoplasm. Another group of methods is based on the analysis of genetic markers in tumours arising in genetically mozaic organisms. The human X-linked glucose 6-phosphate (Glc-6-P) dehydrogenase proved to be a very useful marker for this purpose. As one of the X-chromosomes present in each female somatic cell is inactivated in the course of embryogenesis, normal tissues of a person heterozygous for the Glc-6-P dehydrogenase locus contain a mixture of cells forming one or another of two isozymes controlled by two alleles. When all the cells of neoplastic lesion contain only one form of this enzyme, this indicates the clonal origin of the neoplasm.

The list of human neoplasms that have been shown to be monoclonal includes chronic myeloid leukaemia, Burkitt's lymphomas, lipomas, mel-anomas, leyomyomas of the uterus, most invasive skin papillomas, thyroid carcinoma, and pheochromocytoma. Examples of exceptional polyclonal neoplasms include multiple hereditary neurofibromatosis and hereditary trichoepitheliomas (Fialkow, 1972, 1974, 1976, 1977; Gartler, 1974; Baylin, Gann & Hsu 1976).

Most chemically-induced tumours in chimaeric mice are monoclonal (Iannaccone, Gardner & Harris, 1978) but certain cases of mouse hepatomas are polyclonal (Mintz, 1978).

Possibly, neoplastic nodules containing descendants of several cell clones are formed in tissues with a high probability of neoplastic transformations; in these cases, several clones formed from neighbouring transformed cells may merge into one lesion. Even in cases where the monoclonal origin of a neoplasm has been demonstrated, one cannot say that only one cell has been transformed at the beginning of the growth. In the course of neoplastic growth several clones formed independently may compete with one another so that eventually only one clone will survive.

Nevertheless, the data on the clonal character of various neoplasms are very important as they confirm that the growth of these lesions is a result of continuous proliferation of self-maintaining cell lines; this growth is not accompanied by recruitment of neoplastic cells from non-neoplastic elements of surrounding tissues. The only known exceptions are certain neoplasms induced by RNA-containing viruses. The cells of these neoplasms continuously form infectious virus particles that may transform surrounding normal cells. This type of growth was observed in the foci of fowl sarcomas induced by Rous sarcoma virus (Ponten, 1964; Morgan, 1964) and in spleen colonies generated by Friend virus-infected cells in mice (Steeves *et al.*, 1978).

The concept of the self-maintaining cell line needs some additional comment. The self-maintaining line can be defined as one in which successive generations of cells are identical to those present in the previous generations. By definition, such a line is potentially immortal; it may proliferate for an unlimited number of generations. We do not know whether there are self-maintaining lines of normal somatic cells in multicellular organisms. Certain normal cell lines (e.g. haematopoietic stem cells) can grow *in vivo* for many generations but probably even these lines have a limited life span. In particular, the self-maintaining ability of haematopoietic stem cells may progressively decline in the course of serial transfers from one animal to the next (see review in Lord, 1976). The limited life span of normal fibroblasts in culture will be discussed in the next chapter. Strictly speaking, all lines with a limited life span have no ability for self-maintenance in the sense defined above: cell generations are 'counted' somehow in these lines, that is, successive generations are somewhat different. When these differences are relatively small, one can sometimes disregard them and describe such a line as that 'with limited ability for self-maintenance'. When this term is used one has to remember that it is somewhat self-contradictory. Thus, it is not clear whether any normal cell line *in vivo* has a potentially unlimited life span. In contrast, many neoplastic cell lines are immortal, as shown by the successful serial transplantations of these lines. Immortality is difficult to prove formally for the cells of a neoplasm that has not been serially transplanted or cultured. Nevertheless, it is probable that many human and animal tumours are potentially immortal; that is, contain self-maintaining neoplastic stem cells. These tumours grow progressively without any obvious endpoint: most neoplasms almost never regress spontaneously. However, certain experimentally induced benign tumours, e.g. skin papillomas arising in the course of chemical carcinogenesis, may regress spontaneously (see review in Foulds, 1975). Spontaneous cessation of growth of human neuroblastomas has been observed many times; it is often accompanied by maturation of neoplastic cells into well-differentiated cells of neuronal type (see review in Foulds, 1969). One cannot exclude that stem cells of these neoplasms (and possibly of a number of other benign neoplasias) have a limited ability for self-maintenance, although regressions, of course, can be due also to many causes other than the loss of this characteristic.

Selective growth of neoplastic cells

This property of neoplasms can be formulated as follows: for each neoplasm there is one or several sets of conditions in which a neoplastic cell population grows at a greater rate than the total population of non-neoplastic cells in the area around the neoplastic lesion.

Certain types of non-neoplastic cells located near neoplastic lesions (e.g. endothelium) can proliferate very actively due to stimulatory effects of the neoplastic cells (see chapter 13). Therefore, to reveal the selective growth of a neoplastic population one has to compare the rate of its increase with that of a total non-neoplastic population and not with that of some special sub-set of a normal population. For the same reason, the area around a neoplastic lesion containing non-neoplastic cells should be wide enough, e.g. not less than the diameter of the lesion. Neoplastic cells of primary neoplasms grow selectively in comparison with the cells of the homotypic parent tissue. The cells of invasively growing neoplasms and of metastases grow selectively compared with the non-neoplastic cells of other tissues.

Each tissue of the multicellular organism resides in a certain territory: it is located in a definite way with regard to the other tissues, to blood and lymph vessels, and to intercellular structure such as basal membranes or collagen networks. Selective growth of neoplastic cells may lead to their spread, that is to an increase of territories occupied by these cells. There are two main ways of spreading. (*a*) Neoplastic cells may occupy the pre-existing territory in which normal cells have been dwelling previously. For instance, neoplastic epithelial cells of carcinomas *in situ* progressively occupy territories on the basal membranes, supplanting the normal homotypic epithelial cells. The cells of invasive carcinomas occupy territories of the tissues under the basal membrane. (*b*) Neoplastic cells may induce the formation of new territories by stimulating growth of stromal elements such as blood vessels, collagen-forming fibroblasts, etc.

Both these variants of the spread of neoplasias will be discussed in more detail in chapter 13.

As mentioned above, a neoplastic population may contain stem cells and non-stem cells. At present we can observe only the growth of total populations, as we cannot reliably distinguish stem cells from non-stem cells in any neoplasm. Theoretically, the spread of a neoplastic population should be accompanied by an increase in the number of stem cells. The growth of some neoplasms may become temporarily confined to the non-stem compartment; for instance, the epithelium of papillomas may form a large number of keratinized cells without an increase in the numbers of basal cells. However, in these cases the increase of the total neoplastic population will be stopped sooner or later due to intensified cell death, unless the growth of the stem cells is resumed. Of course, each type of neoplastic cell grows selectively only under certain conditions. When these conditions are changed, the cells may lose their selective advantage, so that the neoplastic population may stop growing and even decrease in size. Long ago, Rous & Kidd (1939, 1941) described so-called conditional neoplasms; that is, neoplasms growing only under certain conditions, e.g. when growth-stimulating agents are used. Among the conditional neoplasms are hormone-dependent tumours growing only in animals with increased blood concentrations of certain hormones.

Certain papillomas of rabbit ear epidermis induced by coal-tar painting grow only while painting is continuous or when a cutaneous wound is made near the tumour; probably, the wound somehow stimulates the neoplastic cells (Rous & Kidd, 1941). Conditional neoplasms stop growing and may partially regress when specific stimulation is stopped, e.g. when the level of specific hormone is decreased in animals with hormone-dependent tumours or when the wound near a papilloma is healed. The growth may be resumed if the stimulation is repeated. Neoplasms growing in animals without any special stimulation are sometimes operationally designated 'autonomous'. Of course, there are no clear-cut boundaries between conditional and autonomous neoplasms.

Deficiency of multicellular structures formed by neoplastic cells

This property of neoplasms can be formulated as follows: populations of neoplastic cells may form multicellular structures which have some deficiencies of organization compared with homotypic normal tissue and organ structures. Deficient organization of neoplastic tissues has various manifestations.

Abnormal maturation

Neoplastic tissue may have altered proportions of various cell types compared with normal tissue. These alterations are most obvious in tissues containing one or several series of maturing cells. For instance, neoplastic tissues of persons with chronic leukaemias contain the same maturation series of haematopoietic cells as normal haematopoietic tissue. However, each variant of leukaemia (myeloid, erythroid, etc.) is characterized by a great excess of cells of one particular maturation series.

Other neoplasms are characterized by the complete disappearance of cell variants representing final stages of maturation. This absence of maturation is, for instance, characteristic of haematopoietic tissues of patients with acute leukaemias. Maturation of basal epidermal cells into flat cells of the upper layers is blocked at the foci of carcinomas *in situ* developing in the uterine cervix.

Abnormal morphogenesis

The patterns of organ and/or tissue structures built by the cells become more simplified. Cells with the most advanced degree of this morphogenetic deficiency lose the ability to build even the most simple tissue-specific structures. Neoplasms with less advanced alterations of morphogenesis form simple tissue-specific structures (alveoles in the tumours of glandular tissue, areas of ossified tissue in bone tumours, etc.) but not the more complex organ-specific structures. Neoplasms with minimal degrees of deficiencies in morphogenesis may contain organoid structures but these are more or less abnormal. For instance, the neoplastic epithelium of organoid mouse mammary tumours builds branching ducts with budding alveoles; however, many of these structures are not connected with each other and a regular system of ducts characteristic of the normal gland is not formed (Foulds, 1956). Several different variants of the deficiences of morphogenesis may be present within the same neoplasm. For instance, in organoid mammary tumours part of the alveolas may have an irregular structure.

Thus, the degree of deficiencies of morphogenesis may vary considerably from one neoplasm to another. However, some degree of these deficiencies is characteristic for each neoplasm.

Pathological differentiation

Are neoplastic tissues able to perform 'pathological differentiations'; that is, are they able to form specialized cell variants or specialized cell products which cannot be formed in progenitor normal tissue? Two phenomena are often thought to be expressions of these 'pathological differentiations': activation of synthesis of embryo-specific proteins in many neoplasms, and production of ectopic hormones by certain human tumours.

Many tumours synthesize proteins that are specific for embryonic tissues but apparently not produced by adult tissues homotypic to the tumours (see reviews in Uriel, 1976b; Fishman, 1976). The discovery by Abelev and collaborators of α-fetoprotein synthesis by the liver tumours gave the first impetus to investigations of this phenomenon. At present, production of α-fetoprotein remains the most thoroughly studied example of embryo-specific protein synthesis by tumour cells (see reviews in Abelev, 1971, 1976, 1978; Sell *et al.*, 1976; Sell & Becker, 1978). α-Fetoprotein is actively synthesized and secreted by embryonic liver cells of all mammals examined including mice, rats, monkeys, and man. This synthesis is almost completely repressed in adult livers but is activated again in many hepatomas. Synthesis of α-fetoprotein can also be temporarily activated in adult liver tissue regenerating after damage. Analysis of this activation suggests that it may be associated with two processes: (*a*) α-fetoprotein may be synthesized by immature cells proliferating in the damaged liver (these immature cells are possibly precursors of mature hepatocytes); (*b*) synthesis may be activated in mature hepatocytes. (As suggested by Abelev (1976), this activation may be a result of alterations of cell–cell contacts.) Thus there is reason to think that activation of α-fetoprotein synthesis in hepatomas may also be due to deficiency of tissue architecture and/or to an increased proportion of immature hepatic cells. The cellular basis of other embryo-specific syntheses has not been examined in detail.

Certain human tumours synthesize hormones that are not characteristic of normal parent, non-endocrine tissues (see reviews in Rees, 1975; Vaitukaitis, 1976; Odell & Wolfsen, 1976; Skrabanek & Powell, 1978). For instance, certain morphological variants of lung tumours (oat cell carcinomas) may synthesize adrenocorticotropin (ACTH) which is normally formed by the cells of anterior pituitary. Analysis of the cellular basis for these 'ectopic' syntheses has not yet been made. In particular, little is known about progenitor cells of hormone-forming tumours. It is possible that ACTH-forming neoplastic lung cells are derived not from 'derepressed' differentiated lung cells but from minority cell types present in the lung. For instance, they could be derived from the pluripotential stem cells, giving rise to various lines of differentiation, or from one of the varieties of special 'endocrine cells' that may be present in small numbers in many normal tissues.

In summary neoplastic tissues are characterized by various types of abnormalities of morphogenesis and differentiation. At present there is no direct evidence proving that any types of neoplastic cells may form 'pathologically differentiated' variants; that is, differentiated variants other than those formed by progenitor normal cells.

Strongly neoplastic and minimally neoplastic clones: clonal pathology?

We have described three main features that are characteristic of neoplasms: the presence of self-maintaining cell lines; ability of these lines to grow selectively; and deficiency of structure formed by the cells of these lines. Combination of these three traits is characteristic only for neoplasms but each can be found in non-neoplastic processes. For instance, reactive proliferations of cells in various organs (hormone-induced hyperplasias, regeneratory hyperplasias, etc.) may lead to considerable alterations of tissue structure (see, for instance, Stewart, 1975). These atypical hyperplasias will, however, remain non-neoplastic as long as they are not accompanied by clonal selective growth of special cell lines.

Grafted haematopoietic stem cells may grow selectively in haematopoietic tissues of pre-irradiated animals but this clonal proliferation is not neoplastic as it eventually leads to formation of normal haematopoietic tissue.

Each typical neoplasm has all the three characteristic traits but concrete expression of these traits can vary considerably from one neoplasm to another; the particular combination of these traits is characteristic for each group of neoplasms and even for each individual neoplasm. For instance, malignant tumours are able, by definition, to spread selectively in non-homotypic tissue by invasion and/or by metastases. Ways of spreading may vary from one neoplasm to another. Certain neoplasms show only local invasion, while others metastasize extensively. Various neoplasms may give metastases with different frequency and, possibly, with different preferential localization.

The degree of atypical morphology may vary considerably, even in malignant tumours of similar origin (see page 6). Carcinomas *in situ* are considerably atypical in morphology but they do not spread outside the basal membrane of the parent epithelium. Leukaemias are neoplasms that spread systemically in their parent haematopoietic tissue; they range from those forming well differentiated tissue with the abnormal proportions of various cells to those forming only undifferentiated cells. Various lines of leukaemic cells grow preferentially in different parts of haematopoietic system (lymph nodes, bone marrow, etc.); certain leukaemic cells may also grow outside the natural territories of haematopoietic cells (in the skin, in the nervous tissue, etc.).

Of special interest are neoplasms with minimal degrees of expression of neoplastic traits. The cells of these lesions do not spread outside the tissue of origin and make only slightly atypical tissue structures. Many of these neoplasms are conditional; that is, they require special stimulation for their growth. For most of these neoplasms it is not clear whether the life span of their cell lines is unlimited as efforts to propagate them serially have not been made, or have been unsuccessful.

If these minimally neoplastic clones grow as distinct nodules in homotypic tissue, they may be diagnosed as benign tumours. In fact, many benign tumours known in human and animal pathology may be properly regarded as minimally neoplastic clones.

Application of modern methods to distinguish clonal and non-clonal proliferates has shown that growth of minimally neoplastic clones may be encountered much more frequently than previously suspected. The athero-sclerotic plaque is the most exciting example of a lesion whose neoplastic nature had been previously unsuspected: recent findings suggest that each plaque may be a result of clonal proliferation of slightly abnormal smooth muscle cells of the vessel wall, that is, a kind of benign tumour (E. P. Benditt & J. M. Benditt, 1973; Pearson *et al.*, 1978).

Certain minimally neoplastic clones may not grow as focal nodules but spread along the pre-existing stromal structures of normal tissues, that is, in a manner similar to that of the carcinoma *in situ* spreading along the basal membrane.

In these cases the macroscopic structure of the organ may remain unchanged and the area occupied by the neoplastic clone would have only slightly atypical microscopic structure and, possibly, some functional abnormalities. So-called 'paroxysmal nocturnal haemoglobinuria' may be an example of this type of neoplastic process. This pathological condition may be due to clonal proliferation of haematopoietic cells within the bone marrow; the main pathological manifestation is production of erythrocytes with increased sensitivity to haemolysis (Fialkow, 1974). Certain other haematological disorders with different clinical manifestations (refractory anaemias, myelo-fibrosis) are also possibly due to proliferation of functionally abnormal clones of haematopoietic cells (Sokal, Michaux & Van den Berghe, 1975; Fialkow, 1977). Probably, proliferation of minimally neoplastic clones may take place not only in the bone marrow but also in many other organs and may lead to various functional abnormalities. Selective growth of these clones may be the basis of a number of immune, endocrine, haematological and metabolic diseases whose neoplastic nature has never been suspected before. Perhaps, the near future will be of the age of 'clonal pathology'; at present, its development is just beginning and it is impossible to predict how far it will progress.

Progenitor cells of neoplasms

Pierce & Cox (1978) likened the atypical structure of neoplastic tissue to a caricature of normal tissue structure. Looking at this caricature we can in most cases recognize what it imitates, that is, we can tell which normal tissue is an analogue of the tissue built by each neoplastic clone. Finding these analogues is one of the main tasks of the histological diagnosis of neoplasms; at present, a well-developed branch of pathology. At the same time, in most cases we are unable to tell which particular cell variant, present in the parent normal tissue, has been the progenitor of the neoplastic clone.

Each normal tissue contains many cell types which are differentiated in various directions and are at various stages of differentiation. Inter-relations between different cell types are known more or less satisfactorily only for the haematopoietic cell system (see review in Chertkov & Friedenstein, 1977; Lajtha *et al.*, 1978). This system contains pluripotent stem cells which are able to maintain themselves for many generations (see p. 5) and may also give rise to committed precursor cells of several types. Each of these precursors (granulocytic, erythroid, megakaryocytic, etc.) can give rise to one or several series of maturing cells.

For instance, granulocytic precursors are polyphyletic as they can produce colonies of granulocytes, or mononuclear cells of the macrophage type. Erythroid precursors are monophyletic as they form only erythroid progeny. Sometimes these cells are designated 'granulocytic' and 'monocytic' stem cells (Moore, Kurland & Broxmeyer, 1976). In each series, several generations of cell division lead to the formation of highly specialized mature cells (erythrocytes, neutrophiles, etc.) which are unable to proliferate ('suicide maturation').

Maturation progress from the pluripotential stem cells to end cells takes not more than 20 cell generations. The structure of other cell populations is much less clear than that of haematopoietic system.

Certain epithelial tissues, e.g. intestinal epithelium or epidermis, contain morphologically undifferentiated cells which give rise to maturing specialized cells; the final cells of these maturation series stop proliferating and die. Some data suggest that the population of morphologically undifferentiated cells in these epithelia is biologically heterogeneous: it may contain pluripotential stem cells and several types of committed precursor cells from which different series of maturation are started (Leblond & Cheng, 1976; Potten, 1976*a*, *b*).

It could be that some other tissues also have a tree-like population structure consisting of pluripotential stem cells and several maturation series with unidirectional movement of cells from one compartment to another (see chapter 2). However, to prove unequivocally that a population has this structure it is essential to develop methods for cloning stem cells and precursor cells. Until now, this has only been done for the haematopoietic system. Therefore, at present it remains possible that certain tissues have population structures different from the tree-like one. For instance, in certain tissues some stages of maturation may remain reversible (retrodifferentiation, see Uriel, 1976*a*) or the whole population may consist of self-maintaining cells with a similar degree of maturation.

The structure of neoplastic populations is not known much better than the structure of normal populations. Only few types of neoplasms have been studied in this respect. One of these neoplasms is human chronic myeloid leukaemia. Neoplastic cells of this leukaemia have a specific chromosomal marker (Philadelphia or Ph′ chromosome). Besides myeloid cells, this marker has been found also in the cells of erythroid and megakaryocytic series. Thus, although most cells of the neoplastic population undergo differentiation in only one direction this population also contains pluripotential stem cells that are able to undergo maturation in several other directions. It is probable that the pluripotential haematopoietic stem cell is a progenitor of the neoplastic clone in chronic myeloid leukaemia.

Mouse teratocarcinoma is another example of a tumour that has been shown to contain pluripotential stem cells. Besides undifferentiated cells these tumours contain cells differentiated in various directions including derivatives of all three embryonic germ layers: ectoderm (e.g. neural tissue, keratinizing epithelium), mesoderm (e.g. muscle, cartilage, bone) and endoderm (glandular epithelium, gut), see reviews in Pierce, 1967, Stevens, 1967; Martin, 1975; Manes, 1976; Jacob, 1977.

The nodules arising in a new host after transplantion of one tumour cell contained the same variety of differentiated cell types as the original tumour (Kleinsmith & Pierce, 1964). Obviously, teratocarcinomas contain pluripotent stem cells that are able to undergo differentiation almost into all cell types present in the adult organism; wide developmental potentialities of these neoplastic stem cells correspond to those of early embryonic cells. In contrast to the normal embryo, differentiated tissues formed in teratocarcinomas are arranged chaotically.

Normal progenitor cells of teratoblastomas are, probably, primordial germ cells or the somatic cells of early embryos.

Human and mouse plasmocytomas can be regarded as an example of neoplasm arising from cells that can be differentiated in only one direction. The cells of each plasmocytoma produce only one type of specific immunoglobulin. Such immunoglobulins may act as transplantation antigens. Mice immunized with purified immunoglobulin produced by a certain strain of plasmocytomas resist further transplantation of the cells of corresponding neoplasm (Lynch *et al.*, 1972). These results suggest that stem cells of these tumours carry specific immunoglobulins on their surfaces. In other words, progenitors of the plasmacytomas probably belong to some sub-sets of B lymphocytes (Potter & Cancro, 1978); that is, these progenitors are differentiated much more than pluripotential stem cells from which chronic myeloid leukaemias arise.

Populations of many other tumours contain morphologically undifferentiated cells and series of cells differentiating in one particular direction. Neoplastic non-stem cells which are at the final stages of maturation lose the ability to proliferate in the same way as analogous normal mature cells. For instance, Pierce (1974) has shown that epidermoid carcinoma cannot be transplanted by keratinized cells present in the tumour. We do not know the structure of subpopulations of the morphologically undifferentiated cells present in these tumours. Some neoplasms may contain pluripotential stem cells while others may arise from more differentiated monopotential cells. It has been suggested that stem cells are the progenitors of malignant tumours, while transformation of cells differentiating in one direction leads to formation of a benign tumour (Pierce, 1974, 1976). As noted above, experimental tests of these hypotheses will become possible only when methods of clonal analysis of various normal populations are better developed. At present, normal progenitors of stem cells of most neoplasms remain unknown. At the same time, development of many neoplasms from pluripotential stem cells with wide developmental potentialities is a possibility that should be taken into account in any analysis of the properties of neoplastic tissue (see, for instance, pp. 8–9).

Progression of neoplasms

Each neoplastic population has a certain genetic stability: a neoplastic cell line may retain its growth requirements for many generations, ability for differentiation, character of spreading, etc. This stability is, however, not absolute and the properties of a neoplastic population can be changed from time to time. Due to these alterations, evolution of each neoplasm is a multistage process which is designated as neoplastic development or neoplastic evolution. Alterations of the properties of neoplasms are designated as progression. Foulds (1969) defines progression as the development of a tumour by way of an irreversible qualitative change in one or more characters of its cells. In accordance with the description of the main traits of neoplasms given above, progression is always due to the stable genetic alteration of the properties of self-maintaining neoplastic stem cell lines. Probably, this alteration does not occur in all neoplastic stem cells simultaneously, but is a result of the alteration of one neoplastic cell; if this new clone has some selective advantage, it gradually replaces the pre-existing neoplastic population (see review in Nowell, 1976). By definition, in the course of progression, new cell variants arise from pre-existing neoplastic cells and not from normal cells present within the neoplasm. In the opposite case, for instance, when a stromal cell of a tumour has become malignant, we are dealing with formation of a new neoplasm within the pre-existing one but not with the progression of pre-existing neoplasm.

A new cell line formed in the course of progression of a neoplasm is usually more strongly neoplastic than the pre-existing cell line. The new line may have a wider range of conditions in which its selective growth takes place. In particular, conditional neoplasms progress to an autonomous state; for instance, a hormone-dependent tumour may become hormone-independent. The routes of cell spread may also become less selective; for instance, carcinoma *in situ* growing only along the basal membrane may undergo progression to invasive carcinoma. The degree of deficiencies of multicellular structures may also increase. For instance, neoplastic stem cells may lose the ability for differentiation into mature cells; the formation of atypical morphological structures may increase. On the basis of studies on mammary tumours in mice, Foulds (1954, 1969) proposed six general rules of progression applicable to various sorts of neoplasms in animals.

Rule I. Progression occurs independently in different tumours of the same animal.

Rule II. Progression occurs independently in different characters of the same tumour.

Rule III. Progression is independent of growth. Progression may occur in latent tumour cells and in tumours whose growth is arrested.

Rule IV. Progression may occur by gradual change or by abrupt steps.

Rule V. Progression follows one of several alternative paths of development.

Rule VI. Progression does not always reach an end-point within the life-time of the host.

It is easy to see that these rules show that progression is not an orderly and predictable sequence of changes; the time and character of the changes may vary widely during the evolution of different neoplasms. Of course, by studying a series of homotypic neoplasms we may learn to predict mean statistical probabilities of various changes in a similar series of neoplasias. However, we can never predict exactly the fate of one neoplastic lesion.

The unpredictability of progression is easy to understand if we assume that each step of progression is a result of the more or less random appearance of a new variant clone that first forms a minority in the population but later may outgrow other cells. From this point of view, Rule IV needs clarification. Probably, at the cellular level, progression never occurs gradually: the appearance of new cell variant is always an abrupt change. Of course, the average properties of the whole neoplastic population may change gradually due to progressive selective growth of new clone. In the course of progression of similar neoplasms, certain cell variants may appear and/or are selected with a high frequency. For instance, chronic myeloid leukaemia may undergo blastic crisis; that is, progression from a relatively benign stage to the terminal stage characterized by growth of morphologically undifferentiated cells that are difficult to control by therapy. All the neoplastic cells at the benign stage of chronic leukemia usually have a single characteristic chromosomal change (see below). During the blastic crisis the neoplastic cells may have various additional chromosomal variations. Although these alterations are very variable, their analysis has shown that they often evolve in certain regular order: alterations of chromosome 17 are usually followed by those of chromosome 8 and alterations of chromosome 19 appear at a later stage (Prigogina & Fleischman, 1975; Prigogina *et al.*, 1978). Thus, despite considerable individual variations, the number of 'alternative paths of development' followed by progression (see Rule V) is not infinite.

With regard to the analysis of phenotypic characteristics for neoplastic cells, the concept of an independent progression of neoplastic characters (Rule II) is of special importance. Foulds (1969) who developed this concept, formulated it as follows: structure and behaviour of tumours are determined by numerous characters that, within wide limits are independently variable, capable of highly varied combinations and assortments and liable to independent progression. The term 'character' used here, according to Foulds, is applicable at all levels of analysis, from general pathological characters (invasiveness, frequency of metastases, atypicalness, etc.) to the presence or absence of individual proteins. Many examples of the independent progression of characters, collected by Foulds and accumulated in the literature following his publications (see review in Medina, 1975), have confirmed the wide applicability of this concept. Many facts summarized in this book also illustrate this rule: in the course of neoplastic evolution *in vitro* various transformed characters can be dissociated from one another (see chapter 2).

The fundamental unsolved problem is that of the universality of the rule of independent progression of characters and there may be two alternative concepts of neoplastic evolution. One of them postulates that the rule of independent progression is universal: there is no single character common to all the neoplasms arising in a certain tissue; independent alterations of various characters are randomly combined in the course of development of each neoplasm. Alternatively, it is possible that one particular 'central' group of cellular characters is always altered in the neoplastic cells, but the degree of this alteration may be different in various neoplastic clones and, in addition, other characters of these clones may be randomly assorted. We will return to the fundamental problem of choice between these possibilities in the final chapter.

The nature of genetic differences between normal and neoplastic cells

Altered phenotypic properties of neoplastic cells are genetically stable, that is, they may be transmitted to progeny cells for many generations. What is the nature of the genetic alterations responsible for these changes of phenotype? Detailed discussion of this problem is outside the scope of this book and here we will outline only briefly several types of genetic changes that may be involved.

For a long time, most hypotheses about the nature of neoplastic transformation have dealt mainly with three types of genetic change: the integration of viral genome into cellular DNA, mutations, and epigenetic changes. Remarkably, during the last decade, each of these has turned out to be heuristically valuable and the search for the three types of genetic change has led to important experimental results.

In particular, investigations of the role of viruses have been very fruitful. Many oncogenic viruses causing different types of neoplasms have been discovered and it has been shown that the development of certain neoplasms is a result of the integration of viral genomes into cellular DNA. Some of the virus-induced neoplastic transformations will be discussed in chapter 2.

The mutational hypothesis has stimulated important discoveries of characteristic chromosomal changes in several types of neoplasia. The most specific of these changes known is that of the Ph′-chromosome in chronic myeloid leukaemia; The Ph′-chromosome is a translocation of part of the long arm of one chromosome of the 22nd pair on to the long arm of chromosome 9, or more rarely, on to other chromosomes. The Ph'-chromosome is present in neoplastic cells of about 90% of patients with chronic myeloid leukaemia. As mentioned above, progression of this disease is manifested by the appearance of new clones with additional chromosomal changes.

The cells of various malignant lymphomas often have rearrangements of chromosome 14 (Fleischman & Prigogina, 1977). Deletion or complete loss of chromosome 22 is characteristic for the cells of human meningiomas (Mark, Levan & Mitelman, 1972; Zankl & Zang, 1972). Certain other chromosomal alterations have been frequently revealed in various types of neoplasms (see reviews in Nowell, 1974; Wolman & Horland, 1975; Rowley, 1975, 1976; Mitelman & Levan, 1976; Wang-Peng, 1977). The regular presence of characteristic chromosomal changes in the cells of certain neoplasms gives reason to conclude that these changes may be responsible for neoplastic transformation but at present there is no direct proof of this.

Many chemicals as well as ionizing and ultraviolet radiation have carcinogenic properties. Most of these agents also have mutagenic activity in various biological systems (review in Montesano, Bartsch & Tomatis, 1976; McCann & Ames, 1976; see also chapter 2).

Humans with genetic deficiencies of their DNA repair systems have a very high incidence of ultraviolet-induced skin cancer (see review in Setlow, 1978). These data indirectly support but do not prove the suggestion that mutagenic effects of chemicals and radiation are the basis of their carcinogenic action.

It has been suggested many times that neoplasms may arise as a result of 'epigenetic changes', or of 'pathological differentiation'; changes that are similar to those occurring in embryogenesis and which result in the formation of cells of different tissue types (see, for instance, Braun, 1974, 1976; Pierce, 1976). Normal changes of this type are genetically stable: tissue-specific features of the cells can be transmitted for many passages in culture. The nature of these 'epigenetic changes' remains at present quite enigmatic so that it is difficult to imagine what sort of evidence might prove or disprove the 'epigenetic' nature of neoplastic transformation.

In connection with the suggestions about the 'epigenetic' nature of neoplastic transformation many experimental efforts to induce reverse transition of neoplastic cells into normal cells have been made.

The most convincing evidence on normalization of neoplastic cells has been obtained in two experimental systems. (*a*) Several varieties of plant tumour tissues (crown gall, Kostoff genetic tumours, habituation), when grafted to cut stem tips of normal plants, gave rise to normally developed shoots that ultimately flowered and set viable seed from which normal plants could be grown (Braun, 1975, 1978; Braun & Wood, 1976). (*b*) The cells of mouse teratocarcinoma transplanted to early mouse embryos gave rise to normal cellular progeny participating in the formation of many different normal mouse tissues. Apparently normal cells carrying the genetic markers of the neoplastic donor cells have been found in many different mouse tissues grown from the implanted embryos (Mintz & Illmensee, 1975; Illmensee & Mintz, 1976; Brinster, 1976). In both systems 'normalized' neoplastic cells gave rise to germ cells carrying specific markers. The plant obtained from tumour cells gave normal seed. Chimaeric male mice mated to non-chimaeric females gave progeny carrying markers of the original teratoma cells; that is, teratoma cells gave rise to normal sperm cells. In summary, the cells of both types of neoplasm became normal when placed in a specific micro-environment and this normalization was stable for many cellular generations.

These experiments are obviously of extreme importance but they cannot be regarded as proof of the epigenetic nature of the differences between the tumour cells and their corresponding normal cells: to obtain these proofs we have to know more about the nature of epigenetic changes occurring in embryogenesis. In particular we do not know whether reversibility is a characteristic feature of these changes. Can normal fibroblasts be transformed into normal epithelial cell or vice versa?

One special possibility should be mentioned with regard to the experiments with tumour cell normalization. One cannot exclude that the cells of certain neoplasms are not mutants and not even cells that have undergone pathologic 'epigenetic' change, but perfectly normal cells placed in an unusual environment. Teratocarcinoma cells may be normal cells of the early embryonic type that grow selectively, subcutaneously or intraperitoneally and do not form a normal structure when placed in this environment, because they do not receive proper stimuli for differentiation and morphogenesis. The same cells returned to the micro-environment of normal embryo tissues would receive proper signals and form normal tissues. In other words, 'misplacement' of certain normal cells may be sufficient for their neoplastic behaviour. This intriguing possibility needs further investigation.

We have outlined above several possible genetic mechanisms of neoplastic transformations. Of course, development of various neoplasms may involve different mechanisms. Genetic changes of different types may occur at various stages of evolution of one neoplasm. Probably, the boundaries between different types of genetic changes are not as sharp as they once have seemed to be. For instance, it had been suggested that alterations accompanying normal and pathological differentiation ('epigenetic changes') may involve integration of new genetic material (protoviruses) transmitted from other cells (Temin, 1971).

At present our concepts about the structure of the eukaryotic genome are rapidly changing due to the enormous growth of new facts and it is probable that the list of possible types of genetic changes of somatic cells will be considerably increased and modified in the near future.

2

Neoplastic transformations in culture

Fibroblasts and epithelial cells in culture

This book is based on the results of experiments made with cultures of fibroblast-like cells* and, to a lesser degree, cultures of epithelial cells. The Evolution and behaviour of fibroblasts in culture are known better than those of other cell types because they are relatively easy to grow *in vitro*. The common feature of both epithelial and fibroblastic cells is their pronounced ability to spread on substrata such as glass or certain plastics.

Morphological differences between epithelia and fibroblast cultures become obvious after spreading of the cells on the substratum (fig. 2.1). The main distinctive feature of epithelial cells is their ability to form coherent sheets consisting of cells firmly linked to each other; in the course of locomotion the marginal cells usually are not detached from the sheet. Fibroblasts are unable to form coherent sheets; in contrast to epithelium, these cells easily break cell–cell contacts and move individually. In mixed cultures, it is not always easy to distinguish an isolated fibroblast from a single substrate-spread epithelial cell that is not in contact with its neighbours. Usually, fibroblasts have several elongated cytoplasmic processes. Depending on the number and size of these processes, the cells may have spindle-like, fan-like or stellate shapes. In contrast, isolated epithelial cells usually do not form elongated processes but have rounded or polygonal contours. Fibroblasts in culture are usually recognized on the basis of the morphological features described above.

Primary fibroblastic cultures may be obtained from mixed embryo tissues or from isolated tissues of embryos and adult animals, e.g. from subcutaneous connective tissues, lung, spleen, bone marrow, blood, etc. Epithelial cultures are obtained from various organs, most often from kidney, liver, and skin. Usually, it is implied that cultures of both these morphological cell types grow from the cells of corresponding tissues present in the explanted populations; that is, epithelial cultures grow from various epithelia while fibroblastic cultures grow from cells of mesenchymal origin, and belong to the mechanocyte group (Willmer, 1965). In most cases this suggestion seems to be reasonable, especially in cases where fibroblastic cultures have been obtained from

---

* For the sake of brevity, fibroblast-like cells are referred to subsequently as fibroblasts.

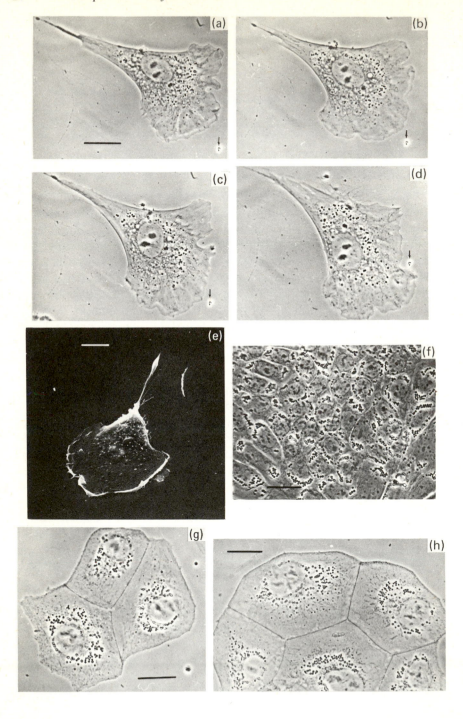

previously isolated connective tissue fragments or epithelial cultures from isolated epithelia.

The origin of the cultures of many specialized epithelia (hepatocytes, epithelia of endocrine glands, epidermis) may be confirmed by the presence of tissue-specific, morphological and biochemical markers (see review in Wigley, 1975). We do not as yet have specific markers for fibroblasts. Fibroblasts often form large amounts of collagens and glucosaminoglycans in culture. These products are similar to those formed by fibroblasts *in vivo*. However, other cell types, including various epithelia (review in Reddi, 1976; Laurent *et al.*, 1978; Sakakibara *et al.*, 1977, 1978; Killen *et al.*, 1978) were also reported to form certain amounts of collagen and of glucosaminoglycans. Different types of collagen are probably formed by epithelia and fibroblasts. In particular, the formation of type IV collagen seems to be characteristic for epithelial cells. We hope that detailed studies of these markers will provide the basis for a more exact identification of cell types.

Epithelia and fibroblasts are two large morphological classes of cultured cells and each may contain many subclasses of cells with different properties. Morphological and biochemical differences between epithelia derived from various tissues are obvious. Fibroblasts from various tissues of the same animal may look similar but, nevertheless, they may have different developmental potentialities. For instance, in the experiments of Friedenstein (1976) fibroblastic clones isolated from the bone marrow and from the spleen were morphologically similar in cell cultures. However, when these clones were implanted in isogenic mice only the cells from bone marrow formed osteoid tissue; spleen-derived fibroblasts were not able to perform osteogenesis. Certain specialized cell types may share many common morphological and behavioural features with fibroblasts or with epithelia. In particular, glial cells in cultures have many morphological similarities with fibroblasts (Ponten, 1975). The endothelial cells from large vessels have many epithelial features. The systematic comparative study of intra-class differences between various species of fibroblast and epithelia remains a task for the future.

Analyses of the cellular origin and population structure of fibroblastic cultures are hindered by the slow progress in the analysis of histogenesis of

Fig. 2.1. Morphology of cultured fibroblasts and epithelial cells. (a–d) A polarized mouse embryo fibroblast moving on glass: (a), zero time; (b), 40 min; (c), 85 min; (d), 130 min. Notice the alterations of cell position with regard to the non-moving particle on the glass (arrow). Phase contrast micrographs of living cultures. (e) Polarized mouse embryo fibroblast by scanning electron microscopy. These and all other SEM pictures were taken from glutaraldehyde-fixed and critical-point-dried cultures. Photographs were taken with a Cambridge Stereoscan-S4 at 10 kV; the tilt angle was 30°. (f) Phase contrast micrograph of normal mouse kidney epithelium, coherent sheet. (g, h) Phase contrast micrograph of the continuous FB line of fetal bovine tracheal cells (see description in Machatkova & Pospisil, 1975). Bars: (a–d) (g–h), 20 μm; (e), 10 μm.

fibroblasts and other mechanocytes *in vivo*. Experiments with radiation chimaeras suggest that mechanocytes seem to form a special cell line (or several lines) *in vivo* different from the line of haematopoietic cells (Friedenstein, 1976). It had been suggested that subendothelial cells in the walls of blood capillaries (so-called pericytes) may be pluripotential elements which may serve as precursors of endothelium, smooth muscle cells, fibroblasts and a variety of other mesenchymal cell types (Rhodin, 1968; Johnson *et al.*, 1973). Franks and collaborators (Franks, Chesterman & Rowlatt, 1970; Franks & Cooper, 1972) suggested that many continuous cell lines arise from vascular endothelium and pericytes. They stressed that tumours formed by various spontaneously transformed mouse cell lines had similar morphology; each tumour had fibrosarcomatous, myxoid, epithelioid and giant cell areas. This mixed pattern may be due to the pericytic origin of these transformed lines; undifferentiated pericytic cells when transplanted *in vivo* might form various more differentiated cell types: fibroblastic cells, muscle cells, endothelioid cells, etc. This suggestion is plausible but it should be stressed that all the hypotheses about the special role of pericytes are based on static morphological pictures and cannot be regarded as proven. Other schemes of hystogenesis of mechanocytes remain possible. For instance, one cannot exclude the possibility that this population does not contain stem cells and irreversibly matured cells but has several reversible directions of cell modulations.

For the same reason, it is not clear whether cultured fibroblast populations are biologically homogeneous or contain cells at various degrees of 'maturations'. Morphologically, most cells in primary fibroblastic cultures and in the cultures of continuous lines are rather similar to each other. However, only a fraction of these cells is clonogenic (Friedenstein, 1976); the same is true for the cells of many continuous lines.

The same difficulties arise when we try to analyse epithelial populations in culture. Considerable indirect evidence suggests that many types of epithelia *in vivo* contain polipotent stem cells that can be differentiated in several directions (see chapter 1). However, there are as yet no reliable methods of distinguishing epithelial stem cells *in vitro* or *in vivo*. Further development of the clonogenic assays will tell us more about the population structure in fibroblastic and epithelial cultures.

Manifestations and stages of cell transformations in culture

In the course of cultivation the properties of fibroblasts and epithelium may undergo alterations and their properties become different from those of primary cultures isolated from normal tissues. Alterations of the culture properties may arise spontaneously or after the treatment of cultures with carcinogenic chemicals, or oncogenic viruses. In many cases the character of both types of alteration is similar. Therefore, the combination of these alterations may be designated by some common term: we will use for this purpose the term 'neoplastic evolution of cultures'. The main manifestations of this evolution include the following cellular changes.

(1) *Development of oncogenicity:* the cultured cells may acquire the ability to form tumours after transplantation to susceptible animals. This alteration will be discussed later in this chapter.

(2) *Morphological changes*, more specifically, deficiency of cell attachment to the substratum (decreased spreading) and to other cells. These changes will be discussed in more detail in chapters 5 and 6.

(3) *Development of an unlimited life span* (*immortalization*). The primary normal cells after a certain number of passages in culture gradually stop multiplying (so-called senescence of the cultured cells). In the course of neoplastic evolution the cells may acquire the ability to grow for an unlimited number of generations. They form potentially immortal lines. Senescence and immortalization will be discussed in the next sections of this chapter.

(4) *Loss of anchorage dependence of growth:* the cells of primary fibroblastic and epithelial cultures in usual conditions do not multiply in suspension; they start proliferation only after spreading on the solid substratum. In the course of neoplastic evolution a considerable fraction of the cell population acquires the ability to form colonies in a suspended state, without attachment to the substratum. To test this ability suspended cells are usually placed in a semi-fluid medium (soft agar or methylcellulose). We will discuss the loss of anchorage dependence in chapter 8.

(5) *Decreased serum dependence*. Normal cells grow only in media containing considerable amounts of serum; in the course of evolution the cell may acquire the ability to proliferate in media with low concentrations of serum unable to support the growth of normal cells. The loss of serum dependence will be discussed in chapter 10.

(6) *Decreased density dependence of growth*. The rate of proliferation of normal cells decreases as the local density of the cell population increases. In the course of evolution this dependence can be altered in several ways: the cells may acquire the ability to grow in an area with a high density of homotypic cells and/or in an area with high density of normal cells. This property will be discussed in the chapter 9.

Other alterations, frequently accompanying neoplastic evolution, include karyotypic changes; increased ability of cells to be agglutinated with plant lectins (see chapter 5), increased resistance to the toxic effects of chemical carcinogens (see this chapter), decreased dependence of proliferation on calcium content in the medium (see chapter 10), and decreased synthesis of fibronectin (see chapter 6).

All these alterations have been studied mostly in the course of evolution of fibroblastic cultures of various mammals and birds. Neoplastic evolution of epithelial cultures is probably accompanied by similar changes but transformations of these cultures have been studied much less thoroughly than those of fibroblasts. In many cases several different alterations listed above may develop simultaneously within the same culture and even within the same clone. However, in other cases the culture may develop only a few of the symptoms characteristic for neoplastic evolution; different symptoms may appear at different times within the same culture. We will discuss the inter-relationships between various alterations accompanying neoplastic evolution in chapter 14.

Here we will define certain terms used in this book. Fibroblastic or epithelial cultures which develop some of the alterations listed above in the points (1)–(6) will be designated 'transformed cultures'. Whenever possible we will try to indicate which particular characters have been transformed. Correspondingly, we will speak about the oncogenically transformed (oncogenic) cultures, morphologically transformed cultures, transformants with loss of anchorage dependence, etc. Often the cells may have several transformed characters simultaneously, in this case we will simply speak about transformed cells. Of course, we will try to specify which particular complex of characters has been altered.

Cultures that do not have any of the transformed characters we will designate as 'normal'. These are the cultures of normal cells during their first passages (primary cultures) and, possibly, also the cultures of diploid human strains with a limited life span. The cultures that have only one or a few transformed characters, while their other characters do not obviously manifest transformations we will designate 'minimally transformed cultures'. The cultures which have most of the transformed characters listed above we will designate 'strongly transformed cultures'. Obviously, the exact list of transformed characters and degree of transformation of each character may be different for each particular transformed culture and there is no clear-cut boundary between minimally and strongly transformed cultures.

In the course of neoplastic evolution of one culture a series of transformations may occur. The normal culture may give rise to a minimally transformed culture; this culture in its turn may give rise to a strongly transformed culture. In other cases, normal culture may give rise directly to strongly transformed cultures. In certain conditions, usually after special selection procedures, strongly transformed cultures may give rise to the culture losing certain specific transformed characters. These cultures are designated 'revertants'.

The terminology related to development of immortal life span in cultures

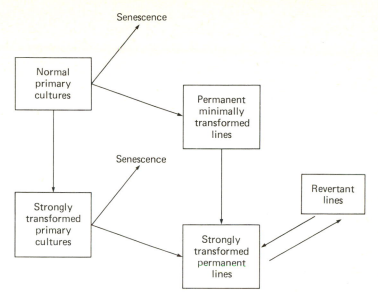

Fig. 2.2. The main stages of neoplastic evolution in cell cultures.

also requires some comment. Cells with a limited life span are designated 'primary cultures', or sometimes, if they have passed several passages *in vitro*, as 'cell strains'. The cultures that have acquired unlimited life span are usually designated 'permanent (continuous) cell lines' or simply 'cell lines'. Immortalization is one of the symptoms of neoplastic evolution; therefore it is not correct to speak about these as normal cell lines, especially, as most lines also have other transformed characters, for instance, karyotypic changes.

In summary, we can distinguish normal primary cultures, strongly transformed primary cultures, minimally transformed cell lines and strongly transformed cell lines. Theoretically, minimally transformed primary cultures can also exist but we are not familiar with any description of a culture of this type. The scheme in fig. 2.2. illustrates, in a very simplified form, interrelationships between these variants. All transformations discussed above are genetically stable; that is, transformed characters are transmitted to the cell progeny. However, in certain special conditions a normal cell may aquire certain features, characteristic of the transformed phenotype (e.g. altered morphology or anchorage independence) but when returned to 'usual' conditions these cells restore their normal phenotype. On the other hand, genetically transformed cells may temporarily acquire a more normal phenotype when placed in some special conditions. We will designate these alterations as 'phenotypic transformations' and 'reversions'; they are discussed in the chapters 6, 11 and 14. Unless otherwise mentioned, the terms 'transformation' and 'reversion' refer to genetically stable changes expressed in usual culture conditions.

In the following section we will discuss in more detail certain variants of cell evolution in culture: cell senescence, spontaneous transformations, transforming effects of chemicals and viruses as well as selection of revertant lines.

Cellular senescence

After an initial period of adaptation to the culture conditions (Phase I), primary cultures of fibroblasts and epithelia may enter into a phase of more or less stable growth (Phase II). This phase is followed by a period of gradual decrease in growth rate and a progressive deterioration of the culture; this period is designated 'Phase III', 'crisis' or 'cellular senescence'.

The boundary between the Phases II and III is not sharp. Duration of the life span (stable growth plus senescence) is apparently dependent on the average number of generations (population doublings) made by the cells in culture rather than on the calendar time spent in culture (see reviews in Hayflick, 1973; Holliday, 1975; Littlefield, 1976; Harley & Goldstein, 1978). This duration varies considerably, depending on the species of animal from which the cultured cells were taken: it is about 50 generations for human embryo fibroblasts (Hayflick & Moorhead, 1961). Skin fibroblasts from the subhuman primates (rhesus macaque and African green monkey) could be cultivated for over 40 generations (Hsu & Cooper, 1974). Maximal growth potential of cloned chick fibroblasts did not exceed 39 generations and only 4% of the clones had this duration of growth, while other cells stopped growth earlier (Beug & Graf, 1977). Mouse embryo fibroblasts have a very short life span: the crisis of these cultures may be evident after five to ten population doublings (Meek, Bowman & Daniel, 1977). Human fibroblasts from various tissues and from persons of different age may have different average durations of life in culture (Hsu & Cooper, 1974). Various cell clones within the same cultures of human fibroblasts may have different life spans (Smith & Hayflick, 1974; Absher, P. M. & Absher, R. G., 1976). The life span also depends considerably on the conditions of cultivation. Of great interest in this connection are the results of Rheinwald & Green (1977). They have found that the culture life time of epidermal cells is increased from 50 to 150 generations by adding to the medium the epidermal growth factor (EGF), a polypeptide mitogen (see chapter 10). Epidermal cells in culture can undergo terminal differentiation (maturation) into cornified cells. EGF increased cell ability to survive subculture and to initiate new colonies. On the basis of these data, Rheinwald & Green (1977) suggested that EGF delays senescence by maintaining the cells in a state further removed from terminal differentiation.

Among the factors reported to increase the life-span of human fibroblasts, hydrocortisone (Grove & Cristofalo, 1976) and intermittent exposure to fluorescent light (Parshad & Sanford, 1977) can be mentioned. The mechanisms of these effects remain obscure.

Development of senescence is accompanied by characteristic alterations of the parameters associated with cell proliferation. Plating efficiency of cells progressively decreases in the course of aging (Couzin, 1978). The fraction of cells entering S phase per unit time also progressively decreases. (Cristofalo & Sharf, 1973). For instance, in culture of human fibroblasts exposed to labelled thymidine for 24 h, the labelling index decreased from 83% after the 25th passage (growth phase) to 37% for the 50th passage (senescence phase). When mitotic cells had been selected from the cultures of both passages and transferred into other flasks duration of the mitotic cycle of these cells (the mean interval till the next mitosis) was similar for growing and senescent cells (Kapp & Klevecz, 1976). These results indicate that a decreased rate of proliferation of senescent cells may be due not to the general increase of duration of the mitotic cycle of all the cells of the culture but to the progressive increase of the fraction of the cells that progress very slowly through G_1 phase (Grove & Cristofalo, 1976) or stop cycling at all (Gelfant & Smith, 1972), while remaining cells continue cycling at the usual rate.

The cells of the non-dividing fraction in senescent cultures are morphologically different from dividing cells in the same cultures: non-dividing cells are larger and occupy a greater area on the substratum and they may have increased activity of lysosomes (see review in Mitsui & Schneider, 1976; Cristofalo, 1977; Couzin, 1978).

The construction of hypothetical schemes of cellular mechanisms limiting life-span has been in progress for many years (see reviews and discussions in Bershadsky & Gelfand, 1970; Holliday, 1975; Cristofalo, 1977; DeMars, 1977; Littlefield, Choy & Epstein, 1977; Sinex, 1977). The main hypotheses may be divided into two large groups.

(1) *Hypotheses postulating the existence of special intracellular mechanisms that causes cell death after a fixed number of generations.* Many concrete schemes of these mechanisms have been proposed. The only known cellular systems with programmed death are series of differentiating tissues such as haematopoietic cells and keratinized cells. These undergo an irreversible maturation process and die after a fixed number of generations. Hypotheses about the similarities between senescence and differentiation have been discussed in the literature (see Cristofalo, 1977). The experiments of Rheinwald & Green (1977) suggesting that EGF increases the life span of epidermal cells by preventing their terminal maturation have been described above. Unfortunately, mechanisms of differentiation are as obscure as those of senescence, so that these hypotheses only postulate similarities between two mysterious phenomena.

(2) *Hypotheses postulating that each round of cell replication is accompanied by a certain number of self-perpetuating errors in synthetic processes.* These errors may be classical mutations or errors in translating mechanisms. It has been suggested that translation errors should be autocatalytic and could be responsible for cell senescence (Orgel, 1963). Possibly, when an accumulated number of mutations or mistranslations per cell reaches a certain critical level, the cell dies. One of the difficulties of these hypotheses is that transformed cells may become immortal although their frequency of mutations and other errors is probably not lower than that of mortal normal cells.

Hypotheses of the first and of the second group are not mutually exclusive but, in fact, may complement each other. Normal cells have the ability to switch themselves from one state to another (e.g. from growing into resting state and/or from undifferentiated into differentiating state) in response to alterations in their environment and, possibly, in response to the alterations of cell metabolism. For instance, various drugs altering different metabolic processes may shift normal cells into the non-growing state (see chapter 10). Also endogenous errors altering cell metabolism (mutations, mistranslations) may increase the probabilities of these shifts. The fraction of cells irreversibly switched into the non-proliferating resting state and/or into the differentiating state may progressively grow in the course of cultivation because the level of errors accumulated by the cells increases. Transformed cells may have a reduced ability for these switches and this may account for their immortality. We will return to this problem again in chapter 14.

Spontaneous transformations
Various cell types

Formation of an immortal cell line is an alternative to cell senescence and culture death (see historical review in Harris, 1974; Ponten, 1976). Formation of a cell line is usually manifested by the appearance of actively growing cells with an abnormal karyotype and often with altered morphology. Formation of a line may become evident during the period of stable growth or during the phase of senescence. Probably, each line is a result of selective proliferation of a clone of altered cells.

Since most experiments with spontaneous transformations have been performed with uncloned mass cultures it has been difficult to determine the time and frequency of appearance of altered cells in culture.

The probability of the formation of a line in the course of continuous cultivation of cells varies greatly depending on the species of animal from which the cultured cells have been taken. This probability is near zero in cultures of human cells: until now, no spontaneously formed line of human fibroblasts or of epithelial cells has been described. The only exception is human lymphoid cells from which a number of lines with unlimited life span have been obtained. This transformation is probably due to the presence in many healthy persons of Epstein–Barr virus potentially oncogenic for lymphoid cells (see review by Ponten, 1976). The chicken is another 'stable' species whose cell cultures are not spontaneously transformed into continuous lines. For instance, Beug & Graf (1977) isolated more than 500 clones of chicken embryo fibroblasts; although some of these clones grew for many generations, they invariably underwent senescence and no established line was obtained.

Mouse cells have the highest probability of spontaneous transformations. In favourable conditions, continuous lines can be obtained from almost every mass culture of mouse embryo fibroblasts (Sanford, 1965; Anderson *et al.*, 1966; Meek *et al.*, 1977). The probability of obtaining a line in experiments with golden hamster fibroblasts and with rat fibroblasts seems to be rather high, although probably lower than with mouse cells. For instance, in the experiments of Kuroki, Miyashita & Yuasa (1975), three cultures of hamster fibroblasts were repeatedly sub-cultivated; two of them eventually stopped proliferation and died while the third one was transformed into a continuous line. Many spontaneously established lines of mouse, hamster and rat cells have been described (see review in Ponten, 1976). Spontaneous formation of transformed cell lines from epithelial rodent cells have also been described; for instance, Colburn *et al.* (1978) obtained them of mouse epidermal cells, Montesano, Saint-Vincent & Tomatis (1973), Montesano *et al.* (1977), Borenfreund *et al.* (1975), Schaeffer & Polifka (1975) from rat liver cells. Possibly there is some reverse correlation between the duration of life-span and frequency of the formation of lines: human cells have a long life-span and low probability of forming lines while the reverse is true for mouse cells. However, the maximal life-span of chicken cells is not longer than that of hamster cells (up to 20–30 generations) while hamster cells seem to have a much higher probability of forming lines.

Alterations of karyotype are a general feature of all spontaneously established lines. Usually, the cells of these lines have heteroploid or pseudodiploid karyotype; numerous chromosomal rearrangements are revealed. Each particular line may retain characteristic peculiarities of karyotype for a considerable number of generations. It is not clear whether different lines obtained from the same source, e.g. different mouse fibroblastic lines, have some common patterns of chromosomal changes.

Minimally transformed lines

Continuous lines may retain a number of normal untransformed characters. One group of these comprises the 3T3 lines (Todaro & Green, 1963; Aaronson & Todaro, 1968; Todaro, Scher & Smith, 1971). The designation '3T3' refers to the sub-cultivation regimen: transfer at 3-day intervals with a relatively low inoculum of 3×10^5 cells per 60 mm dish. It has been suggested that frequent transfers and a low inoculum minimize cell–cell contact and prevent the selection of transformed lines with a lowered density dependence of growth.

In fact, the cells of these lines usually stop growing at a much lower population density than primary normal cells (see chapter 9). The 3T3 lines established from random-bred Swiss mouse embryos and from inbred BALB/c mouse embryos (Todaro & Green, 1963; Aaronson & Todaro, 1968) are among the most frequently used culture lines. Lines similar to 3T3 have been also obtained from rat embryos (Kimura, Itagaki & Summers, 1975) and from hamster embryos (Kuroki *et al.*, 1975). Like all other lines, 3T3 cells have aneupoloid karyotypes. They are well spread on the substratum, do not grow in low serum media and give a very low proportion of colonies in semi-solid media. These cells do not give colonies when implanted to syngenic animals by the usual methods. However, when certain special methods of transplantation are used 3T3 cells may give rise to tumours; these tumours are probably not formed from the 'usual' 3T3 cells but from their strongly transformed variants (see later).

The histogenesis of 3T3 cells is not clear. They are regarded either as fibroblast-type cells or, on the basis of their morphology, as the cells of endothelial origin (Porter, Todaro & Fonte, 1973*b*). Certain sub-lines of 3T3 may give rise to cells similar to the elements of adipose tissue (Green & Kehinde, 1975, 1976; Kuri-Harcuch, Wise & Green, 1976).

As we discussed above the inter-relationships between fibroblasts and other cellular types *in vivo* are not clear; it cannot be excluded that endothelial cells, adipose cells and other elements may arise from common precursors. If 3T3 lines were derived from these hypothetical precursor cells, they could acquire a morphology resembling that of various cell types. Needless to say, at present all these explanations remain speculative.

Besides 3T3 lines there are many other minimally transformed lines. The BHK 2/line described by Stoker & Macpherson (1964) has been widely used. This is a line of fibroblast-like cells obtained from baby hamster kidney. Several other minimally transformed lines deserve mention: the 1OT 1/2 C3H fibroblastic line derived from the embryo of C3H strain mice (Reznikoff, Brankow & Heidelberger 1973*a*; Reznikoff *et al.*, 1973*b*); the NRK line derived from normal rat kidney (Duc-Nguyen, Rosenblum & Zeigel, 1966) containing well-spread fibroblastic and epithelioid cells, and a series of cloned lines of cells from LEW/CB rat embryos (Vesely *et al.*, 1978).

Strongly transformed lines

Spontaneous transformations of fibroblastic and epithelial cells, especially in rodent cultures, may also lead to the formation of strongly transformed lines that are obviously oncogenic and that possess most of the transformed characters listed earlier. There are two main variants of these lines.

(1) *Strongly transformed lines can be obtained directly from primary cultures without obvious intermediate stages.* Certain regimens of sub-cultivation seem to select for cells with strongly transformed characters. Todaro & Green (1963) and Aaronson & Todaro (1968) transferred mouse embryo cultures repeatedly at various intervals and by various inocula. When the cells were transferred by small inocula, minimally transformed 3T3 lines were obtained while transfer by larger inocula led to formation of strongly transformed lines, such as the 3T12 line obtained by transfer at 3-day intervals using inocula containing 12×10^5 cells per 60 mm dish. In this case, the conditions were probably favourable for the selection of cells that could grow at high population densities.

(2) *Strongly transformed lines may be formed as a result of an additional spontaneous transformation of a minimally transformed line.* This additional transformation may take place at various times after the establishment of the line. It may be observed not only in mass cultures but also in cloned cultures. When one clone obtained from a certain cell line of mouse fibroblasts was divided into several subclones, additional transformations occurred at different times in the various subclones (Sanford, Likely & Earle, 1954; Sanford, 1958). These results indicate that strongly transformed variants appear *de novo* in the course of cultivation of minimally transformed lines.

Mechanisms of spontaneous transformations

We do not know the genetic mechanisms that are responsible for either the appearance of the immortal cells that initiate cell lines or for subsequent additional transformations. Efforts to reveal any role of endogenous viruses in spontaneous transformation have not given positive results (Avery & Levy, 1978). In the experiments recently described by Spandidos and Siminovitch (1978a) primary senescent cultures of the Chinese hamster lung were incubated with purified metaphase chromosomes isolated from spontaneously transformed lines of the ovary (CHO line). After this treatment, actively growing colonies having the transformed morphology were formed in cultures with a frequency of about one in 10^5 of the incubated cells. Although CHO cells are tumourigenic and formed colonies in agar, these two characters were not transferred to normal primary cells. In other experiments Spandidos & Siminovitch (1977) used not primary cells but several permanent minimally transformed lines (BHK and several Chinese hamster lines, including one obtained by chromosome transfer in the experiments quoted above. The cells were incubated with isolated CHO chromosomes which induced formation of colonies consisting of cells that were able to grow in suspension and were tumourigenic. Thus, chromosomes from strongly transformed cells were reported to induce a transition from normal cells to minimally transformed cells as well as a transition from minimally transformed to strongly transformed cells. This chromosome material was not, however, able to induce a one-step transition from normal to strongly transformed cells. At present all these data still await confirmation.

Action of chemical carcinogens on cultured cells
Metabolism of carcinogens

Active carcinogens belong to several different groups of chemical compounds such as polycyclic hydrocarbons, nitrosamines, and aromatic amines.

Chemical carcinogens produce several types of effects in cell cultures: toxic effects, cell transformations and mutagenic effects. To produce these effects most carcinogens require metabolic activation. Initial carcinogens are metabolized in the cell to ultimate carcinogens; that is, to derivatives that are responsible for biological effects. This metabolism is catalysed by a series of closely associated cellular enzymes.

A common property of the molecules of various ultimate carcinogens is their electrophilic character. These molecules contain relatively electron-deficient atoms that easily react with nucleophilic sites in other molecules. Nucleophilic sites are abundant in DNA, RNA and proteins and ultimate carcinogens of different types are bound to these macromolecules in the cell (see review in Weisburger & Williams, 1975; Miller, 1978). It is currently believed that most biological effects of carcinogens are due to reactions of their ultimate derivatives with these macromolecules.

The first steps of metabolism of most carcinogens are catalysed by microsomal mono-oxygenases. These multicomponent enzyme complexes are located in the endoplasmic reticulum. One of their characteristic components is a haemoprotein, cytochrome *P*-450; there are probably many species of cytochrome *P*-450 differing in substrate specificity. Mono-oxygenases are responsible for the oxidation of a wide variety of endogenous and exogenous lipophilic compounds including many carcinogens, drugs and steroids. Action of these enzymes leads to the introduction of polar groups (such as hydroxyl) into the parent molecules. At the next stage these oxidized products are further metabolized by other enzymes which catalyse their hydration or conjugation to other compounds. The result of this metabolism is the transformation of lipophilic compounds into polar, water-soluble molecules that can be removed from the cells and from the body. However, as mentioned above, certain metabolites formed at intermediate steps have high chemical reactivity and may become covalently bound to macromolecules. Several exceptional carcinogens are biologically active without previous enzymatic activation. For instance, ethylnitrosourea may spontaneously decompose giving a water-soluble ultimate carcinogen an ethyldiazonium ion.

Cultured normal primary cells retain the ability to metabolize different carcinogens. Metabolism of one particular group of carcinogens, polycyclic aromatic hydrocarbons (PAH), has been studied in detail in cultured cells of different types (Gelboin, Wiebel & Kinoshita, 1974; Gelboin *et al.*, 1976; Sims, 1976; Brookes, 1977; De Pierre & Ernster, 1978). Monooxigenases transform PAH into epoxides and another enzyme (or enzymes), hydratase, converts epoxides into dihydrodiols.

As a result of the many enzymatic and non-enzymatic transformations an array of intermediate metabolites is formed but only one or a few of them act as ultimate carcinogens binding to macromolecules. Owing to the high reactivity of these active compounds their identification has proved to be a difficult task. Recent studies implicate 7β, 8α-dihydroxy-9α, 10α-epoxy- 7, 8, 9, 10-tetrahydrobenzo(a)pyrene (Huberman *et al.*, 1976; Weinstein *et al.*, 1976; Levin *et al.*, 1977).

As discussed later, certain neoplastic cells lose the ability to metabolize PAH; accordingly, PAH do not produce any biological effects in these cells. However, when non-metabolizing cells are co-cultivated with normal metabolizing cells, carcinogens produce biological effects (toxic or mutagenic action) in the cells of both types. Cell–cell contact is essential for this transmission (Mittelman, Sharovskaja & Vasiliev, 1972; Huberman & Sachs, 1974; Newbold *et al.*, 1977). Probably, active metabolites of carcinogens formed in the metabolizing cells migrate through cell–cell contacts into non-metabolizing cells and are bound there by macromolecules. There is probably a certain cell specificity in the metabolic activation of carcinogens by different cell types. For instance, rat hepatocytes metabolized in culture both benzo(a)pyrene and aflotoxin B_1, a liver specific carcinogen, while rat fibroblasts metabolized only benzo(a)pyrene but not aflatoxin B_1 (Langenbach *et al.*, 1978).

Toxic and mutagenic effects of carcinogens

Manifestations of these effects in cultures are manifold. Typical toxic effects of PAH in sensitive cultures of rodent fibroblasts include progressive inhibition of the rate of DNA synthesis and decreased mitotic activity. Analysis of the effects of PAH on the proliferation of mouse and rat fibroblasts have shown that these substances do not block cells in a particular phase of mitotic cycle but slow down their passage through all the phases. While carcinogens inhibit replicative synthesis of DNA, they may also induce repair synthesis. This synthesis is probably a response to DNA damage by active metabolites of carcinogens (see review by Stich *et al.*, 1976).

Cultured fibroblasts damaged by carcinogens are unable to grow but they nevertheless remain attached to the substratum for many days; these cells retain the ability to move and to phagocytose (Andrianov *et al.*, 1971). The carcinogen-treated cells acquire certain features similar to transformed cells: they have decreased ability to spread on the substratum and increased ability for agglutination with plant lectins (Vasiliev, 1976). The biological significance of these 'quasi-transformations' of non-proliferating cells damaged by carcinogens is not clear. Human fibroblasts are much less sensitive to the toxic effects of PAH than rodent fibroblasts. The intensity of PAH metabolism in cultures of human fibroblasts is also lower than in those of rodent cells (Belitzky *et al.*, 1970).

Mutagenic effects of carcinogens have been observed in experiments with different types of normal rodent cells and in permanent lines. Mutations of different genetic loci have been observed, including resistance to ouabain, and 8-azoguanine amongst others (Huberman & Sachs, 1976; Newbold *et al.*, 1977). Recently, induction of mutations in human fibroblasts by benzo(a)-pyrene and urethane have been reported (Spandidos & Siminovitch, 1978*b*). Carcinogens may also induce microscopically visible chromosome damage in different cells (see review in Pogosianz, 1973; Brookes, 1977).

Transforming action of carcinogens

Chemical carcinogens induce transformation both in primary cultures and in cultures of permanent lines (see reviews by Heidelberger, 1973; Heidelberger & Boshell, 1975; Kuroki, 1975). The transforming action of PAH on primary cultures was first observed by Berwald & Sachs (1963) in experiments with hamster embryo fibroblasts. Since then, hamster fibroblasts have been widely used in transformation experiments. Di Paolo and collaborators (Di Paolo, Donovan & Nelson, 1969; Di Paolo, Nelson & Donovan, 1972*a;* Casto, Janosko & Di Paolo, 1977) and other investigators (see review in Kuroki, 1975; Dunkel, Wolff & Pienta, 1977) have shown that many different chemical types of carcinogens induce transformation in this system; the transforming properties of various chemicals were well correlated with carcinogenicity *in vivo* (Purchase *et al.*, 1976). Transformation of hamster cultures may also be produced by ionizing radiation (Borek, Hall & Rossi, 1978). Usually, transformed colonies were diagnosed by their abnormal morphology; at least some of these morphologically transformed clones were shown to form colonies in agar and to produce tumours after implantation in hamsters. There was usually a considerable time interval (up to several weeks) between the incubation of cells with carcinogen and the appearance of morphologically transformed colonies. Development of anchorage independence may be observed later than morphological transformation (Barrett & Ts'o, 1978*b*). The rate of morphological transformation was not higher than a few per cent;

often it was less than 1% of clone-forming cells. For instance, Casto *et al.* (1977) obtained about 4–5 transformed foci per 10^5 hamster cells surviving the toxic effects of chemicals. Only the colonies with obvious morphological alterations were diagnosed as transformed ones; therefore it was possible that the appearance of a certain number of other less strongly altered cells passed unnoticed. Transformations by various chemicals were also obtained in rats (Sekely *et al.*, 1973; Rhim & Huebner, 1973) and in Chinese hamster cells (Kirkland, 1976). Strongly transformed lines of embryo mouse fibroblasts treated with methylcholanthrene were obtained in the classical experiments of Earle (1943). The most famous of these lines is that of *L* fibroblasts, widely used in many laboratories. Many sub-lines and clones of *L* fibroblasts have been obtained; they differ from each other in oncogenicity and other properties. Because of the high level of spontaneous transformations in mouse primary cultures, these cells are rarely used for the evaluation of transforming effects of carcinogens.

Primary cultures of human fibroblasts seem to be much more resistant to the transforming action of carcinogens than rodent cultures. For instance, Spandidos & Siminovitch (1978*b*) were unable to induce morphological transformation in cultures of human lung cells using benzo(a)pyrene or urethane as the carcinogens although both these substances induced various mutations (resistance to 6-thioguanine, resistance to ouabain, ability to use fructose) in the same cells. Freedman & Shin (1977) treated primary cultures of human fibroblasts with *N*-methyl-*N'*-nitro-*N*-nitrosoguanidine; this carcinogen does not require metabolic activation. Carcinogen-treated cultures gave colonies in methylcellulose with a frequency of one in 10^5; in control, non-treated cultures the frequency was lower than one in 10^8. Cloned cultures of the treated cells isolated from methylcellulose colonies during subsequent passages formed colonies in methylcellulose with the same frequency of one in 10^5. These cultures did not give tumours in nude mice and became senescent at the same time as control cultures. Thus, in these experiments the carcinogen probably induced the formation of peculiar, minimally transformed cultures that apparently had only one abnormal character: a somewhat increased frequency of colonies in a semi-solid medium. Using a special experimental protocol (treatment of synchronized cultures with carcinogen during the S phase of the cell cycle) Milo & Di Paolo (1978) recently reported the successful transformation of human skin fibroblasts by six different carcinogens including aflotoxin B_1 and 4-nitroquinoline oxide.

Additional transformations of rodent cells by various chemical carcinogens have been observed in experiments with various permanent lines: mouse 3T3-type (Di Paolo *et al.*, 1972*b*; Mareel *et al.*, 1975; Kakunaga, 1975), BHK cells (Fradkin *et al.*, 1975; Altanerova, 1975; Ishii *et al.*, 1977), rat cell line H43 (Kouri *et al.*, 1975), various lines of rat hepatic cells (see review in Williams, 1976; Bridges, 1976), M2 clone of fibroblasts from the mouse prostate (Marquardt *et al.*, 1974; Marquardt, Grover & Sims, 1976; Marquardt *et al.*, 1977). One particular line widely used in experiments of this type is the mouse fibroblastic line C3H 1OT 1/2 which apparently has an almost normal morphology (Reznikoff *et al.*, 1973a, b). It is not tumourigenic when implanted by the usual methods and gives a low percentage of spontaneous additional transformations. The cells of this line were reported to be transformed by ultraviolet irradiation (Mondal & Heidelberger, 1976), X-rays (Terzaghi & Little, 1976; Bronty-Boyé & Little, 1977) and by many chemical carcinogens including polycyclic hydrocarbons (Reznikoff *et al.*, 1973a, b), cigarette smoke condensate (Benedict *et al.*, 1975), and halogenated pyrimidine nucleosides (Jones *et al.*, 1976a). Transformed colonies of the C3H 1OT 1/2 line are characterized by a strongly abnormal morphology; many of the morphologically transformed colonies are also tumourigenic and able to grow in semi-solid media. The frequency of transformed colonies in carcinogen-treated cultures rarely exceeded 1–2%.

Also, a human permanent line derived from cultured osteosarcoma cells, in contrast to primary human cells, was found to be susceptible to transformation by 7,12-dimethyl-benzo(a)anthracene (Rhim *et al.*, 1975). Before carcinogen treatment this line formed colonies in agar and was aneuploid. After treatment it became oncogenic for nude mice and altered its morphology; piled-up foci appeared in the culture. One might say that in this case a strongly transformed line gave rise to a very strongly transformed one.

In-vivo chemical carcinogenesis is probably a multistage process. There is a long latent period between the initial application of the carcinogen and the appearance of tumours; the action of certain carcinogens, so-called initiating agents, may be markedly enhanced by subsequent application of so-called promoting agents. Promoting agents *per se* have little, if any, carcinogenic activity (Friedewald & Rous, 1944; Berenblum, 1941, 1974, 1978; Berenblum & Shubik, 1949). In experiments with epidermal carcinogenesis in mice, a single application of polycyclic hydrocarbons is often used as an initiating treatment, while promotion is achieved by repeated applications of croton oil or of its active constituent, 12-O-tetradecanoyl-phorbol-13-acetate (TPA). TPA was found to accelerate transformation initiated by polycylic hydrocarbons in rat cells (Lasne, Gentil & Chouroulinkov, 1974, 1977) and to enhance the transforming effect of polycyclic hydrocarbons in C3H 1OT 1/2 cells (Mondal, Brankow & Heidelberger, 1976). Analysis of the processes involved in initiation and promotion of transformations in culture seems to be an important field for future studies.

Another special approach to the analysis of the effects of chemical carcinogens is the so-called in-vivo–in-vitro carcinogenesis. In experiments of this type an animal is treated with carcinogen and then its cells are cultivated *in vitro*. In these conditions the initial stages of the effect of the carcinogen (metabolism, binding to macromolecules) take place in normal tissue and not in the special environment of cell culture. It is suggested that in these conditions it is more easy to obtain and observe transformation of cells that may be putative progenitors of carcinogen-induced tumours *in vivo*. An example of this approach is provided by the experiments of Laerum & Rajewsky (1975) and Laerum *et al.* (1977). They studied the action of ethylnitrosourea on the brain of rat embryos. If this carcinogen was given to pregnant females, their progeny developed brain tumours after approximately 200 days. Brain cell cultures obtained from control embryos contained glia-like cells that died after about 40 days. In contrast, the cultures isolated from the brains of carcinogen-treated embryos contained glial cells that could be grown continuously; after 200 days of growth these cultures already contained cells that were able to form colonies in agar and tumours when transplanted in rats. Hard & Borland (1975, 1977) observed an accumulation of morphologically altered mesenchymal and epithelial cells in cultures of rat kidney explanted after treatment with dimethyl-nitrosamine. Control cultures senesced after four passages while treated cultures gave rise to morphologically transformed continuous lines.

At present, many investigators favour the possibility that genetic changes responsible for chemical transformation are identical to mutations. Experimental facts which are regarded as supporting this point of view include: (*a*) ultimate carcinogens are bound to DNA (see above), (*b*) many carcinogens induce not only transformations but different mutations in cultured cells (see above). However, the frequencies of carcinogen-induced mutations of any single locus were found to be significantly lower than those of morphological transformations in the same cultures (e.g. Huberman, Mager & Sachs, 1976; Barrett & Ts'o, 1978*a*). For instance, in the experiments of Huberman *et al.* (1976) benzo(a)pyrene induced mutations (ouabain resistance) in primary hamster cell cultures with a frequency of about 200 per 10^6 colony-forming cells and morphological transformation with a frequency about 4500 per 10^6 cells. Bouck & Di Mayorca (1976) found in BHK cells that carcinogens induced three variants of transformation to anchorage independence: (*a*) formation of cells growing in suspension at 32 °C and at 38 °C, (*b*) formation of cells growing in suspension at 32 °C but not at 38 °C (heat-sensitive clones), (*c*) formation of cells growing in suspension at 38 °C but not at 32 °C (cold-sensitive clones). The total frequency of transformed clones was 10^{-4} per survivor cell and about half the clones were either cold-sensitive or heat-sensitive. Induction of variants with different temperature sensitivities of the same character is easily explained by the suggestion that carcinogen may induce different mutations of one gene or several genes controlling this character. Temperature-sensitive variants of rat liver epithelial cells transformed by *N*-acetoxyacetylaminofluorene were also obtained by Yamaguchi & Weinstein (1975). It is easy to see that all the evidence in favour of the mutational nature of carcinogen-induced transformations is at present indirect.

At its best, this evidence shows that DNA changes are involved in these transformations. We do not know whether these changes involve one or many genes, or what the inter-relationships are between classical mutations, alterations of endogenous viral genomes, possible 'epigenetic' changes of DNA, etc.

The experiments summarized above suggest that each particular transformation induced by a chemical carcinogen is one stage in a complex multi-step process. For instance, morphological chemical transformation of a primary culture or of cells *in vivo* may be followed by subsequent (spontaneous?) changes during repeated passages in culture. Spontaneous transformations leading to the establishment of permanent lines may be followed by additional transformations by carcinogens. There are probably many different combinations of the various transformation steps.

Resistance of transformed cells to the toxic effects of carcinogens

Haddow (1938) first suggested that the cells of neoplasms induced by carcinogenic hydrocarbons may be more resistant to the toxic effects of these compounds than normal parent cells. Many years later this suggestion was confirmed experimentally: it was shown that cultured cells of rat sarcomas induced by 7,12-dimethyl-benzo(a)-anthracene (DMBA) were much more resistant to its toxic effects than normal rat fibroblasts (Starikova & Vasiliev, 1962). In subsequent years it has been found that many strongly transformed rodent fibroblastic cell lines are more resistant to the toxic effects of carcinogenic hydrocarbons than parent normal fibroblasts (Alfred *et al.*, 1964; Diamond, Defendi & Brookes, 1967, 1968; Irlin *et al.*, 1968; review in Andrianov *et al.*, 1971). Cultures of mouse hepatomas (Vasiliev & Guelstein, 1963) and those of premalignant hyperplastic liver nodules (Laishes, Roberts & Farber, 1978) have proved to be more resistant to the toxic effects of hepato-carcinogens than normal cultures of hepatocytes. Alterations of resistance observed in the course of neoplastic transformation may be very large: 25 ng/ml of DMBA completely stopped proliferation of normal rat fibroblasts while concentrations 100-fold higher, suspended in the medium, had no effect on the growth of rat sarcomatous cells (Starikova & Vasiliev, 1962). Resistance to carcinogenic hydrocarbons is developed not only in the course of transformations induced by chemical carcinogens but also in the course of spontaneous transformations and those induced by viruses. For instance, the cells of sarcomas induced in rats by plastic films were found to be as resistant to the toxic effects of DMBA as the cells of sarcomas induced by the hydrocarbon itself (Starikova & Vasiliev, 1962). Resistance to carcinogenic hydrocarbons may vary in different cell lines having other transformed characters: certain hamster lines were found to retain considerable sensitivity to PAH, although this sensitivity was still lower than that of primary fibroblasts (Zavadina & Khesina, 1971). Additional transformation of one carcinogen-sensitive hamster line by SV-40 virus led to a considerable additional increase of cell resistance to the toxic action of DMBA (Irlin & Vasiliev, 1968).

The resistance of transformed cells to chemical carcinogens is due to a deficiency of the enzymatic systems which metabolize these carcinogens and convert them into active compounds; the permeability of the cells to the carcinogens is not changed (Andrianov *et al.*, 1967; Diamond, Defendi & Brookes, 1967; Diamond, 1971).

The cells of a number of transformed lines lose completely the ability to metabolize polycyclic hydrocarbons. For instance, a million normal mouse or hamster fibroblasts may metabolize 1 mg benzo(a)pyrene added into the flask during 2–3 days. In contrast, a similar number of strongly transformed cells, e.g. L or BHK cells, will not metabolize benzo(a)pyrene, so that all of the initially added hydrocarbon can be extracted from the culture after several days of incubation (Andrianov *et al.*, 1967). Similarly, when labelled carcinogenic hydrocarbons are added to the medium of these resistant cultures, the label is not bound to DNA or proteins (Andrianov *et al.*, 1967). Enzymes metabolizing polycyclic hydrocarbons are not found in the extracts of resistant cells or their activity is considerably diminished (Gelboin & Wiebel, 1971). The lines that retain their sensitivity to carcinogenic hydrocarbons usually also retain the ability to metabolize these hydrocarbons (Zavadina & Khesina, 1971).

Can the transforming effects of carcinogens just described be due to the selection of pre-formed cell variants resistant to the toxic effects of these chemicals? A number of facts contradict this suggestion. For instance, in many experiments transformations in culture are obtained by using the carcinogens at low concentrations, which do not have considerable toxic effects. Carcinogens often increase considerably the relative number of transformed clones per total number of seeded cells; this effect cannot be due to selection of pre-existing variants. Apparently, carcinogens induce the formation of transformed cell variants. Of course, in certain experimental conditions these compounds *in vivo* and *in vitro* may also have an additional selecting effect.

The deficiency of carcinogen-metabolizing systems associated with an increased cell resistance is not a specific property of cells transformed by carcinogens. This property also develops spontaneously or arises regularly in the course of neoplastic evolution induced by viruses. Thus, it can be regarded as one of the general symptoms of neoplastic cell transformation. We do not yet know all the normal functions of microsomal oxidative systems participating in the metabolism of carcinogens; therefore, it is difficult to say what alterations of cell behaviour may result from a deficiency of these systems. These enzymatic systems are responsible for the metabolic transformation not only of carcinogens but also of many other foreign xenobiotic substances and possibly of certain cell-made compounds such as steroids. Therefore alterations of these systems may be associated in some way with abnormalities of cell–environment interactions characteristic for transformed cells.

Transforming action of SV-40 and polyoma viruses
Oncogenic papova viruses

DNA-containing viruses shown to have cell transforming properties in fibroblastic and epithelial cultures belong to three different groups: papova viruses, adenoviruses and herpes viruses. Most studies concerned with the properties of these transformed cultures have been performed with two viruses belonging to the papova group: SV-40 and polyoma virus. Therefore, we will discuss here only the transformations induced by these two viruses.

Polyoma and SV-40 viruses have many similar properties. Their genomes contain a region of DNA coding for early proteins and another region coding for the late proteins (see reviews in Dulbecco, 1973, 1976; Rapp & Westmoreland, 1976; Reddy *et al.*, 1978). Early proteins, in contrast to the late ones, are not incorporated into virions. Infection of a cell by SV-40 and polyoma viruses may follow one of two alternative courses: permissive or non-permissive. In the course of permissive infection, expression of the 'early' region is followed by the replication of viral DNA, expression of the 'late' genes and formation of infectious virions accompanied by cell degeneration. In the course of non-permissive infection the 'early' genes may be expressed but the products of the 'late' genes and infectious virions are not formed and the cell may carry the viral genome in an integrated state for many generations. Only non-permissive infection can lead to transformation.

The type of infection depends on the animal species from which the cells are derived. Polyoma virus gives permissive infection in most cells of mouse cultures; only small fraction of cells in these cultures is infected non-permissively and may become transformed. SV-40 virus gives typical permissive infection in cultures of green monkey cells. Both SV-40 and polyoma viruses give non-permissive infection in cultures of golden hamster cells. The SV-40 virus infection of human cultures is permissive in a certain fraction of cells but remains non-permissive in many other cells of the same culture.

Viral transformations

SV-40 and polyoma viruses may transform non-permissively infected cells of primary cultures or of permanent lines. The efficiency of transformation is usually low: even at a very high multiplicity of infection, the fraction of stably transformed cells is usually not more than a few per cent; the proportion of morphological transformations in primary cultures is often lower than 1%.

These figures are based on the counts of colonies with considerable morphological changes. Possibly, many colonies with minimal morphological alterations are also present in infected cultures but are not regarded as transformed in these tests, so that total number of morphologically transformed colonies may be higher than that usually given by investigators. The same may be true with regard to the frequency of transformation of growth properties (see below).

The time interval between infection and the formation of cell colonies with stably transformed characters is usually not less than several weeks. During this interval, virus infection may induce unscheduled synthesis of cellular DNA in the infected cells. For instance, in the experiments of Lehman & Defendi (1970), 20–30% of Chinese hamster cells infected with SV-40 virus underwent DNA synthesis without subsequent mitosis, so that at 48 hours after infection the fraction of polyploid cells was considerably increased. Polyploidy increased in subsequent cycles and many other chromosomal aberrations were also observed (Lehman, 1974; Zuna & Lehman, 1977) Non-permissive polyoma infection may also lead to chromosomal alterations but they are less pronounced and less frequent than in SV-40 virus infected cells.

Stoker (1968) described the abortive transformation of BHK cells by polyoma virus. Soon after infection many cells temporarily acquired morphological changes and anchorage independence but in the course of subsequent subcultivation most cells lost these properties and only a few remained stably transformed. The mechanism of this interesting phenomenon is not yet clear.

When stably transformed clones are formed in virus-infected cultures, their properties vary considerably from colony to colony. For instance, Risser & Pollack (1974) analysed the growth properties of 40 randomly selected colonies of 3T3 mouse cells infected with SV-40 virus. It was found that these clones could be classed into four categories: (*a*) cells indistinguishable from 3T3 cells (five clones), (*b*) cells with a low serum requirement and somewhat higher saturation densities than 3T3 cells (minimal transformants, 15 clones); however, like 3T3 cells, these clones were unable to form colonies in methylcellulose and on the monolayer of 3T3 cells, (*c*) cells with a low serum requirement, the ability to grow on the normal monolayer and in methylcellulose (standard transformants, 10 clones), (*d*) cells with growth properties intermediate between 3T3 cells and standard transformants (10 clones). Thus, expression of transformation in different clones obtained from the same culture showed a wide spectrum of variation. Clones with different degrees of transformed properties after SV-40 virus infection have also been described by Todaro *et al.* (1971). Vogt & Dulbecco (1963) described two morphological variants of hamster cell colonies transformed by polyoma virus: thin colonies consisting of relatively well-spread fibroblasts and thick colonies consisting of multi-layered cells.

The morphology of transformed colonies depends also on the nature of the infecting virus: in cultures of golden hamster cells, polyoma virus often induces formation of colonies consisting of elongated fibroblastic cells, while SV-40 virus often induces formation of colonies consisting of polygonal 'epithelioid' cells without significant elongation (Defendi, 1966).

Virus-transformed cells may undergo additional transformations in the course of subsequent cultivation. For instance, thin morphological colonies were reported to give rise to thick colonies (Vogt & Dulbecco, 1963). Transformed primary human cells with a limited growth potential may enter crisis in subsequent passages; only some of the cells recover from this crisis and give rise to a transformed permanent line with unlimited growth potential (Koprowski *et al.*, 1962; Shein *et al.*, 1964; Defendi, 1966; Fogh, 1971; Oshima *et al.*, 1977).

The genetic basis of virus-induced transformations

Virus-transformed cells always contain integrated copies of the viral genome and most transformed lines have multiple copies of it. Recently, an established rat embryo line containing only one copy of the SV-40 virus genome per cell was described (Botchan, Topp & Sambrook, 1976).

Viral genomes in transformed cells may be integrated into specific chromosomes. By hydridization experiments, functioning viral genomes in several lines of cells transformed with SV-40 virus were found to be associated with chromosomes 7, 8 or 17 (Croce & Koprowski, 1974; Croce, 1977; Kucherlapati *et al.*, 1978).

Only the 'early' region (A gene) of the viral genome is active in transformed cells. This region codes for several proteins one of which is the T antigen, a protein of 90–100 kilodaltons which is present in the nuclei of transformed cells. A 15–20 kilodalton protein, (t antigen) is also found in SV-40 virus transformed cells. T antigen is immunologically related to t antigen and has a number of peptide sequences in common (Paucha *et al.*, 1978). It has been suggested that certain parts of the t- and T-proteins are coded by identical sequences of the A region of the SV-40 virus genome while other parts of the molecules are coded by two, non-identical sequences of this region (Volckaert, van der Voorde & Fiers, 1978; Crawford *et al.*, 1978). The polyoma virus 'early' region also codes for several proteins: the large T protein and for proteins of lower molecular weight (Schaffhausen, Silver & Benjamin, 1978).

Virus-transformed cells express a tumour-specific transplantation antigen (TSTA) at their surfaces. The exact relationship between this antigen and the 'T' proteins is not yet clear.

Numerous temperature-sensitive mutants of the 'early' regions of polyoma and SV-40 viruses have been obtained and many cell lines transformed with these mutant viruses have been described. The expressions of many transformed characters have proved to be temperature sensitive in certain lines of this type.

For instance, mouse neuroglial cells transformed by the ts-A mutant of SV-40 virus restored serum dependence, anchorage dependence and density dependence after transfer to a non-permissive temperature (Anderson & Martın, 1976). However, in other lines transformed with ts-A mutants of SV-40 virus and polyoma viruses not all the transformed characters disappeared at non-permissive temperatures (Di Mayorka *et al.*, 1969; Dulbecco & Eckhart, 1970; Kimura & Itagaki, 1975).

Several cell clones obtained from one culture and transformed with the same ts-A mutant virus may have different temperature sensitivities of their transformed characters. For instance, in the experiments of Tenen *et al.* (1977) three clones of Chinese hamster lung cells transformed with the ts-A mutant of SV-40 virus were obtained. Two transformed characters were assessed: loss of density dependence (ability to grow on the surface of a normal cell layer), and loss of anchorage dependence (ability to grow in agar). In two lines both these characters were temperature sensitive; that is, were expressed at permissive (33 °C) but not at non-permissive (39 °C) temperature. However, in the third line only anchorage independence was temperature sensitive, while the loss of density dependence was expressed at both temperatures. Seif & Cuzin (1977) infected a permanent line of the 3T3 type with a ts-A mutant of polyoma virus. Subsequent cultivation of infected cells on plastic led to the development of several lines that expressed several transformed characters (ability to form colonies in agar, high saturation density, ability to grow in low serum) only at permissive temperature but not at the non-permissive one. Parallel cultivation of similarly infected cells in agar led to selection of several lines that retained all these transformed characters even at the non-permissive temperature.

It is not clear which factors determine this variability of temperature sensitivities in various lines transformed by the same ts-A virus mutant. Additional cellular changes arising after primary transformation and changing the expression of the viral genome may be important in this respect. Virus-transformed cells may be prone to these changes, as SV-40 virus infection was found to cause chromosomal alterations (see above) and gene mutations (Marshak, Varshaver & Shapiro, 1975). The method of isolation of the colonies transformed by SV-40 virus was found to affect its karyotype (Clark & Pateman, 1978).

Recent discoveries of several protein products coded by various parts of the 'early' region of SV-40 and polyoma viruses (see above) have prompted investigations of the relative roles of each of these products in transformation. Bouck *et al.* (1978) have studied the transforming ability of the viable mutants of SV-40 virus with deletions in different regions of the viral genome. Only mutants with deletions located in the proximal part of the 'early' region (0.59–0.54 map co-ordinates) were defective in the ability to produce anchorage independent transformants. The regions of genome deleted in these mutants probably code for t antigen, but not for T antigen. Schlegel & Benjamin (1978) studied the effect of an hr-t mutant of polyoma virus on rat fibroblasts. This mutant had a defect in the proximal portion of the 'early' region: it coded for apparently normal large T protein but not for early protein species of lower molecular weight. It was found that hr-t mutants were defective in the ability to cause abortive morphological transformation but partially retained the ability to induce DNA synthesis in infected cultures. Thus, these preliminary results suggest that 'smaller' species of 'T' proteins are of special importance for transformation. Further studies are needed for to show whether T antigen is also involved in the control of the transformed phenotype.

In summary, cell transformation after infection with papova virus is a relatively rare event that probably requires some special variety of cell–virus interaction; e.g. integration of the viral genome into certain regions of cellular DNA. Transformation may lead to the formation of many different cell variants and this variability may be increased by additional cellular changes. Maintenance of the transformed state seems to depend on the activity of certain parts of the viral genome; the precise nature of one or several virus-coded proteins essential for the maintenance of the transformed phenotype still requires elucidation.

The transforming action of RNA-containing viruses

All RNA-containing oncogenic viruses belong to the group of oncorna or retroviruses. Their virions contain a characteristic enzyme, reverse transcriptase. When a virus particle penetrates the cell, this enzyme catalises the synthesis of virus-specific DNA provirus which may become integrated into the cell genome (see reviews in Temin, 1976, 1977a, b; Baltimore, 1976; Todaro, 1975, 1978). Typical oncorna viruses having strong transforming properties in fibroblastic cultures are Rous sarcoma viruses (RSV) and mouse sarcoma viruses (MSV, Moloney and Kirsten variants). These viruses are the most efficient transforming agents known at present. For instance, infection of chick fibroblasts by RSV may lead to morphological transformation of all cultured cells 2–3 days after infection; transformed cells may also acquire the ability to grow in agar and in low-serum media and may lose density-dependent inhibition. Under optimal conditions, the probability of any one virus particle causing transformation is 0.1–0.5 for RSV compared with 10^{-5} for DNA-containing viruses (Temin, 1977a). Probably, various clones from one culture transformed by RSV will have somewhat different degrees of expression of transformed characters, but systematic comparative studies of this type have not yet been made. The character of morphological alterations of chick cells depends on the nature of the transforming virus: certain genetic variants of RSV induce the formation of transformed foci consisting mainly of rounded, almost spherical cells while other virus strains may induce foci with elongated fusiform cells (Temin, 1960, 1961; Diglio & Dougherty, 1977). Transformation of chick cells with RSV does not confer immortality on these cells. After a number of passages they enter the crisis stage and die, like non-infected cells. However, certain RSV-transformed chick cells may undergo additional transformation and give rise to permanent lines (Dinowitz, 1977). RSV transforms not only fibroblasts but also a number of other cell types, e.g. neuroretinal cells, myoblasts, pigmented epithelium of the eye. Certain strains of RSV may transform not only chicken cells but also the cells of duck, quail and of other birds as well as those of primary or permanent cultures of the cells of different mammals. Usually, the efficiency of transformation of these mammalian cells is not as high as that of chick cells. For instance, in the experiments of Kuwata *et al.* (1976), morphologically transformed foci did not appear in the cultures of human fibroblasts until three weeks after RSV infection. Mouse sarcoma viruses transform mouse fibroblasts as effectively as RSV transforms chicken cells. Certain variants of MSV were also reported to transform permanent mouse lines of the 3T3 type, primary and permanent cultures of rat and hamster cells, and human cells. Here again, in certain systems, MSV transformation is not as rapid and efficient as that observed in the cultures of mouse fibroblasts. For instance, when a rat bone cell culture had been infected with MSV, the foci of morphologically transformed and oncogenic cells were formed only three to four months later (Kano-Tanaka *et al.*, 1976).

All cells transformed by RNA-containing viruses contain DNA copies of the viral genomes integrated into their DNA. Possibly, transformation requires integration into some specific site of the cell genome but this question is not yet answered. In contrast to the cells transformed by DNA viruses, those transformed by oncorna viruses may form infectious virus particles which bud from the cell surface and are shed into the medium. The replication of virus is not correlated with the maintenance of the transformed state: certain oncorna viruses (transformation-defective viruses) may be replicated in the infected cells but do not induce transformation while infection with certain other viruses may lead to formation of stably transformed cells that do not produce infectious virus.

Genetic analysis of RSV and of MSV showed that the genomes contain four genes (Coffin, 1976; Van Zaane & Bloemers, 1978). Three of these genes code for proteins contained in the virions; these genes are essential for virus replication and for infectivity of virions but not for the maintenance of the transformed state. In contrast, the fourth gene called *src* is essential for the maintenance of the transformed state but not for replication of the virus particles.

(1) Virus strains with a part of the *src* gene deleted are able to infect cells and are replicated in these cells but lose their transforming properties.

(2) Many ts mutations of the *src* gene have been obtained. In cells transformed with the viruses carrying these mutations, most manifestations of transformation are expressed only at permissive temperatures and disappear reversibly after transfer to the non-permissive temperature.

A residual expression of some transformed characters e.g. anchorage independence at the non-permissive temperature has been observed in cells transformed with certain ts mutants of Rous sarcoma virus. This phenomenon may be a result of the 'leakiness' of the mutants, that is, of the incomplete inactivation of the product of the mutant gene at the non-permissive temperature (see review in Wyke, 1975).

The properties of product(s) coded by the *src* genes are currently under active investigation. Translation of this part of the viral RNA in cell-free systems leads to the formation of a protein of about 60 kilodaltons (kd). A transformation-specific antigen with similar molecular size was found in virus-transformed chick cells. Methionine-containing tryptic peptides from the 60 kd proteins obtained from translation *in vitro* and from immunoprecipitation of the antigen of transformed cells were identical (Purchio *et al.*, 1978). Probably, this 60 kd protein is the product of the *src* gene. DNA sequences similar to the *src* gene of exogenous viruses are contained in the normal genomes of non-infected chick cells; correspondingly, sequences similar to those of the *src* gene of MSV are found in the normal mouse DNA. A normal cell protein similar but not identical to the viral transforming gene product was found in non-infected cells (Collett *et al.*, 1979). The exact relationship between this normal protein and viral protein as well as their role in viral and non-viral transformation remain to be elucidated.

Revertant cell lines

Revertants are genetically stable variants obtained from transformed lines but differing from them in reduced expression of one or several transformed characters. Special selection procedures are used to obtain revertants that have lost certain transformed characters.

(1) Selection for morphological reversion is performed by cloning the transformed lines and observing under the microscope the variant 'flat' colonies which are cells that are better spread on the substratum than the cells in other 'usual' transformed colonies (Macpherson, 1965, 1971; Fishinger *et al.*, 1972; Stephenson, Reynolds & Aaronson, 1973; Boettiger, 1974; Groneberg *et al.*, 1978). Morphological revertants have also been selected by shaking poorly attached transformed cells from the substratum: revertant cells are better attached and removed by shaking (Nomura, Dunn & Fishinger, 1973). Pre-incubation of cultures in a medium with very high serum content (80%) may facilitate the selective detachment of transformed cells:

cells resistant to detachment have revertant properties (Bradley & Culp, 1977).

(2) Serum revertants can be obtained by placing a transformed line into a medium with a low serum concentration in the presence of an agent which selectively kills proliferating cells (e.g. fluorodesoxyuridine or other thymidine analogues, colchicine). Transformed cells with a serum requirement continue to proliferate under these conditions and are selectively killed by toxic agents. In contrast, revertant cells stop proliferation in the low serum medium and are therefore not killed by the toxic agents. A similar procedure has been used for the selection of 'density revertants' – cells which are unable to proliferate in dense cultures; and also of 'anchorage revertants' – cells which are unable to grow in suspension (Pollack *et al.*, 1968; Pollack, Wolman & Vogel, 1970; Culp *et al.*, 1971; Mondal *et al.*, 1971).

(3) Selection for resistance to the toxic effect of the plant lectin, concanavalin A (Culp & Black, 1972; Ozanne & Vogel, 1974), as well as selection for the ability to grow on a glutaraldehyde-fixed normal monolayer (Rabinowitz & Sachs, 1968; Rosenberg *et al.*, 1975) have also been used for selection of revertant cell lines. The rationale of these two methods is not clear.

Usually, the characters restored in revertants are intrinsically similar to the corresponding characters of normal cells but in certain cases they may be only superficially similar (see chapter 10). In these cases 'false reversion' takes place: the cell apparently does not restore the normal controlling mechanism but develops another mechanism which stops proliferation.

In the papers listed above, revertants selected for one particular untransformed character usually restored several other normal characters that had not been selected for. For instance, morphological revertants were usually observed to have more normal characters associated with growth control (anchorage dependence, etc.). Density revertants and, especially, anchorage revertants often had more normal morphology than the parent transformed cells as well as reduced oncogenicity. However, reversion of the various characters was not always co-ordinated. For instance, density revertants selected using fluorodesoxyuridine retained the ability to grow in a low serum medium. However, colchicine-selected density-revertant sublines also restored serum and anchorage dependence (Vogel, Risser & Pollack, 1973).

The genetic nature of the alterations leading to reversions has been studied mainly in revertant lines obtained from virus-transformed cells. In revertant lines obtained from RSV- and MSV-transformed cells the viral genome was found to be lost from some lines (Cho *et al.*, 1976; Frankel *et al.*, 1976) but in others, virus-specific DNA was retained (Deng *et al.*, 1974; Boettiger, 1974).

Several types of genetic change have also been observed in various revertant lines compared with parent lines transformed with DNA viruses.

(1) *Viral DNA is lost from the cellular genome*. This type of change was observed in one of the three classes of revertants obtained from a rat line transformed with SV-40 virus, containing one copy of viral genome (Steinberg *et al.*, 1978).

(2) *The viral DNA is partially deleted*.

(3) *The viral DNA is retained but the ability to synthesize the virus-specific product, T antigen, is lost*. Revertants of groups (2) and (3) were also obtained from the rat cell line with a single viral DNA copy (Steinberg *et al.*, 1978).

(4) *The viral DNA is retained and the revertant cells continue to synthesize T protein*. That is, the state of the viral genome in revertant cells cannot be distinguished from that in the parent transformed lines. Most revertants obtained from transformed lines carrying multiple copies of SV-40 or polyoma virus genomes belong to this group (e.g. Pollack *et al.*, 1968; Culp & Black, 1972). Possibly, this group of revertants is formed as a result of some changes of the cellular genome that suppress the effect of the viral products. Revertants of this group are often karyotypically different from parent transformed cells. In particular, the mean number of chromosomes often considerably increases in revertant cells (Pollack, Wolman & Vogel, 1970; Hitotsumachi, Rabinowitz & Sachs 1971; Bloch-Stacher & Sachs, 1977).

Cultures of tumour cells

Many lines of neoplastic cells derived from animal and human neoplasms have been described (e.g. see Fogh, 1975; Owens *et al.*, 1976; Smith *et al.*, 1976). The establishment of these lines has sometimes been complicated by the overgrowth of tumour cells by non-neoplastic stromal cells present in the original material. Laboratory contamination of lines by HeLa cells has also been a frequent complication in these experiments. Lines of tumour cells can obviously be used alongside cells transformed in culture in investigations of the nature of transformed characters. Selection of normal cells for pairing with each tumour line may meet difficulties. Certain tumour cell lines do not have all their transformed characters at the start of cultivation. These lines may undergo additional transformation in culture, arising spontaneously or induced by various carcinogens and viruses. For instance, a line of human osteosarcoma cells was not oncogenic in immunosuppressed hamsters until an additional transformation was induced by MSV (Jones *et al.*, 1975).

Cultures of cells from persons with genetic defects

Cells from apparently normal tissues of persons with certain inborn genetic defects (hereditary adenomatosis of the colon and rectum, ACR) may reveal a number of transformed characters when brought into culture. ACR is a hereditary disease in which numerous adenomatous polyps develop from the mucosa of the large intestine and rectum; subsequent frequent malignization of these polyps has been observed. Other extra-colonic changes including fibroblastic desmoid tumours may develop in certain cases. ACR is believed to be carried by an autosomal dominant gene; other genes may modify its expression. Kopelovich and collaborators (Kopelovich, Pfeffer & Lipkin, 1976; Kopelovich, Conlon & Pfeffer, 1977; Pfeffer *et al.*, 1976; Pfeffer & Kopelovich, 1977; Kopelovich, 1977) found that cultured human skin fibroblasts derived from apparently normal cutaneous biopsies of persons with ACR had a number of properties different from those of the skin from normal individuals or from individuals with non-hereditary colon adenocarcinoma. (*a*) In contrast to normal fibroblasts, those from persons with ACR had formed multiple areas of criss-crossed arrays and multi-layering. Another symptom of the transformed morphology of the ACR fibroblasts was the disorganization of bundles of actin microfilaments, a characteristic feature of poorly spread transformed fibroblasts (see chapter 5). (*b*) In contrast to normal fibroblasts, ACR cells grew in a medium with a low serum content (1 %); that is, they had a decreased serum requirement. (*c*) Fibroblasts from patients with ACR infected with MSV formed transformed foci about 100–1000 times more often than similarly infected normal cells.

Fibroblasts derived from some asymptomatic children of patients with ACR have properties similar to those of ACR fibroblasts while fibroblasts from other children are indistinguishable from those of normal persons. This dichotomy in properties of ACR progeny fibroblasts is supposed to correspond to the presence or absence of a gene responsible for this syndrome. Thus, the cells from patients with ACR have at least two transformed characters: altered morphology and a decreased serum requirement. These cells apparently do not have other transformed characters such as anchorage independence and oncogenicity. Additional virus-induced transformation converts ACR cells into strongly transformed cells that are able to grow in suspension and to form tumours in athymic mice.

As noted above, one distinctive feature of minimally transformed ACR cells is their increased propensity to undergo additional transformations after MSV virus infection. Cells from individuals with certain other genetic diseases are also observed to have greater frequencies of transformation with SV-40 virus than fibroblasts from normal persons. In particular, an increased frequency of virus-induced transformations was observed in the cultures from patients with ataxia-telangiectasia (Webb, Harnden & Harding, 1977), xeroderma pigmentosum (Key & Todaro, 1974), and von Recklinghauses's neurofibromatosis (Riccardi, 1977). It would obviously be important to examine cultures from individuals with these and other genetic defects for the possible presence of various transformed characters. Experiments with ACR-derived fibroblasts show that minimally transformed cells may behave as apparently normal cells *in vivo*. In fact, normal connective tissue of the patients with ACR is probably made of these cells, whose abnormal properties are revealed only after their transfer into the special environment of cell culture.

Oncogenicity of cultured cells
Assessment of oncogenicity

As we discussed above, cultured cells may acquire a number of new, transformed characters either spontaneously or after the action of viruses and chemicals. One of these characters is oncogenicity; that is, the ability of cells to grow as tumours when transplanted to animals. The development of oncogenicity certifies that cells *in vitro* can undergo changes similar to those occurring in the course of carcinogenesis *in vivo*. Development of oncogenicity may be one special alteration that bears no relation to other transformations occurring in culture, i.e. to morphological transformation and to transform-ation of growth properties. Alternatively oncogenicity may be one manifesta-tion of a series of changes which go in one direction and are mutually inter-related; that is, it may be one of the many manifestations of neoplastic evolution occurring in culture. One possible approach to this problem is an analysis of correlations between oncogenicity and other transformed charac-ters. Here, we will discuss these correlations (see earlier review in Ponten, 1976).

When cultured cells implanted in an animal have formed a progressively growing tumour nodule at the site of implantation, the result of the assessment of oncogenicity is obviously positive. An ideal experiment should also include a histological examination of the tumour as well as a comparison of its chromosomes and other markers with those of the implanted cells. These additional studies are especially valuable in cases where we wish to distinguish the tumour growing from the implanted cells from that arising from the host cells under the action of an inducing agent (e.g. a virus) contained in the implanted material. If a single tumour had appeared a long time after the implantation, it would be important to exclude the formation of a spontaneous neoplasm from the host cells. If small, slowly growing nodules are formed at the site of implantation, histological examination will distinguish true tumours from reactive granulomas. In certain cases, examination of tumours shows that they have grown from the implanted cells but the properties of the tumour cell populations are stably different from those of the cultured cells before implantation. In these cases it would be correct to conclude that only some special variants of the implanted cells are oncogenic; these variants might have been present initially in the implanted population or might have arisen after implantation. When these simple controls and limitations are taken into consideration, positive results of transplantation undoubtedly prove the oncogenicity of the implanted cells. A negative result of trans-plantation is much more difficult to interpret, as it may be due to certain limitations of the assay methods. The immunological incompatibility of implanted cells with the host is the most obvious of these limitations. When cultured cells have been originally taken from an inbred animal the negative result of implantation can still be due to secondary antigenic changes of the cells developing spontaneously or as a result of experimental treatments. For instance, in the experiments of Boon & Kellerman (1977) 12 out of 55 clones isolated from cultures of mouse teratocarcinomas lost the ability to form the tumours in syngenic mice after treatment with the mutagen, N-methyl-N'-nitro-N-nitrosoguanidine. However, these clones formed tumours as readily as the original cells when injected into pre-irradiated mice with suppressed immune reactions. Viral transformation may confer on cells an additional antigenicity which may interfere with their growth in animals. For instance, Mora *et al.* (1977), using mouse cell lines which had been spontaneously transformed and were highly oncogenic, found that when the cells were additionally transformed with SV-40 virus their oncogenicity in syngenic mice considerably decreased. The average number of tumour cells required to produce tumours in 50% of the animals (TD_{50}) increased about 100-fold after additional transformation.

Experiments of this type confirm that the probability of obtaining a positive result after transplantation of cultured cells increases if the immunological reactions of the host are decreased. There are several possible ways to prevent or to suppress the immunological reactivity of the host: (*a*) cell implantation into 'immunologically privileged' sites where immune reactions develop with difficulty (e.g. the anterior eye chamber), (*b*) cell implantation to a host with its immune reactivity suppressed either by total irradiation or by immuno-suppressive drugs, (*c*) cell implantation to animals with inborn genetic defects of immune reactivity. The animals most often used for this purpose are nude athymic mice which lack the ability to develop certain forms of immune reactions. During recent years cell implantation to athymic mice has become the favourite method for the assessment of oncogenicity, especially, in experiments with human cells. In the experiments of Fogh, Fogh & Orfeo (1977), 127 human tumour cell lines formed tumours after implantation to nude mice.

None of the methods listed above prevents all types of immune reaction completely, so that here again one needs to be cautious in the interpretation of negative results.

Another group of factors potentially affecting the results of oncogenicity tests is that connected with local conditions at the site of cell implantation. Certain cell variants form tumours only if they find an optimal micro-environment at the site of implantation. This was demonstrated in the experiments of Boone (1975, 1976). He used 'minimally transformed' cell lines (3T3, 1OT 1/2) that do not form tumours in syngenic mice after implantation of suspended cells detached from the substratum. Boone attached these cells in culture to the surface of glass beads or to polycarbonate platelets and then implanted them subcutaneously in syngenic mice. In these conditions, certain variants of the implanted cells formed tumours at the site of implantation. The role of the implanted foreign body is not clear: its surface might serve as a substratum for the mutiplication of anchorage-independent cells or it might stimulate inflammation and stromal proliferation.

The site of implantation may also affect the result of implantation. For instance, in mice the frequency of positive takes and the rates of tumour growth are two to four times higher when the tumour cells are inoculated subcutaneously into more cranial regions of the trunk compared with more caudal regions (Auerbach, Morrissey & Sidky, 1978).

The growth of transplanted tumour cells can be stimulated by factors inducing inflammation at the site of implantation (Vaitkevicius, Sugimoto & Brennan, 1962) as well as by simultaneous implantations of large numbers of lethally irradiated homotypic tumour cells (Revesz, 1958; El Mishad *et al.*, 1975), of nitrosourea-sterilized tumour cells (Dykes, Griswold & Schabel, 1978) or of living syngenic normal cells (Vasiliev, 1962). It would be interesting to test the effects of these modifying factors on the results of implantation of various cultured lines. Examination of the effects of local factors on the results of tumour implantation is at its very beginning and one hopes that eventually methods of local conditioning for different cell types will be developed, enabling the testing of cultured cells for various types of 'conditional oncogenicity'.

One of the obvious factors affecting the results of oncogenicity tests is the number of implanted cells. The critical number of tumour cells leading to positive 'takes' is different for various malignant cell lines and for various conditions of implantation; usually, this minimum is considerably higher than one cell. The same is true for cultured cells. To quote but a few examples, Winterbourne & Mora (1977) compared the spontaneously transformed mouse fibroblast 201 line and two clones derived from this line. The number of cells that produced tumours in 50% of the injected syngenic animals was $10^{2 \cdot 6}$ for cells of the parent line, and $10^{6 \cdot 4}$ and $10^{4 \cdot 8}$ for the two different clones. In the experiments of Koprowsky & Croce (1977), certain strains of human fibroblasts transformed with SV-40 virus gave tumours only when 10^8 cells had been implanted to athymic nude mice. The same was true for certain human tumour lines tested by Fogh *et al.* (1977). It should be noted that in most experiments of various authors not more than 10^6–10^7 cells per animal were injected in the oncogenicity tests. Why should such a large number of tumour cells be required to obtain a positive tumour 'take'? These cells may co-operatively form a favourable micro-environment for themselves e.g. induce stromal growth. Another possible explanation is that only a small fraction of the implanted cells is clonogenic; that is, may give rise to a tumour nodule. Certain other, almost trivial, factors affecting the state of implanted cells may also interfere with the results of animal implantations. For instance, mycoplasma infection of cultured cells was reported to reduce considerably their oncogenicity (van Diggelen, Shin & Phillips, 1977). In summary, the validity of negative results of small implantations of cultured cells should be interpreted with caution: this validity obviously increases when a variety of the methods is used for the assessment of oncogenicity of each tested line.

Correlations of oncogenicity with other cell properties

To study the correlations of transformed characters several cell lines of similar origin are usually compared. Morphological alterations of the cultured cells are found to be well correlated with the development of oncogenicity in certain systems. The clearest demonstration of this correlation has been obtained by Sanford and collaborators who compared the properties of many sublines of mouse fibroblasts at various stages of spontaneous transformation. Cytological assessment by light microscopy of fixed and stained cultures showed that oncogenic and non-oncogenic cultures could be distinguished on the basis of a complex of their morphological properties; that is, a trained cytologist examining only the stained culture could diagnose its oncogenicity or non-oncogenicity (Barker & Sanford, 1970; Sanford *et al.*, 1970; Sanford, Handleman & Jones, 1977). Later it was shown that the oncogenicity of cell lines derived from hepatocytes can be also diagnosed cytologically (Montesano *et al.*, 1977). In their preliminary experiments with fibroblasts, Sanford and collaborators regarded 'retracted cytoplasm' – decreased cell spreading on the substratum – as one of the significant morphological features of oncogenic cultures. Later the same group (Fox *et al.*, 1977) used a quantitative method for determination of the area of lamellar cytoplasm; diminishment of this area corresponds to decreased cell spreading on the substratum (see chapter 5). These measurements confirmed that oncogenic cultures of spontaneously transformed mouse fibroblasts had a smaller area of lamellar cytoplasm than paired non-oncogenic cultures of the same origin.

Another manifestation of morphological transformation associated with decreased spreading is disorganisation of microfilament bundles. This alteration was found to be correlated with anchorage independence and oncogenicity in a number of virally transformed lines (Shin *et al.*, 1975) and spontaneously transformed lines (Tucker, Sanford & Frankel, 1978). Several exceptional tumourigenic lines with abundant bundles were described by Celis *et al.*, (1978). In experiments with a large series of cell lines of diverse origin, Wright *et al.* (1977) observed a correlation of oncogenicity with the ability of cells to form aggregates in suspension. This test probably also reveals some changes of morphogenetic properties of the cells. The relation between alterations affecting the ability of cells to aggregate and those leading to decreased spreading remains obscure.

At present we do not know of any oncogenic line of fibroblasts which could be described as morphologically identical to the primary parent culture. But not all morphologically altered colonies, with abnormal cell distribution and with abnormal spreading, are found to be oncogenic (see, for instance, Spandidos & Siminovitch, 1978a). One should note also that morphological alterations of cultures are assessed rather subjectively in most experiments. Possibly, the application of a system of more objective tests would reveal that only certain degrees of morphological alterations are associated with oncogenicity.

Among the transformed growth properties, anchorage independence seems to be the best correlated with the oncogenicity: most lines that do not form tumours in animals have a low ability to form colonies in in semi-solid media, while most oncogenic lines have high plating efficiencies in these conditions. For example, in the experiments of the Sanford group (Tucker *et al.*, 1977) mouse and rat fibroblastic lines that did not form tumours in animals remained anchorage-dependent: they produced no colonies in agarose. Spontaneously transformed oncogenic descendants of the same lines formed colonies with various efficiencies. Shin *et al.* (1975) isolated non-selectively a number of clones from mouse 3T3 cells and rat fibroblasts infected with SV-40 virus. Six clones that proved to be anchorage dependent were not oncogenic in nude mice; in contrast, each of the three anchorage-independent clones was tumuourigenic. When anchorage independent sublines were selected from an anchorage-dependent line transformed with SV-40 virus, they became oncogenic. Conversely, the selection for tumourigenic cells *in vivo* from anchorage-dependent cells resulted in isolation of an anchorage-independent subline. Comparative examinations of other groups of mutually related cell lines also revealed correlations between oncogenicity and anchorage independence: this correlation was observed in experiments with a series of liver-derived cell lines (Montesano *et al.*, 1977), of epidermal cell lines (Colburn *et al.*, 1978) of chemically transformed derivatives of the C3H 1OT 1/2 line (Jones *et al.*, 1976b) and of a series of hybrid cells between a hepatoma line and L line of mouse fibroblasts (Lyons & Thompson, 1976).

The experiments of Boone and his group (Boone, 1975; Boone & Jacobs, 1976; Boone *et al.*, 1976; Yoshida *et al.*, 1976; Paranjpe, Eaton & Boone, 1978) deserve special comment. As mentioned above, these authors had shown that anchorage-dependent permanent cell lines Balb/3T3 and C3H 1OT 1/2, which were not oncogenic in standard tests, gave tumours if the cells were attached to glass beads and implanted to syngenic animals. Analysis of the properties of the tumours showed that they were formed from the implanted cells. At the same time, each tumour nodule formed after implantation had individual transplantation antigens and a unique karyotype. When the cells of these nodules were brought into culture they proved to be anchorage-independent, in contrast to the parent cells before implantation.

Obviously, the tumours had arisen not from the 'usual' cells of the permanent line, but from some special strongly transformed variants of these cells that were present in the original population or were induced by the *in vivo* environment. Here, as in the experiments of Shin *et al.* (1975), selection for high oncogenicity led also to simultaneous selection for anchorage independence. There are, however, a number of exceptions to the rule that anchorage independent cells are oncogenic and *vice versa*. Stiles *et al.* (1975) did not obtain tumours in nude mice from several human lines transformed with SV-40 virus. As shown by Koprowsky & Croce (1977), several of the lines used by Stiles *et al.* may give tumour nodules in nude mice if very high numbers of cells are implanted. Gammon & Isselbacher (1976) obtained a variant of polyoma-transformed BHK cells that had lost the ability to grow in agar: it gave 0.3% colonies as compared with 25% given by the parent cells. These variant cells remained highly oncogenic. When variant cells were brought back into culture after an *in vivo* passage, they still retained anchorage dependence. In the experiments of Stanbridge & Wilkinson (1978) a series of cell hybrids between a highly oncogenic human HeLa line and non-oncogenic human fibroblasts had been obtained. None of the three hybrid cell lines was oncogenic in nude mice; however, all three were anchorage independent.

The evidence for correlation of oncogenicity with other altered growth properties is not very convincing or negative. In particular, in the experiments of Shin *et al.* (1975) decreased serum dependence was characteristic of a number of virus-transformed clones that were not oncogenic in the usual tests.

Sanford *et al.* (1977*a*) also did not observe a correlation between serum dependence and oncogenicity of their spontaneously transformed mouse fibroblast lines. Acquisition of an unlimited life span also may occur much earlier than acquisition of oncogenicity: as we have seen above, many 'minimally transformed' permanent lines are not oncogenic in the usual tests.

To summarize, development of oncogenicity in cultures seems to be best correlated with decreased anchorage dependence and with certain morphological alterations. All these three characters have been repeatedly observed to develop in co-ordinated fashion in the course of neoplastic evolution. However, the linkage between these characters is not absolute: in certain cases a dissociation of oncogenicity from anchorage independence and from morphological changes has been observed to occur in the course of cultivation or after cell hybridization.

In our opinion these apparently contradictory observations suggest that the unknown intracellular changes which are immediately responsible for morphological alterations, for anchorage independence and for oncogenicity, are not identical. At the same time, in the course of neoplastic transformations, the changes of these three types may develop as various secondary manifestations of the alteration of one 'central' cellular character. In other words, one hypothetical 'central' change may lead to several 'peripheral' changes, one of them being responsible for oncogenicity, and others for anchorage independence and morphological transformation. Diverse additional genetic alterations may suppress independently each of the 'peripheral' changes without suppressing other 'peripheral' change or 'central' change(s). This secondary suppression occurring in the course of cultivation or of hybridization can be the cause of dissociation between different transformed characters. We will return to this problem in chapter 14.

Conclusion

Fibroblasts and epithelial cells brought into culture may undergo a series of genetically stable alterations which arise spontaneously or are induced by external agents. This evolution of cellular properties *in vitro* has many similarities to the evolution of neoplastic cells *in vivo*.

(1) As a result of evolution in culture cells may aquire the ability to form tumours *in vivo*.

(2) As a result of evolution *in vitro*, cells may acquire the ability to grow selectively in conditions in which normal cells do not proliferate, e.g. in suspension, in a low serum medium, and in high density areas. This ability of transformed cells for selective growth *in vitro* is similar to the ability of neoplastic cells for selective growth *in vivo*.

(3) As a result of evolution *in vitro*, cells begin to form abnormal multicellular structures (morphological transformation); this alteration is similar to the alteration of the ability to form organized multicellular structures which is characteristic for neoplastic cells *in vivo*.

(4) Most agents that induce transformations *in vitro* (oncogenic viruses and carcinogenic chemicals) are identical to these agents that induce formation of neoplasms *in vivo*. Neoplastic evolution *in vitro*, like its analogue *in vivo*, is a process which has many steps and innumerable variants.

Transitions from one stage to another may probably be caused by genetic changes of a diverse nature. At advanced stages of neoplastic evolution cells acquire a strongly transformed phenotype that includes a complex of transformed characters. Some of these characters, namely anchorage independence and certain morphological changes, seem to be correlated with the development of oncogenicity. Certain other characters, for instance, serum dependence or infinite life-span do not show a good correlation with oncogenicity. Nevertheless, these characters may also be manifestations of important steps in neoplastic evolution; for instance, they may be characteristic of 'pre-malignant' and/or conditionally neoplastic cells that are able to form tumours only in a special environment. As yet we have no adequate in-vivo tests to reveal altered biological potentials of 'conditionally neoplastic' cells.

These considerations suggest that studies of all the manifestations of cell transformations *in vitro* may be relevant to, and important for, the better understanding of neoplastic evolution *in vivo*. In the following two parts of this book we will describe and discuss in more detail two main groups of transformation accompanying neoplastic evolution *in vitro:* morphological transformation and transformation of growth control.

Part II

Morphogenesis in normal and transformed cultures

3

Cell structures and reactions associated with morphogenesis

Introduction

Alterations of cell shape are observed in the course of many diverse processes such as spreading and locomotion on the substratum, endocytosis, formation of cell–cell contacts capping of surface receptors, etc. Analysis of these alterations suggests that they consist of various combinations of a few types of cellular processes designated 'basic morphogenetic reactions' (Vasiliev & Gelfand, 1977). Pseudopodial reactions form the main group of these basic morphogenetic reactions. In this chapter we will describe several types of pseudopodial reactions and discuss their possible mechanisms. We will also characterize other basic morphogenetic processes controlling the distribution of pseudopodia in various parts of cell surface: stabilization processes and contact paralysis. Discussion of these morphogenetic processes will be preceded by a brief review of the structure of the cell surface and the cytoskeleton.

Cell surface and cytoskeleton
Cortical layer and cytoskeleton

The surface (periphery) of animal cells consists of two main parts: the cortical layer and the plasma membrane. The cortical layer is a zone of cytoplasm located under the plasma membrane containing only fibrils but not vesicular organelles or ribosomes. The thickness of the cortical layer varies considerably in different parts of culture cells; usually it is about 0.1–0.5 μm. Cultured non-muscle cells contain three main types of fibrils (fig. 3.1): microfilaments (4–6 nm in diameter), intermediate filaments (7–12 nm in diameter) and microtubules (about 25 nm in diameter). The various fibrillar elements are sometimes collectively designated the 'cytoskeleton'. Of all the cytoskeleton elements, microfilaments can be regarded as the most characteristic components of the cortical layer, but intermediate filaments and microtubules may also be present.

Electron microscopic examination of sections of cultured normal fibroblasts suggests that most cellular microfilaments are located in the cortical layer. Microfilaments may form linear bundles of various diameters and lengths. The

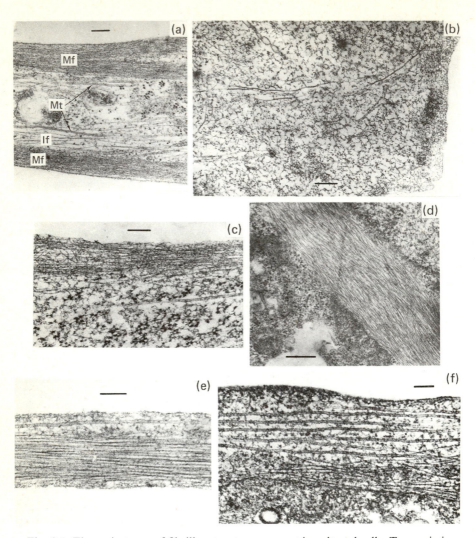

Fig. 3.1. The main types of fibrillar structures present in cultured cells. Transmission electron micrographs of sections parallel to the plane of the substratum. (a) Micro-filament bundles (Mf), intermediate filaments (If) and microtubules (Mt) in the tail part of a polarized fibroblast; (b) matrix cortical layer in the anterior lamella of polarized fibroblasts; (c) microfilament bundle and microtubule near the parallel edge of a polarized normal mouse fibroblast; (d) bundle of intermediate filaments in a colcemid-treated transformed L fibroblast; (e) intermediate filaments in the cytoplasmic strand of a fibroblast treated with cytochalasin B; (f) microtubules in the cytoplasmic strand of a fibroblast treated with cytochalasin B. Bars: (a, c, e, f), 0.2 μm; (b, d), 0.5 μm.

microfilaments in each bundle are approximately parallel to each other. These bundles are also called the 'stress fibres' or 'actin cables'. Microfilaments may also form various configurations in which they are less closely packed and have much more varied orientations than in the bundles. These apparently less organized configurations are designated 'matrices', 'webs', 'meshworks' or 'lattices' (Buckley & Porter, 1967; Wessels *et al.*, 1971; Wessels, Spooner & Luduena, 1973; Goldman *et al.*, 1973; Goldman, Schloss & Starger, 1976*a*; Allison, 1973; Temmink & Spiele, 1978).

Both lattice and bundle microfilaments contain polymerized actin as shown by their ability to bind heavy meromyosin molecules (Ishikawa, Bischoff & Holtzer, 1969; Small & Celis, 1978*b*) and anti-actin antibodies (Lazarides & Weber, 1974; Lazarides, 1976*a*, *b*; Goldman *et al.*, 1976*a*). Different microfilaments within the same bundle of cultured CHO cells have been found to have antiparallel polarities, as indicated by the direction of the arrow-head complexes formed by microfilaments with heavy meromyosin (Begg, Rodewald & Rebhun, 1978). Bundles of microfilaments with uniform polarity have been found in certain other cells such as platelets, intestinal cells, and sea urchin coelomocytes (Mooseker, 1976; Nachmias & Asch, 1976; Edds, 1977).

Actin is one of the main proteins of non-muscle cells: the actin content of chicken fibroblasts has been estimated to be 6–8% of the total proteins (Bray & Thomas, 1975; Anderson, 1976). Actins isolated from various cells closely resemble each other but have certain chemical differences. Two forms of actin (β and γ), separable by isoelectric focussing, were found in various non-muscle cells; a third form (α-actin) is present in skeletal muscle (Lu & Elzinga, 1976; Whalen, Butler-Browne & Gros, 1976; Storti & Rich, 1976; Gallagher, Detwiler & Stracher, 1976; Rubinstein & Spudich, 1977). Only a part of the actin is polymerized; another part exists in cells in a non-polymerized form (Bray & Thomas, 1976). Possibly, non-polymerized actin forms complexes with other proteins; one such protein (profilin) has been isolated (Carlsson *et al.*, 1977). Polymerized actin can also interact with a number of proteins (see review in Korn, 1978) including those listed as follows.

(1) *Myosin* is able to form actomyosin complexes with actin. Fibroblast myosin is probably different from the myosins of striated muscle and of smooth muscle (Pollard *et al.*, 1976; Yerna *et al.*, 1978). In contrast to muscle cells no special myosin filaments are found in cultured fibroblasts.

(2) *Tropomyosin* consists of two subunits with a total molecular weight about 70 000. Tropomyosins from non-muscle cells are probably similar but not identical to muscle tropomyosin (see review in Hitchcock, 1977; Bretscher & Weber, 1978).

(3) *α-Actinin* consists of two subunits each with a molecular weight of about 95 000. This protein is present in the Z-discs of striated muscle cells and in

the dense bodies of smooth muscle cells. Both types of structure serve as attachments for the ends of actin microfilaments. Probably, actinin plays a similar role in fibroblasts (Schollmayer *et al.*, 1976; Lazarides, 1976*a*, *b*, *c*).

(4) *Filamin*, with a molecular weight of about 250000, has been isolated from smooth muscle but is present also in fibroblasts (Wang, Ash & Singer, 1975; Shizuta *et al.*, 1976).

(5) *A high molecular weight protein* (actin-binding protein) has been isolated from macrophages and forms gels with actin (Stossel & Hartwig, 1975, 1976). Probably this protein is different from filamin (Hartwig & Stossel, 1977) but this question is not yet resolved. An actin-binding protein with similar but not identical properties has been isolated from BHK-21 fibroblasts (Schloss & Goldman, 1978).

(6) *Other proteins* interacting with actin have been isolated, e.g. spectrin from erythrocytes, gelactins from amoebae, proteins from crab sperm, etc. (Tilney, 1976*a*, *b*; Hitchcock, 1977; Korn, 1978). Possibly, cultured tissue cells also contain a number of yet unidentified actin-binding proteins.

It is probable that interactions with various proteins may regulate the state of actin in cells: these proteins may inhibit or promote polymerization of actin and they may also affect the distribution of actin microfilaments as well as their contractile properties.

Immunofluorescence studies show that distribution of proteins interacting with actin is different in various parts of the cortical layer. Besides actin, fully assembled microfilament bundles contain myosin, tropomyosin and actinin (Lazarides, 1976*a*, *b*). In the matrix of the cortical layer, in certain areas of the cell where active formation and contraction of pseudopodia takes place (see below), actin and filamin are present but the amounts of myosin are diminished and tropomyosin diminished or absent (Heggeness, Wang & Singer, 1977; Lazarides, 1976*c*).

The structures formed by actin and other proteins in the cortical layer are able to contract. ATP can induce contraction of glycerol-treated fibroblasts ('cell models') that retain most of their cytoskeletal structures (Hoffman-Berling, 1954; Arronet, 1971; Vasiliev *et al.*, 1975*c*). By laser microbeam dissection Isenberg *et al.* (1976) isolated individual bundles of microfilaments from cultured cells and demonstrated their contractility. Isolated cytoplasm from various non-muscle cells is another experimental system convenient for studies of assembly and movements of actin-containing structures. Polymerization of actin microfilaments, leading to formation of gel-like structures, has been observed in various preparations of isolated cytoplasm (Spudich & Cooke, 1975; Kane, 1976; Wohlfarth-Bottermann & Isenberg, 1976). Contraction and other movements of isolated cytoplasm depend on the presence of ATP and divalent cations (Izzard, C. S. & Izzard, S. L., 1975; Pollard, 1976; Taylor, 1976; Condeelis *et al.*, 1976). The calcium concentration is probably one important regulator of the state of actin-containing structures in non-muscle cells. Several forms of calcium regulation of contraction have been observed in various cells.

(1) *Actin-linked regulation.* The troponin complex bound to actin via tropomyosin can make actin–myosin interactions sensitive to calcium. As mentioned above, tropomyosins have been isolated from fibroblasts and other non-muscle cells. Troponin-like proteins have also been obtained from various non-muscle cells but their properties and functions are not yet clear (see review in Hitchcock, 1977, Korn, 1978).

(2) *Myosin-linked regulation.* Calcium activates the enzyme which phosphorylates myosin (myosin light chain kinase); phosphorylation of myosin catalysed by this enzyme seems to be essential for the activation of myosin ATPase by actin and, therefore, for contraction in smooth muscle and non-muscle cells (Adelstein, 1978; Yerna & Goldman, 1978). A protein similar to muscle troponin C (calcium-binding modulator protein, calcium dependent regulator protein) has been isolated from many non-muscle cells (see review in Hitchcock, 1977). This protein participates in many cellular activities; in particular, it may act as one of the subunits of the myosin light chain kinase mentioned above (Yerna & Goldman, 1978).

In summary, the structure and properties of the cortical layer in various cell parts may be highly differentiated due to variations of proportions of different proteins and various configurations of the actin microfilaments.

The cortical layer is also very dynamic; it may change cell shape and provide the motile force for translocating the cell on its substratum. Alterations of the shape and local structure of the cortical layer may be achieved in several ways. One way is the local polymerization and depolymerization of actin microfilaments. Depending on the presence of proteins or changes in local conditions microfilament conformations can be formed or destroyed in a very short time. Another way to change the structure of the cortical layer is to move microfilaments relative to each other: these movements may lead to contraction or to streaming of certain parts of cortical layer. The exact phenomenology and mechanisms of changes accompanying the various alterations of the cortical layer remain to be studied.

Intermediate filaments (intermediate-sized filaments, 100 Å filaments) are present in cells of many types including fibroblasts, nerve cells (neurofilaments), epithelial cells (tonofilaments), smooth and striated muscle cells. Proteins with molecular weights of about 55–60000 seem to be the major structural components of these filaments. Proteins of intermediate filaments from various cells are antigenically different; probably there are several chemically distinct classes of these filaments (Davison, Hong & Cooke, 1977; Lazarides, 1978; Starger *et al.*, 1978; Bennett *et al.*, 1978; Fellini, Bennett & Holtzer, 1978; Gipson & Anderson, 1978). Buckley *et al.* (1978) have reported that specific proteins of intermediate filaments surround an actin-containing core; this intriguing finding awaits confirmation. Immunofluorescence reveals a complex network of intermediate filaments in cultured fibroblasts and epithelial cells: they are often concentrated around the nucleus and radiate into the peripheral regions (Blose, Shelanski & Chacko, 1977; Osborn, Franke & Weber, 1977; Gordon, Bushell & Burridge, 1978; Franke *et al.*, 1978*b;* Small & Celis, 1978*a*). The function of these filaments remains unknown. Possibly, they participate in the formation of a tensile structural framework of cells.

Besides these fibrillar structures cells may also contain a 'microtrabecular lattice' pervading all parts of the cytoplasm. Porter and collaborators (Buckley & Porter, 1975; Porter, 1976; Wolosewick & Porter, 1976) revealed this three-dimensional lattice of thin (3–6 nm) trabeculae by high voltage electron microscopy of whole cultured fibroblasts. The chemical composition of this labile lattice, its relation to other fibrils and its functional significance remain unknown. It was suggested (Porter, 1976) that microtrabeculae form a highly asymmetric gel-like structure that may be responsible for 'keeping in place' other fibrillar structures and cell organelles: it may also participate in various intracellular movements.

Cytochalasins, especially cytochalasin B, have been widely used to affect the cortical layer more or less selectively. Cytochalasins reversibly alter cell shape (fig. 3.2) and inhibit many forms of cell motility (Carter, 1967*b;* Sanger & Holtzer, 1972; Boyde, Bailey & Vesely, 1974). These drugs rapidly and selectively destroy most lattice microfilaments and microtrabeculae, as well as many, but usually not all, the microfilament bundles (Wessels *et al.*, 1971, 1973; Fonte *et al.*, 1978). The microfilamentous material collected into a star-like aggregate is often seen in fibroblasts treated with cytochalasin B (Weber *et al.*, 1976). Intermediate filaments and microtubules are not destroyed by cytochalasins. The molecular targets for cytochalasins have not yet been identified; probably, they are cortical proteins (Lin, D. C. & Lin, S., 1978). Cytochalasin B has been found to decrease the viscosity of actin-containing gels; possibly this drug interferes with the interaction of actin microfilaments with each other or with other proteins present in the gels (Hartwig & Stossel, 1978; MacLean *et al.*, 1978). Glycerol-extracted 'models' prepared from fibroblasts treated with cytochalasin B before extraction, in contrasts to 'models' of control cells, did not contract in solutions containing ATP. But cytochalasin B added to the incubation medium after glycerol extraction did not inhibit ATP-induced contraction (Vasiliev *et al.*, 1975*c*).

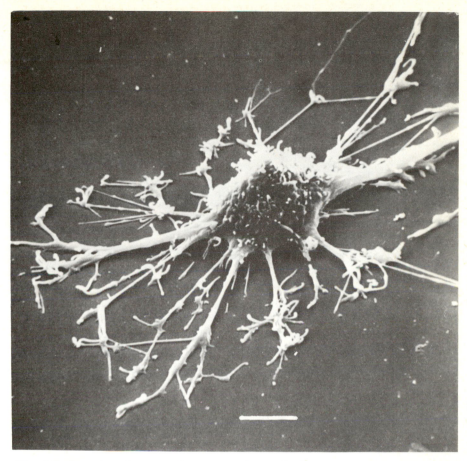

Fig. 3.2. Scanning electron micrograph of an arborized, cytochalasin B-treated mouse fibroblast 24 h after seeding. Cells were incubated for 1 h with cytochalasin B (10 μg/ml). Bar: 5 μm.

These results suggest that the effect of cytochalasin on the contractility of the cortical layer is not direct: some stage of the action of this drug takes place only in living cells but not in glycerol-extracted ones. Cytochalasin B inhibits sugar and nucleoside transport through the cell membrane but this effect is not related to its action on the cortical layer and cell motility. Dihydrocyto-chalasin B, which differs from cytochalasin B by the absence of a double bond, has no effect on membrane transport at concentrations at which it affects cell motility and morphology in the same way as its parent compound (Lin, Lin & Flanagan, 1978). Despite unknown mechanisms of action, cytochalasins remain attractive experimental tools because their cellular effects are both marked and reversible.

Plasma membrane

A large number of reviews on the structure of the plasma membrane are available (Weiss, 1967; Curtis, 1967; Robbins & Nicolson, 1975; Bretscher & Raff, 1975; Nicolson, 1976a, b) and we will review here only few points relevant for further discussion. The membrane lipids are of three chemical classes: phospholipids (many different chemical types), neutral lipids (mainly cholesterol) and glycolipids (many different chemical types). All membrane lipids are amphipathic molecules, having a hydrophilic (polar) and hydrophobic (non-polar) portion. The inner and outer leaflets of the bilayer seem to have significantly different lipid compositions.

Some proteins (so called peripheral or extrinsic proteins) can be easily eluted from membrane preparations by minimal treatments such as altering the ionic strength of the medium. Other proteins (integral or intrinsic proteins) extend into, and sometimes across, the lipid bilayer and can be solubilized only by treatments that disrupt the bilayer, e.g. detergents such as sodium dodecyl sulphate. Electrophoresis in polyacrylamide gels containing this detergent have revealed about 15 major protein bands in the membranes of human erythrocytes. The variety of protein types in the membranes of nucleated cells is probably considerably larger. Intrinsic membrane proteins are asymmetrically oriented across the bilayer. Many, although possibly not all intrinsic proteins span the bilayer, that is, are exposed on both sides of the membrane (Walsh & Crumpton, 1977; Rothstein *et al.*, 1978; Guidotti, 1977). Some of these transmembrane proteins have most of their mass and functional properties in the aqueous environment outside the cytoplasm; the short intramembrane portion anchors these proteins to the membrane. In contrast, other proteins have the major part of their mass in the cytoplasm and only a small part of the molecule is exposed to the external side of the membrane. Some of the membrane proteins have enzymatic properties e.g. adenylate cyclase. A number of proteins are involved in transport of ions and nutrients across the lipid bilayer, e.g. Na^+, K^+-dependent ATPase, or the anion exchange complex of the human erythrocyte.

Certain membrane lipids and proteins (glycolipids and glycoproteins) have oligosaccharide chains covalently attached to them. These chains are probably located exclusively on the external side of the membrane (see review in Virtanen, Miettinen & Wartiovaara, 1978). About ten distinct types of monosaccharides are found in glycolipids and glycoproteins. Different combinations of these monosaccharides give rise to a considerable diversity and specificity of oligosaccharide chains. Oligosaccharide side chains of membrane molecules form the carbohydrate-rich zone on the external side of the plasma membrane. This zone ('glycocalyx') may also include peripheral membrane proteins as well as proteins and proteoglucans absorbed from the external environment. It is difficult to draw a sharp distinction between peripheral molecules of the cell surface and those of the immediate cellular micro-environment.

External cell membrane structures undergo considerable turnover: some of the pre-existing structures move to the interior of the cell and are replaced by new exterior structures. The main movement of cell membrane structures is probably endocytosis. Calculations of the rate of pinocytosis have shown that macrophages and mouse L fibroblasts move the equivalent of their cell surface area to the interior every 33 and 125 min, respectively (Steinman, Brodie & Cohn, 1976). This surface material obviously has to be replaced by an equal or larger (in growing cells) amount of exterior membrane components. Probably, the transfer of membrane vesicles formed in the Golgi apparatus to the external surface is the principal way of doing this. By fusion with the external membrane materials, these vesicles may be incorporated into the cell membrane (see e.g. Whaley, Dauwalder & Kephart, 1972). Another possible way of membrane renewal is the incorporation of isolated molecules into the membrane at its internal side (see Harris, 1976a). We do not know whether material moved to the exterior initially has some preferential location in the membrane. Membrane turnover may serve as a repair mechanism replacing altered membrane components e.g. glycoproteins with enzyme-modified carbohydrate chains (Buck & Warren, 1976). In addition to the constant turnover of components between the cell membrane and the cell interior there is, probably, a considerable exchange of molecules between the cell surface and the environment. Shedding of the cell surface macromolecules into tissue culture medium has been observed (Kapeller *et al.*, 1973, 1976). The shed membrane macromolecules as well as other proteins present in the medium can be re-absorbed at the surface. The nature of the processes involved in this shedding and reabsorption remain unknown.

A membrane receptor may be operationally defined as an integral component of the membrane that binds, more or less specifically, certain molecules present in the external environment: these molecules are designated 'ligands'. Any externally exposed membrane component may be regarded as a receptor if we can find a ligand for it, e.g. if we have prepared an antibody against it. The reason for using the term 'receptor' in such a broad sense is that it describes all types of membrane molecules that can initiate significant cellular alterations by reacting with corresponding ligands. For instance, antibodies to different externally exposed membrane components can induce capping reactions (see later) as well as proliferation of cells (chapter 12). Several other examples of mammalian cell surface receptors are given below.

(1) *Receptors for lectins.* Lectins are proteins that react specifically with end groups of carbohydrate chains. Membrane glycoproteins and glycolipids having a corresponding group act as receptors for these ligands (see reviews in Sharon & Lis, 1972; Nicolson, 1974). Many types of lectins specific for

various types of monosaccharide groups have been isolated from plants. The most well-known is concanavalin A (Con A) which binds glycosyl and mannosyl groups.

(2) *Receptors for carbohydrate end groups.* These are lectin molecules present in the plasma membrane. The most studied receptor of this type is the hepatic binding protein present in the membrane of rabbit hepatocytes; it binds glycoproteins containing carbohydrates with galactose end groups (Ashwell & Morell, 1974; Simpson, Thorne & Loh, 1978). In this case the lectin acts as a receptor and a carbohydrate chain as the ligand. This situation is the reverse of (1) above.

(3) *Receptors of sensitized lymphoid cells binding corresponding antigens.* The best characterized molecules of this type are immunoglobulin receptors present at the surface of B-lymphocytes (see review in Brondz & Rocklin, 1978).

(4) *Receptors for peptide hormones and neurotransmitters* (see review in Kahn, 1976). Many types of these receptors are being actively studied, especially acetylcholine receptors isolated from cholinergic synaptic membranes, insulin receptors present on the membranes of many cell types, thyrotropin receptors present on thyroid cells. Most receptors of this group are proteins, although possibly certain glycolipids (gangliosides) may also act as receptors for peptide hormones (Fishman & Brady, 1976).

(5) *Receptors for bacterial toxins.* In particular, cholera toxin, a ligand binding to specific surface receptors present in many cell types, was identified as specific ganglioside G_{M1} (Fishman & Brady, 1976; Yamakagawa & Nagai, 1978; O'Keefe & Cuatrecasas, 1978).

Several types of membrane components may serve as receptors for one ligand. For instance, many types of glycoproteins having corresponding end groups in their carbohydrate chains may act as receptors for concanavalin A. The ligand molecule can be monovalent or polyvalent, that is, it can react with one or several molecules of membrane receptors simultaneously. Ligands can be either solvated in the extracellular medium or immobilized by chemical attachment to some surface. Interaction of membrane receptors with immobilized ligands can result in cell attachment to the ligand-carrying surface.

Distribution and mobility of plasma membrane components

In physiological conditions lipids forming the membrane bilayer are predominantly in a liquid-crystalline state. This means that individual lipid molecules easily exchange places with their neighbours within their own monolayer but rarely migrate from one monolayer to another. The rate of diffusion, possibly, is different in various parts of the same lipid layer. In particular, molecules surrounding certain membrane proteins (annular lipids) may be partially immobilized. As the temperature is lowered, the lipid layers decrease their fluidity and eventually undergo a phase transition from a liquid-crystalline to a crystalline state. The temperature of transition depends on the chemical composition of the lipid layer. Alterations in fluidity of the lipid layer produced by temperature shifts or by changes of lipid composition probably affect all types of membrane-associated functions; the exact consequences of these changes in the plasma membrane and their biological role remain to be studied.

In physiological conditions not only lipid molecules of the membrane but also many intrinsic proteins have considerable mobility. This mobility of various membrane components has been demonstrated by several groups of experiments. (*a*) When human and mouse cells are fused by Sendai virus, different histocompatibility antigents present on their surfaces (human HLA antigens and mouse H-2 antigens) first remain segregated in different parts of the surface of the formed heterokaryon but gradually intermix; 20–30 min are required for complete intermixing to occur in about half of the heterokaryons at 37 °C (Frye & Edidin, 1970; Edidin & Wei, 1977*a*, *b*). (*b*) Incubation of cells with labelled polyvalent ligands may induce redistribution of the receptors for these ligands: they are assembled into microscopically visible patches (see later). (*c*) In 'bleaching' experiments the surface receptors are first labelled by a fluorescent molecule. This label is either attached directly to membrane proteins or, more often, the cell is incubated with the pre-labelled ligand to a certain receptor. The fluorescence of a small spot on the cell surface is then bleached by a laser beam. The fluorescence gradually returns to the bleached area probably due to the lateral movement of fluorescent receptors from the non-bleached areas of the membrane (Edidin, Zagyansky & Lardner, 1976; Jacobson, Wu & Poste, 1976; Zagyansky & Edidin, 1976; Axelrod *et al.*, 1976; Schlessinger, Webb & Elson, 1976; Schlessinger *et al.*, 1977*a*; Wolf *et al.*, 1977; Fowler & Branton, 1977). (*d*) Application of a constant electric field of 1–10 V/cm causes a gradual redistribution of concanavalin A receptors on the surface of living muscle cells: these receptors become more concentrated on the side of the cell facing negative electrode. Possibly, this redistribution is a result of the electrophoresis of the charged receptor molecules within the plane of the membrane (Poo & Robinson, 1977; Poo, Poo & Lam, 1978).

These experiments show that many receptors are able to move in the lipid layer. However, they do not prove that this movement is due to non-restricted, free diffusion of molecules. In fact, the observed rates of receptor movements in many experiments were 10–100 times slower than the calculated rates for free diffusion of proteins. These results were obtained both in the bleaching experiments and in the experiments with heterokaryons quoted above. Bleaching experiments showed that certain fractions of concanavalin A receptors and of other receptors are immobile. Movement of receptors in plasma membrane was inhibited by cytochalasin B which had no effect on diffusion in an artificial lipid layer.

Thus, free diffusion of proteins in the plasma membrane may be restricted by several factors. Another approach to the same problem is an analysis of the distribution of various protein components within the membrane. Light microscopic and electron microscopic examination of the distribution of labelled receptors in pre-fixed membranes shows in many cases their apparently random uniform arrangement (see review in Nicolson, 1976).

Freeze-fracture electron microscopy of the plasma membrane reveals intramembrane particles which are believed to be proteins: these particles usually have a random arrangement (see reviews in Branton, 1969; Weiss, Goodenough & Goodenough, 1977; Verkleij & Ververgaert, 1978). However, in certain cases the distribution of membrane components may become non-random. This is usually observed in specialized cell–cell contact structures of various types. For instance, intramembrane particles form organized polygonal arrays in the areas of gap junctions (Gilula, 1974). Organized arrays of intramembrane particles ('necklaces' and other variants) are also seen in cilia and flagella (Gilula & Satir, 1972; Weiss *et al.*, 1977). In cultured fibroblasts, receptors for plasma low density lipoprotein have been observed to be concentrated within large pits (0.1–0.5 μm in diameter) on the cell surface (Orci *et al.*, 1978).

Cells without stable surface specializations may also have non-uniform distribution of certain membrane components. For instance, surface immuno-globulin molecules of mouse spleen lymphocytes were found in a higher concentration on the surface of microvilli compared with the cell body (de Petris, 1978*a*). When rabbit thymocytes formed protrusions (uropods), concanavalin A receptors and various surface antigens, recognized by anti-thymocyte serum, became depleted from the surface of these outgrowths and concentrated over other parts of the cell body (de Petris, 1978*b*).

Theoretically, restriction of receptor movement may be due to several different mechanisms: anchoring of receptors to some structure on the external side of the membrane, anchoring of receptors to some cortical structures under the membrane, or aggregation of receptors within the membrane. In most cases receptors in non-specialized areas retain considerable mobility so that anchoring, if any, has to be reversible. It is usually assumed that apparently random movements of surface receptors are a result of their free diffusion in the membrane, while extramembrane factors only restrict this movement. However, an alternative explanation is also possible: these receptors may move not by free diffusion but because they are pulled in different directions by some cortical structures to which they become reversibly attached. As yet, we have no ways of finding out which particular mechanism induces or restricts the movements of membrane components in intact cells.

We have seen that membrane components probably interact in many ways with the components of the underlying cortical layer. The molecular basis of this interaction is not clear: there is no convincing evidence that any characteristic cortical protein (e.g. actin or myosin) is exposed at the external side of the membrane; more probably, microfilaments may become attached to some membrane components at the internal side of the membrane. Several membrane proteins may be linked to a single microfilament (Loor, 1976). It was suggested that α-actinin may serve as intermediate element in this linkage.

Pseudopodial attachment reactions

In this section we will describe and discuss pseudopodial reactions: the formation and attachment of pseudopodia. We regard reactions of this group as a basic mechanism that is used in various morphogenetic processes. We will describe these pseudopodial reactions taking as an example the processes responsible for cell spreading on artificial substrata. We will show that the surface of pseudopodia has special properties different from those of other cell parts. We will then compare pseudopodial reactions induced by the substratum with those induced by cell-attached particles and by soluble ligands. Finally, we will discuss possible mechanisms of pseudopodial reactions.

Cell spreading as an example of multiple pseudopodial reactions

Isolated fibroblasts and epithelial cells suspended in a fluid medium usually have an almost spherical shape. When these cells make contact with an 'adhesive', rigid flat surface (e.g. glass, plastics, metals) these cells rapidly spread on the substratum; that is, they acquire a flattened shape and form numerous contact sites with the substratum (fig. 3.3). The morphology of spreading of normal embryo fibroblasts has been studied by many investigators (Taylor, 1961; Witkowski & Brighton, 1971; Domnina *et al.*, 1972; Rajaraman *et al.*, 1974; Bragina, Vasiliev & Gelfand, 1976). These studies have shown that several stages of spreading can be distinguished: initial attachment, radial spreading and polarization.

At the stage of initial attachment, the cells become attached to the substratum but do not form new pseudopodia. The lower surface of the cell contacting the substratum becomes flattened.

The beginning of radial spreading is manifested by the formation of pseudopodial protrusions around the spherical cell body. Microcinemato-

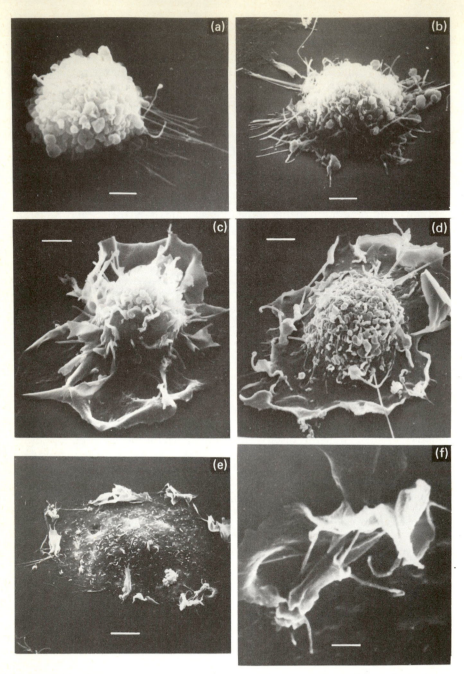

Fig. 3.3. Stages of spreading of normal mouse fibroblasts on a glass surface as seen by scanning electron microscopy. (a) Unspread cell with filopodia attached to the substratum. (b) Numerous filopodia and blebs at the cell periphery. (c, d) Formation of a rim of lamellar cytoplasm. (e) Cell with a rim of lamellar cytoplasm and a flattened central part. Ruffles near the external edge. (f) Magnification of one of the ruffles seen in (e). Times after seeding: (a), 10 min; (b, c), 30 min; (d), 1 h; (e, f), 2 h; Bars: (a), 2 μm; (b–d), 5 μm; (e), 10 μm; (f), 1 μm.

graphic studies show that pseudopodia are continually formed at the cell periphery; they are either attached to the substratum or are withdrawn. In fixed preparations examined by scanning or transmission electron microscopy one can see at this stage different numbers of pseudopodia attached to the substratum. There are two main morphological variants of these pseudopodia: (*a*) cylindrical filopodia about 0.2–0.5 μm in diameter and up to 10–20 μm long, (*b*) flattened lamellipodia about 0.1–0.5 μm thick and 2–5 μm wide. Large ovoid or cylindrical processes about 1–2 μm in diameter are called lobopodia (blebs); the blebs attached to the substratum are not often seen during the standard course of spreading but they may become numerous after special pre-treatment of cells e.g. after repeated washing in saline (Albrecht-Buehler & Lancaster, 1976). Morphological forms of pseudopodia that are intermediate between these types are also seen in spreading cells. Filopodia are predominant at the beginning of radial spreading, lamellipodia become predominant at the later phase. A thin cytoplasmic plate firmly adherent to the substratum (lamellar cytoplasm) is gradually formed from numerous attached pseudopodia at the cell periphery. After that, formation of pseudopodia becomes confined to the outer edge of the lamellar cytoplasm. Attachment of pseudopodia gradually leads to the extension of the ring of lamellar cytoplasm. Simultaneously, the central part of the cell body is gradually flattened until the entire cell acquires the flattened shape.

Until the end of the radial attachment stage, all the external edge of the cell remains active. At the next stage of cell polarization, this activity becomes confined to certain parts of the edge, while the other parts become stable and stop extending pseudopodia. A polarized cell, because of division of its edge into stable and active zones, becomes able to move directionally on the substratum. In the moving cell, the widest active zone (leading edge) and the largest area of lamellar cytoplasm (anterior lamella) are usually located at the anterior of the cell. Besides lamellipodia that are parallel to the substratum, vertical cytoplasmic folds (ruffles) are continually formed at the active edges; these ruffles then move along the upper surface towards the central part of the cell. Ingram (1969) obtained side-view photographs of the lamellipodia formed at the leading edge of fibroblasts. He showed that rapid (about 4–5 μm/min) extension of a lamellipodium in the direction parallel to the substratum is followed by its contraction. Ingram suggested that contraction may follow several different courses. If contraction is stronger on the upper surface of the lamellipodium, it will lift upwards and will be transformed into a ruffle. An equally strong contraction of both surface will lead to retraction. If contraction of the lower surface is stronger, the pseudopodium will curl towards the substratum and eventually make contact with it.

Thus, examination of the morphology of spreading shows that this process consists of a great number of pseudopodial reactions, of the extension and attachment of pseudopodia. A special cell part (lamellar cytoplasm) is formed

from the attached pseudopodia. Locomotion of the substrate-attached fibroblast is also accompanied, and probably caused, by the polarized formation
and attachment of pseudopodia at the leading cell edge.

Special properties of the surface of pseudopodia and lamellar cytoplasm

The surfaces of pseudopodia and lamellar cytoplasm formed in the course of
spreading have a number of properties in common which are different from
the properties of the surface of other cell parts or the body of unspread cells
and of the endoplasm of the spread cells. These special properties include:
(*a*) the ability to form focal sites of cell–substratum attachment with
accompanying bundles of microfilaments; (*b*) the ability to develop centripetal
tension in the cytoplasm; (*c*) the ability to attach to various inert particles
and move them centripetally along the surface; (*d*) the ability to move various
membrane receptors, cross-linked by external ligands, centripetally along the
surface. We will now discuss each of these properties in more detail.

Formation of focal attachment sites. The focal sites of attachment to the
substratum (plaques, focal contacts) formed by fibroblasts in the course of
spreading have the following morphological features (Abercrombie, Heaysman
& Pegrum, 1971; Brunk *et al.*, 1971; Revel & Wolken, 1973; Harris, 1973*a;*
Revel, 1974): close apposition of the membrane to the substratum (the mean
distance does not exceed 10–15 nm); increased electron density of the
cytoplasm near the membrane; and attachment of a bundle of microfilaments
(fig. 3.4). Interference–reflection microscopy shows that each focal contact in
the spread fibroblast is 2–10 μm long, by about 0.25–0.5 μm wide (Curtis,
1964; Izzard & Lochner, 1976). These focal contacts are absent at the initial
stage of spreading. The first attachment sites with accompanying microfilament
bundles are formed at the beginning of radial spreading; at this stage they
are localized only in filopodia but not in other cell parts (Bragina *et al.*, 1976).
At the more advanced stage of spreading numerous attachment sites and
bundles are seen in the lamellar cytoplasm; some of the bundles extend also
into the cortical layer of the central cell part. A polarized fibroblast moving
on the substratum also forms new attachment sites at the active cell edge
(Abercrombie, *et al.*, 1971).

As shown by Heath & Dunn (1978), each distal end of a microfilament
bundle present in a spread chick fibroblast is associated with a focal contact
and, conversely, each focal contact is associated with the bundle. Thus, focal
attachment sites are formed only by pseudopodial surfaces; formation of
most, possibly all, of the microfilament bundles present in the spread cell
seems to be initiated at the attachment site.

Focal contacts are not the only structures involved in the cell–substratum
attachment of fibroblasts and other cultured cells. Interference–reflection

(a)

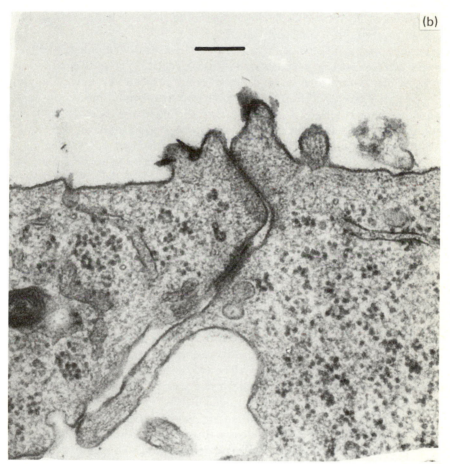

(b)

Fig. 3.4. Cell contacts with the substratum and transmission electron micrographs of sections perpendicular to the substratum. (a) Cell–substratum attachment sites near the anterior edge of a polarized mouse embryo fibroblast. (b) Cell–cell contacts between lateral surfaces of epithelial cells of the MPTR line. Bars: 0.5 μm.

microscopy and electron microscopy reveal besides focal contacts, so-called 'close contacts' at the lower surface of the lamellar cytoplasm of fibroblasts. These are wide areas representing about 30 nm separation from the substratum (Izzard & Lochner, 1976; Heath & Dunn, 1978). Close contacts are not connected to microfilament bundles but may be associated with a dense meshwork of microfilaments (Heath & Dunn, 1978). Certain substratum-attached cells, such as polymorphonuclears or macrophages, have close contact areas but no focal contacts (Abercrombie, Dunn & Heath, 1977). It is not clear whether formation of these contacts is associated with pseudopodial activity but this seems probable, because these contacts, like focal contacts, are preferentially located at the lower surface of the lamellar cytoplasm, but not on the surface of other cell parts.

Besides their distinctive ability to form cell–substratum contacts, pseudopodial surfaces are also able to attach various particles. This is especially clear in experiments with substratum-attached cultures of kidney epithelium and certain other epithelia. The cells in these cultures form coherent sheets in which the upper surfaces of the central cells are inactive, that is, do not form pseudopodia. In these sheets are formed pseudopodia only on the lateral surfaces of the marginal cells that are free of cell–cell contacts. Correspondingly, various types of particles (carmine, carbon, etc.) are attached preferentially to these active areas of the marginal cells (Di Pasquale & Bell, 1974; Vasiliev *et al.*, 1975*a*).

Development of tension. Extension of a pseudopodium is probably always followed by the development of centripetal tension within the cytoplasm of this pseudopodium. The tension may be responsible for the retraction of unattached pseudopodia, upward bending of the pseudopodia and subsequent movement of the ruffle on the upper cell surface. The tension continues to exist after the attachment of the pseudopodium to another surface: when cell–substratum contact is destroyed by any treatment, e.g. mechanical, contraction of the cytoplasm near the contact takes place immediately. Cells spread on fibrin fibres are able to distort this meshwork (Harris, 1973*c*). This observation shows that tension persists within the attached cell. We do not know whether the degree of tension is similar in an unattached pseudopodium, in attached pseudopodium and in a detached pseudopodium. Pseudopodia and lamellar cytoplasm do not contain large vesicular organelles that can be visualized by phase contrast microscopy. Possibly, centripetal tension acts not only on the contact structure but also on vesicles which happen to be present within the cytoplasm of these cell parts. As a result, these organelles are displaced into more central parts of the cell.

Centripetal movement of particles. Various types of particles (carbon, colloidal gold, anion exchange resin, carmine, etc.), attached to the upper or lower surfaces of pseudopodia and lamellar cytoplasm, start to move centripetally along that surface. This movement was observed and studied by Abercrombie and his group (Abercrombie, Heaysman & Pegrum, 1970*c*; Harris & Dunn, 1972; Harris, 1973*c*) and later by other investigators (Di Pasquale & Bell, 1974; Albrecht-Buehler & Lancaster, 1976). All the morphological types of pseudopodia formed at the early stage of cell spreading have the ability to move particles of colloidal gold (Albrecht-Buhler & Lancaster, 1976). On the surface of polarized, spread chick fibroblasts the particles moved with an initial velocity of about 3–4 μm/min, and gradually this movement slowed down and stopped (Harris, 1973*c*). Our observations show that when a particle is moving on the surface of lamellar cytoplasm its directional movement usually stops when it reaches the boundary of the endoplasm. Particles located on the surface of the endoplasm perform only random displacements.

Ability to clear the surface of cross-linked surface receptors. Incubation of cells with a medium containing a soluble multivalent ligand to certain surface receptor may induce redistribution of the corresponding receptors (see review in De Petris, 1977, 1978*c*). This redistribution has two components.

(1) *Patching.* Receptors that are more or less diffusely distributed on the surface of an intact cell become assembled into microscopically visible clusters or patches; it is usually assumed that patching is due to cross-linking of receptors by the ligand.

(2) *Clearing (capping).* Patched receptors move in the plane of the membrane so that they are selectively removed from certain parts of the surface. In particular, in the experiments of several authors (Abercrombie, Heaysman & Pegrum, 1972; Edidin & Weiss, 1972; Vasiliev *et al.*, 1976; Brown & Revel, 1976), when the substratum-spread fibroblasts or epithelial cells were incubated with certain ligands, patches were formed over the whole cell surface, but only the surface of pseudopodia and lamellar cytoplasm was subsequently cleared of patched receptors. For instance, in our experiments (fig. 3.5) the cleared zone first appeared near the active peripheral edges and gradually progressed centripetally. After 30–60 min the entire surface of the lamellar cytoplasm became cleared of the patched receptors but the cleared zone did not progress beyond the boundary between the lamellar cytoplasm and endoplasm. Suspended fibroblasts that had no attached pseudopodia developed patches but not cleared surface zones. When epithelial islands were incubated with ligands, only the outer active zone of marginal cells became cleared; patching but not clearing was observed on the inactive surface of the central cells of the islands.

This type of clearing was observed in experiments with normal mouse

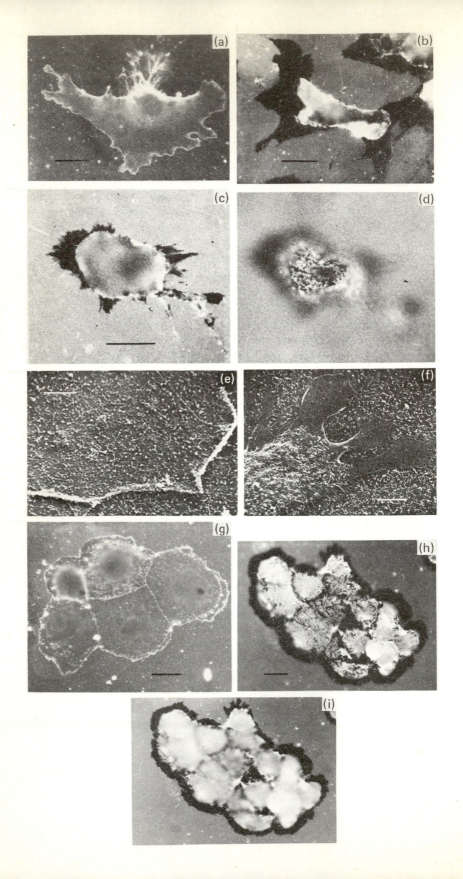

fibroblasts, transformed mouse L fibroblasts, or kidney epithelial cells, treated with concanavalin A. Similar clearing was also observed in experiments with mouse fibroblasts and epithelial cells treated with cationized ferritin which presumably cross-links negatively charged surface groups. The binding of ligands to surface receptors of substratum-attached cells has not been followed in all cases by patching and clearing: in certain published experiments, the distribution of receptors remained random on untransformed cells (e.g. Ukena *et al.*, 1974). An interesting variant of ligand-induced redistribution was described by Ash & Singer (1976) and by Ash *et al.* (1977). In their experiments human fibroblasts and rat kidney line (NRK) cells were incubated with different ligands (concanavalin A or antibodies to three different membrane proteins). In each case, corresponding receptors became patched and in a certain percentage of cells the patches became aligned over microfilament bundles present in the lamellar areas of cytoplasm. We also have observed linear arrays of concanavalin A patches in certain mouse fibroblasts pre-incubated with this ligand for a short time. Possibly, these arrays are formed at the early stage of redistribution preceeding clearing. It is not clear why in many systems incubation of the substratum-attached cells

---

Fig. 3.5. Redistribution of concanavalin A receptors on the surface of cultured cells after incubation with the corresponding ligand. The receptors were revealed by an indirect immunofluorescence method (a–d, g–i) or by scanning electron microscopy with phage T4 particles as markers (e, f). For a description of the methods used see Vasiliev *et al.* (1976). (a) Normal mouse fibroblast pre-fixed before incubation with concanavalin A showing diffuse distribution of receptors on all parts of the cell surface. The surface of the substratum around the cell is also fluorescent because of adsorption of serum proteins reacting with concanavalin A. (b) Normal mouse fibroblast incubated in the medium with concanavalin A (25 μg/ml) for 20 min and then in the medium without concanavalin A for 60 min before fixation. Concanavalin A receptors were removed from the lamellar cytoplasm (dark zones). (c, d) Transformed mouse L fibroblast incubated for 20 min with concanavalin A and for another 60 min without. Photograph focussed at the small peripheral lamellar areas: (c), reveals the removal of concanavalin A receptors from these areas; (d) photograph of the same cell focussed on the surface of its central part reveals numerous patches of concanavalin A receptors. (e, f) Scanning electron micrographs of lamellar areas of normal fibroblasts pre-fixed before incubation with concanavalin A (e), or incubated with concanavalin A and then fixed (f). The distribution of concanavalin A receptors is revealed by marker phage particles: (e), random distribution of receptors on the surface of a pre-fixed cell; (f) receptors were cleared from the surface of peripheral cytoplasmatic outgrowths (smooth zones) after incubation with ligand. (g–i) Redistribution of receptors on the surface of cells of an MPTR line, forming epithelial islands: (g) pre-fixed cells, receptors are distributed diffusely; (h, i), the island was pre-incubated with concanavalin A before fixation. The photograph focussed on the peripheral free edges reveals the clearing of their surfaces (h). A photograph of the same field of view focussed on the surface of the central part of receptors (i) shows that the surface areas near lateral cell–cell contacts are not cleared or cleared zones are very narrow. Bars: (a–c), 20 μm; (e), 1 μm; (f), 2 μm; (g–i), 20 μm.

with ligands leads to complete clearing of receptors from certain areas of the surface while in other conditions redistribution of receptors is less pronounced. The valency of the ligands and density of receptors over the cell surface may possibly affect the final result. For our present discussion, it is significant that in all cases in which the clearing of cross-linked receptors took place at all, the only cleared parts of the surface of substratum-attached cells were pseudopodia and lamellar cytoplasm.

Properties of the substratum essential for spreading

In the course of spreading the cell surface is attached to some ligands present on the surface of the substratum. What is the nature of the ligands and receptors participating in this reaction? The only case in which attachment involves well-known receptors and ligands is that of lymphocytes carrying receptors to F_c fragments of immunoglobulin molecules. These lymphocytes are able to spread on surfaces covered with immobilized antigen–antibody complexes; that is, on surfaces with the exposed immunoglobulin molecules (Alexander & Henkart, 1976). It has been reported that fibroblasts and other tissue cells, in contrast to lymphocytes, do not attach to surfaces covered by immunoglobulins (Giaever & Ward, 1978). This avoidance, if confirmed, will be a surprising exception. Generally, tissue cells are not very selective in their choice of substratum; they are able to spread on many types of solid surfaces, including metals, glass, plastics and layers of different proteins. The final degree of spreading depends on the nature of the substratum but the factors involved are not fully understood. Empirically, it was found that certain surfaces, for instance, polylysine-coated surfaces (McKeehan & Ham, 1976) are especially favourable for spreading. Polylysine-coated surfaces carry cationically charged groups. In other conditions, the anionic charge of the substratum was found to enhance spreading: the attachment and spreading of mouse fibroblasts on a polystyrene surface in a serum-free medium was improved when the anionic charge of the surface was increased by sulphonation to a certain optimal level (Maroudas, 1977). Possibly, an optimal polarity of the surface is essential for spreading; this polarity can be provided either by negatively charged or by positively charged groups.

The substratum-absorbed proteins derived from serum or secreted by cells can promote spreading (see review in Grinnell, 1978). Two particular groups of proteins should be mentioned in this connection: collagen and fibronectins. Collagen fibres form the natural substratum to which fibroblasts are attached *in vivo*. Collagens are secreted by the cells in culture (Gay *et al.*, 1976; Bornstein & Ash, 1977); these fibrils probably form an intermediate layer between the artificial substratum and the cell surface. Fibronectins (LETS-proteins,* cold insoluble globulins) are of high molecular weight (about

---

* LETS, Large external transformation sensitive.

200000–250000) glycoproteins (see reviews in Hynes, 1976; Yamada & Olden, 1978; Vaheri & Mosher, 1978). These molecules may undergo polymerization into dimers and larger aggregates (Keski-Oja, 1976; Ali & Hynes, 1978b). Closely related varieties of these proteins are present in plasma and synthesized by normal fibroblasts, glial cells, endothelial cells (Birdwell, Gospodarowicz & Nicolson, 1978) and by certain epithelial cells (Chen *et al.*, 1977b). Fibronectins are bound by collagens and fibrin (Engvall, Ruoslahti & Miller, 1978). Polymerized cellular and plasma fibronectins can form fibrillar arrays at the cell surface: these fibrils are located externally from the cell membrane. The morphological distribution of fibronectin corresponds to that of collagen fibrils (Bornstein & Ash, 1977). Probably these proteins interact in a pericellular matrix.

The involvement of fibronectins in cell spreading on the substratum is indicated by several groups of facts. (*a*) Plasma or cell-derived fibronectins attached to plastic or collagen-coated substrata promote cell attachment and spreading. Serum factors essential for spreading are probably identical to serum fibronectin (Pearlstein, 1976; Grinnell, 1976; Grinnell & Hays, 1978a). (*b*) At least some of the fibronectin fibrils are usually located at the cell–substratum interface of the spread cells (Stenman, Wartiovaara & Vaheri, 1977; Mautner & Hynes, 1977; Hedman, Vaheri & Wartiovaara, 1978; Yamada, 1978). (*c*) Fibronectin added into the medium of normal cultures improves cell migration on the substratum (Ali & Hynes, 1978a). (*d*) Fibronectin added to cultures of badly attached transformed cells improves their spreading (see chapter 6).

In summary, these facts suggest that the cell membrane may become attached to diverse 'non-specific' solid substrata via intermediate molecules of fibronectins and collagens. Glycosaminoglycans contained in the serum or secreted by the cells also possibly serve as intermediate molecules. The inter-relations between all these molecules are not clear. It is possible that in certain cases the 'attachment sandwich' may consist of the substratum, collagen, fibronectin and the cell membrane. We do not know which membrane receptors interact with fibronectin or collagens. The exact localization of fibronectins and collagens in relation to focal and close cell–substratum attachment sites also remains to be elucidated.

Fibronectin-mediated attachment is probably a common mechanism but, of course, not the only one possible. We have mentioned already the attachment of lymphocytes to immunoglobulin-covered surfaces. Attachment of cells to surfaces precovered by concanavalin A was observed by several authors (Edelman, Rutishauser & Millette, 1971; Seglen & Fossa, 1978); it is probably mediated by specific receptors. In the experiments of Grinell and Hays (1978b) the substrata were pre-coated with ligands directed against various receptors of cell surface: concanavalin A, cationized ferritin, and antibodies against BHK plasma membrane. BHK cell spreading on these substrata was similar to that on a substratum pre-coated with fibronectin.

Cultured tumour cells that have surface receptors for certain hormones or toxins can be selectively attached to the glass or sepharose bead surfaces carrying immobilized specific ligands; for instance, rat adrenal tumour cells were bound to immobilized corticotropin (Venter, Venter & Kaplan, 1976). In principle, the cell can probably attach itself to any surface for which it has receptors and different receptors may be used by the same cell for attachment to various surfaces.

As discussed above, attached cells exert tension on contact structures. To withstand this tension the surface of the substratum should have sufficient mechanical rigidity (Maroudas, 1973a). One particular example demonstrating the role of this rigidity in experiments testing the adhesiveness of various artificial lipid films for fibroblasts (Ivanova & Margolis, 1973; Margolis *et al.*, 1978). A comparative study of the films made from different lipids showed that their adhesiveness is correlated with the fluidity of the films: all films in a crystalline state at 37 °C were adhesive for fibroblasts while all films in a liquid-crystalline state at this temperature were non-adhesive. Probably, molecules of a liquid-crystalline film are too easily displaced by the contacting pseudopodia and therefore the attachment is not stable.

Other examples of pseudopodial reactions

Phagocytosis. Phagocytosis and spreading are two variants of the same process. Spreading can be regarded as an effort by the cell to phagocytose an infinitely large foreign body. Phagocytosis can be regarded as the formation of pseudopodia near the site of attachment of the microscopical particle and eventually leading to complete internalization of this particle. Solid particles of various kinds, not covered with specific ligands, can be phagocytosed by polynuclears, macrophages and fibroblasts. Possibly, their attachment occurs through the interaction of the particle with the same unidentified surface receptors that mediate cell–substratum attachment. Pre-covering of the particle or of the cell with concanavalin A and other lectins may enhance their attachment and phagocytosis (Goldman, Sharon & Lotan, 1976; Goldman, 1977). Attachment may be also enhanced by the pre-covering of particles with certain protein molecules (e.g. immunoglobulin) to which the macrophage surface has specific receptors. Apparently, in these cases specific receptors of the corresponding ligands participate in attachment. Cell membrane attached to the particle probably contains only certain types of receptors and other receptors are excluded from this area. For example, the internalization of particles did not lead to decrease of the membrane transport sites for amino acids remaining on the cell surface (Tsan & Berlin, 1971) – possibly, these sites were excluded from the phagocytosized membrane. Phagocytosis, like spreading, can be accompanied by morphological alterations of the cortical layer. Development of the bundles of microfilaments near the site of attachment of particles at the first stage of phagocytosis has been observed in experiments with macrophages (Reaven & Axiline, 1973). Berlin & Oliver (1978) observed the formation of a dense network of microfilaments in pseudopodia formed in the course of phagocytosis in leucocytes; this mesh disappeared after internalization of the particles.

Reactions induced by soluble ligands. As described above, surface receptors cross-linked by molecules of soluble ligands can be cleared from the surface of pre-formed pseudopodia of substratum-attached fibroblasts and epithelial cells; soluble ligands themselves can also induce formation of pseudopodia. One type of pseudopodial reaction induced by these ligands in various cells is endocytosis leading to internalization of patched receptors. The multivalency of these ligands has been found to be essential for the induction of endocytosis (Goldman, Sharon & Lotan, 1976). Patching of receptors in suspended lymphocytes may also be followed by capping; that is, by directional movement of cross-linked receptors into one part of cell surface. As shown by de Petris (1975) a large cytoplasmic protrusion is formed on one pole of the cell in the course of capping. The surface of this protrusion, in contrast to that of other cell parts, is free from patched receptors. This protrusion gradually increases until almost all the cell organelles are translocated into its cytoplasm. Thus, capping can possibly be regarded as a special type of pseudopodial reaction that can be induced in suspended cells by soluble ligands.

In our unpublished experiments, suspended lymphocytes and fibroblasts were incubated with concanavalin A followed by anti-concanavalin A antibody. Patching of corresponding receptors was observed on the cells of both types; however, capping was observed only on lymphocytes but not on fibroblasts. The cause of this difference is not clear.

Capping, like other pseudopodial reactions, was reported to be accompanied by redistribution of actin and myosin as revealed by immuno-morphological methods (Toh & Hard, 1977; Schreiner *et al.*, 1977; Gabbiani *et al.*, 1977). Alterations in the distribution of microfilament bundles in the course of capping were also revealed by electron microscopy (Clark & Albertini, 1976; Albertini & Anderson, 1977). According to Berlin & Oliver (1978), protuberances formed in the course of capping and pseudopodia formed during phagocytosis in leucocytes have a similar ultrastructure: neither formation contains microtubules but both have numerous microfilaments.

Capping can usually be induced only by polyvalent ligands; monovalent ligands do not induce capping, although certain exceptions have been described (Stackpole *et al.*, 1974). The cross-linking of many types of different receptors induces capping; the list of these receptors includes not only membrane proteins but also ganglioside receptors of cholera toxin (Revesz & Greaves, 1975).

Capping can be followed by endocytosis or by 'shedding' of cross-linked receptors from the membrane into the surrounding medium (Karnovsky, Unanue & Leventhal, 1972; Unanue, Perkins & Karnovsky, 1972). The nature of 'shedding' is not clear.

Possible mechanisms of pseudopodial reactions

We have seen that cell contact with the substratum, a particle, or a soluble ligand, can induce the formation of pseudopodia which are able to attach themselves to different surfaces. The molecular mechanisms for the different stages of these pseudopodial reactions remain unknown. In this section we will briefly discuss several specific questions arising with regard to these reactions.

What induces the formation of pseudopodia? Different factors inducing the formation of pseudopodia (the solid surfaces of particles and substrata; molecules of various soluble ligands) interact with different surface receptors; this is especially clear in experiments with phagocytosis, endocytosis and capping (just described). One common link in the action of all these factors may be the immobilization of a group of surface receptors by contact with solid surfaces or by cross-linking caused by soluble ligands. Possibly, the formation of a locally immobilized group of surface components may be a signal for further change in the membrane or in the cortical layer.

Another membrane-related phenomenon, exocytosis in mast cells, was also found to be induced by cross-linking of certain membrane receptors. Binding of monomeric immunoglobulin E molecules to mast cell membrane receptors did not induce the release of granules while dimers and trimers of the same molecule produced effective degranulation (Segal, Taurog & Metzger, 1977).

How are pseudopodia extended? Obviously, this extension is a result of local movements of some components of the cortical layer but the exact nature of these movements is not known. Among the several possible mechanisms discussed in the literature are the following. (a) Actin microfilaments may slide relative to each other as a result of their interaction with myosin molecules (Bray, 1973; Huxley, 1976). (b) Extension may be due to local polymerization of microfilaments from monomeric actin (Tilney *et al.* 1973). (c) Local depolymerization of pre-existing microfilaments may produce a 'hole' in the cortical layer. Because of hydrostatic pressure existing in the cell interior, cytoplasm would start to flow through this 'hole' and produce a local extension (Harris, 1973c).

These suggestions are not mutually exclusive: all three mechanisms may participate in the advancement of pseudopodia. Small, Isenberg & Celis, (1978) have shown that microfilaments of the bundles and meshwork in the advancing lamella of a human fibroblast have uniform polarity: arrowheads formed by binding of myosin subfragment-I to these microfilaments are always directed towards the cell body. The direction of polymerization of actin microfilaments in cell-free systems is usually opposite to that of arrowheads; therefore, Small *et al.* suggest that actin microfilaments in the advancing lamella are always polymerized in the forward direction: from the cell body to the edge of the lamella. This result indirectly confirms the role of actin polymerization in the extension of pseudopodia.

The membrane which covers extended pseudopodia can be formed either from pre-existing material coming from other parts of the plasma membrane or by new material coming from the cell interior. We do not yet have reliable methods that can distinguish between these two mechanisms. For instance, the chemical composition of the membrane around latex particles phago-cytosed by mouse fibroblasts was reported to be different from the plasma membrane of the same cells. Labelled precursors of proteins, sugars and lipids were found to be differentially incorporated into the plasma membrane and into the membranes of phagosomes (Vicker, 1977). It was argued that these results suggest the formation of phagocytic vacuoles from new membrane material of unique composition. However, it remains possible that the phagocytic membrane is made from pre-existing plasma membrane but certain components are excluded from the area attached to the surface of the phagocytosed particle.

Why do particles and cross-linked receptors move centripetally on the surface of pseudopodia? This movement can be explained by the existence of a continuous flow of surface material in the plane of the membrane from the periphery to the centre: particles and receptor patches are carried away by this flow. An alternative hypothesis postulates that only immobilized groups of receptors are translocated centripetally by some special mechanism. The surface flow hypothesis (Abercrombie *et al.*, 1970c; Harris, 1973c, 1976a) suggests that movement of new surface material to the exterior takes place at the sites of formation of pseudopodia (see above); excess of this material flows back to the surface of central parts of the cell where it is moved to the interior by some unknown mechanism. Eventually material moves again from the interior to the exterior at the active part of the cell surface; thus, recycling of surface material takes place in pseudopodia-forming cells. All the components of the membrane (Harris, 1976a) or only the lipids (Bretscher, 1976) may participate in this recycling. As mentioned above, at present we have no way of proving or disproving the existence of surface flow of this type. Another group of hypotheses (De Petris & Raff, 1973; Ash *et al.*, 1977) postulates the existence of a special mechanism translocating only immobilized groups of receptors; that is, cross-linked receptors or receptors attached to the surface of a particle. It is suggested that these groups of receptors become attached (anchored) to some cortical structures which exert centripetal tension upon them. Association of receptor patches in lymphocytes with actin- and myosin-containing areas observed by immunofluorescence can be regarded as indirect evidence for a linkage between patched receptors and cortical structures. This suggestion also explains the formation of linear arrays of patched receptors of different types over microfilament bundles in spread fibroblasts (see above). It is more difficult to explain these phenomena on the basis of the surface-flow hypothesis. Clearing of patched receptors leads to temporary disappearance of this class of receptors from the cell surface: when cells preincubated with a certain ligand are fixed and then reincubated with this ligand, corresponding receptors are not found in the cleared zone; new receptors appear in this zone only after several hours (Vasiliev *et al.*, 1976). This fact is also more difficult to explain by surface flow, than by anchoring of patched receptors. Thus, at present anchoring seems to be a more plausible explanation of centripetal movements. The exact mechanism of anchorage remains obscure. The difficult point here is: why does formation of patches of many different types of surface receptors lead to a similar result; that is, to their anchorage by cortical components? Ash *et al.* (1977) and Bourguignon & Singer (1977) suggest that actin is associated with the cytoplasmic surface of the plasma membrane by attachment of the ubiquitous membrane integral protein X; when other integral proteins are induced to form clusters, they become bound to protein X and hence to actin. Another possible suggestion is that the membrane contains some component preventing the anchorage of cortical structures. This anchorage-inhibiting component may be distributed randomly in the intact membrane, and patching may lead to exclusion of this component from certain zones of the membrane and thereby permit anchoring. At present we have no evidence which favours either of these schemes.

How are focal attachment sites formed? The process of formation of attachment sites is probably similar to patching in the sense that it also involves local accumulation of certain membrane components. These receptors can be various well-known membrane components – e.g. the receptors involved when lymphocytes spread on immunoglobulin-coated surfaces – or unknown receptors participating in the spreading of cells on various surfaces. Selective accumulation of certain membrane components was observed in various types of focal cell–cell contacts. As discussed above, this accumulation of receptors may initiate their anchoring to some cortical components. Anchoring may be essential for stabilization of attachment sites: membrane receptors attached to another surface at the external side of the fluid membrane can easily be displaced by any mechanical stress unless they are anchored at the internal side of the membrane.

Formation of microfilament bundles associated with attachment sites indicates that anchoring of the membrane receptors leads to extensive alteration of the structure of the cortical layer. Possibly, anchoring somehow induces polymerization of actin microfilaments or their redistribution in the cortical layer. Accumulation of actinin at the attachment site under the membrane has been observed (Lazarides & Burridge, 1975; Scholmeyer *et al.*, 1976). Possibly, actinin is accumulated at the first stage of anchoring of the membrane receptors, while microfilament bundles are bound to actinin-containing structures at the following stage. Actin microfilaments are bound by actinin-containing Z-plates in striated muscle. Membrane–microfilament connections via actinin-containing structures were also described in the intestinal microvilli (Mooseker, 1976).

As discussed above, the polarity of actin microfilaments in the anterior lamella of the fibroblast suggests that polymerization of actin is directed from the cell body to the tip of an advancing pseudopodium. The distal end of a newly formed microfilament may establish a connection with the structures formed at the cytoplasmic side of the attached membrane area. When the contractile cytoplasm of *Physarum* containing a meshwork of microfilaments undergoes isometric contraction, parallel alignment of microfilaments takes place (Fleischer & Wohlfarth-Botterman, 1975). In a similar way, isometric contraction of the cytoplasm of an attached pseudopodium may lead to parallel alignment of the membrane-attached microfilaments; that is, to formation of microfilament bundles (Abercrombie *et al.*, 1977). Other proteins may become associated with newly formed bundles at a later stage. For instance, tropomyosin was absent in bundles formed in human lung cells during the first hour of spreading but several hours later tropomyosin became associated with them (Lazarides, 1976c). Both the bundles and the meshwork of microfilaments may be responsible for the tension acting on the attachment sites. The nature of this tension is probably the same as the centripetal force which translocates patched receptors and particles. In this connection it is interesting to note that normal mouse fibroblasts spreading in a medium containing cytochalasin B retain the ability to form cylindrical pseudopodia and to attach these pseudopodia to the substratum (Goldman & Knipe, 1973). However, as shown in fig. 3.6, the attachment sites of these cells, in contrast to normal attachment sites, are not linked to microfilament bundles. The pseudopodia of cytochalasin-incubated cells, in contrast to normal ones, are not contracted after detachment; patched concanavalin A receptors are not cleared from the surface of these pseudopodia. These experiments indicate that the formation of focal attachment sites can be dissociated from subsequent reactions leading to formation of microfilament bundles, development of tension, and of the clearing ability. Only this later stage is inhibited by cytochalasin B.

As described above, the ability to form focal attachments and to move patched receptors and particles centripetally are characteristic properties of pseudopodial surfaces, and this distinguishes them from the other 'inactive' parts of the cell surface. What is the basis of this difference? The considerations

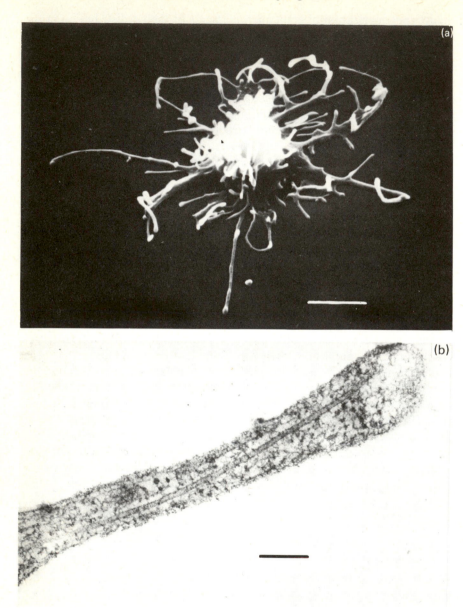

Fig. 3.6. Spreading of normal mouse fibroblasts in a medium containing cytochalasin B. (a) Scanning electron micrograph of attached filopodia extended by a fibroblast. (b) Electron micrograph showing a section of a microtubule in the cytoplasm of a filopodium; microfilament bundles are absent. Bars: (a), 5 μm; (b), 0.2 μm.

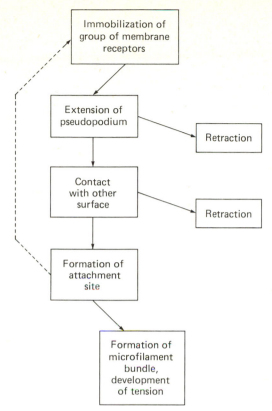

Fig. 3.7. Possible stages of the pseudopodial attachment reaction.

discussed above make it probable that the ability to anchor immobilized receptors may be characteristic only the pseudopodial surface. We do not know what the molecular basis is for this distinctive ability. Possibly, it may be related to peculiarities of the chemical composition of the cortical layer in these zones.

Concluding remarks. Formation of a group of immobilized receptors can be regarded as an initial event of the pseudopodial reaction (fig. 3.7). Formation of this group of receptors may induce some cortical changes leading to extension of the pseudopodium. If an initially formed or other group of immobilized receptors is present at the pseudopodial surface it may induce another set of cortical changes leading to anchorage of the surface and association with microfilaments. Depending on the reaction conditions, this process will lead either to the formation of an attachment site or to the centripetal movement of immobilized receptors. Needless to say, this scheme should only be regarded as a working hypothesis.

Stabilization of surface activities: the role of cytoplasmic microtubules

We have discussed pseudopodial reactions of the cell surface, which are induced by substrata, particles and soluble ligands. Cells can co-ordinate pseudopodial reactions taking place at various parts of the surface; they can regulate the ability of various parts of the surface to perform these reactions. The system of cytoplasmic microtubules is responsible for many aspects of this co-ordination of surface activities. Here we will briefly describe the organization of this system and the effects of its destruction on the course of pseudopodial reactions.

Organization of the microtubular system in interphase tissue cells

Microtubules are cylindrical structures with an outer diameter of about 22–25 nm. They are composed of a special protein called tubulin: a dimer consisting of α and β monomers. Besides tubulin, microtubules possibly contain certain other proteins including the so-called τ-proteins (Murphy & Borisy, 1975; Connolly *et al.*, 1977).

Microtubules have an intrinsic polarity: all microtubules contained within the interphase tissue cells (fig. 3.8) radiate from a few centres known as microtubule-organizing structures. These structures correspond to centrioles or basal bodies and are often associated with a short or long cilium. The growth of microtubules is polar and proceeds from the centre to the periphery of the cell (Osborn & Weber, 1976*a, b;* Frankel, 1976; Weber *et al.*, 1978). In cell-free systems microtubules can be polymerized from the solution of tubulin without the addition of any microtubule-associated proteins or microtubule-organizing structures (Herzog & Weber, 1977); however, in tissue cells formation of microtubules without an association with centres has not been observed.

Electron microscopy often reveals thin transverse bridges connecting the surfaces of microtubules with each other, with vesicular organelles, or with the plasma membrane. The chemistry and function of these bridges are not clear. Microtubule-associated proteins can increase the viscosity of mixtures of actin microfilaments and microtubules (Griffith & Pollard, 1978). Possibly these proteins mediate actin–tubulin interactions.

Certain drugs (so-called antitubulins) selectively affect the microtubular system. This group of drugs includes colchicine, colcemid, vinblastine, and podophyllotoxin. These drugs reversibly destroy cytoplasmic microtubules in tissue cells but the central structure remains unaffected. After removal of the drug from the medium the microtubular system is restored by polar growth from the centres. Colchicine binds to tubulin dimers in solution. These colchicine–tubulin complexes are then bound by the ends of the assembling microtubules and prevent their further growth (Margolis & Wilson, 1977).

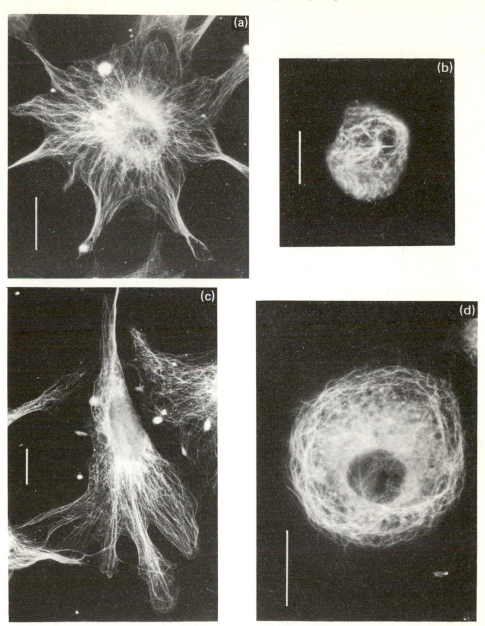

Fig. 3.8. Microtubules in spreading fibroblasts as revealed by indirect immuno-fluorescence. (a) A suspended cell; (b) peripheral circular bundle of microtubules in a radially spread cell 3 h after seeding; (c) microtubules radiating from the centre into peripheral outgrowths of a star-like cell at the beginning of polarization 5 h after seeding; (d) microtubules radiating into the anterior lamella and into the tail part of a polarized fibroblast 24 h after seeding. Bars: (a), 10 μm; (b–d), 20 μm.

The mechanism of action of the drugs is not quite clear. Probably, a dynamic equilibrium between the polymerization and depolymerization of microtubules exists within the cells and by inhibiting polymerization antitubulins shift this equilibrium in favour of depolymerization. Lumicolchicine, an inactive analogue of colchicine which is not bound to tubulin, is often used in control experiments in order to distinguish specific effects of colchicine from non-specific side effects. Experiments with antitubulins have provided a number of facts about the possible roles of microtubules in different cellular processes. We will describe the effects of these drugs on various processes involving pseudopodial reactions (spreading, phagocytosis, capping, and endocytosis).

Effects of antitubulins on spreading

These effects have been studied in experiments in which suspended fibroblasts from mice were allowed to spread on the substratum in media containing colcemid or other antitubulins in concentrations sufficient to destroy most microtubules (Ivanova *et al.*, 1976; Vasiliev & Gelfand, 1976). The cells spreading in these media, like control cells, start to form pseudopodia which attach to the substratum. However, the size and shape of individual pseudopodia become much more variable in the drug-containing media and their distribution along the cell perimeter becomes highly irregular. Numerous reversals of spreading are observed: the cells become partially detached from the substratum and begin to spread again. Most cells reach the well-spread stage only after several such unsuccessful efforts. Once reached, this state can be preserved indefinitely. Most cells reach the well-spread state in a colcemid-containing medium after 6–8 h as compared with 0.5–1.0 h in the control medium. In contrast to control cells, well-spread fibroblasts kept in medium containing antitubulins are unable to undergo polarization; that is, they are unable to stop pseudopodial activity over large areas of their edges. Cells that have already been polarized return to a non-polarized state when they are transferred into a medium containing antitubulins (Vasiliev *et al.*, 1970; Gail & Boone, 1971*a*; Bhisey & Freed, 1971). All, or almost all, the edge of these cells remains active indefinitely; therefore the cell is unable to perform directional translocation on the substratum. Correspondingly, the lamellar cytoplasm of these cells is not divided into several discrete zones but forms a ring around the central cell part (fig. 3.9). These antitubulin-treated cells, like controls, contain numerous focal contacts (Lloyd *et al.*, 1977). They also retain the ability to move centripetally attached particles and cross-linked receptors along their upper surface (Harris, 1973*c*; Domnina *et al.*, 1977). All these alterations are reversible: the normal state of the cells is restored after removal of colcemid or colchicine from the medium. These alterations are not produced by lumicolchicine. The data suggest that alteration of pseudopodial activity is a direct or indirect result of the destruction of the microtubules.

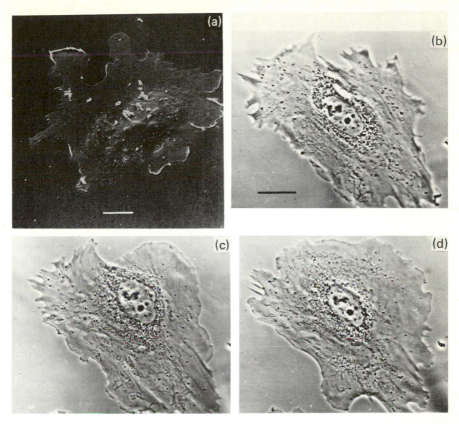

Fig. 3.9. Normal mouse fibroblasts incubated for 24 h in a medium containing: (a) colcemid (0.1 μg/ml) or (b–d) colchicine (0.2 μg/ml). (a) Scanning electron micrograph; lamellar cytoplasm is distributed around the whole cell perimeter. (b–d) A series of phase contrast micrographs of one cell at zero time, 45 min and 75 min, respectively. Extension and withdrawal of pseudopodia occur around the whole cell perimeter (cf. Fig. 2.1). Bars: (a), 10 μm; (b–d), 20 μm.

Effects of antitubulins on phagocytosis and reactions with soluble ligands

Cells of different types treated with antitubulin drugs retain the ability to phagocytoses (Moizhess, 1969). As reported by Ukena & Berlin (1972), colchicine interferes with the redistribution of membrane components accompanying phagocytosis. In colchicine-treated cells, in contrast to normal ones phagocytosis of particles leads to a decrease in the number of amino acid transport sites remaining on the cell surface. Colchicine promotes endocytosis induced by the soluble ligand, concanavalin A (Goldman, 1976).

The reported effect of antitubulins on the formation of caps by suspended cells varies. An increase in the number of cap-forming leucocytes and other cells was reported by several authors (Ukena *et al.*, 1974; Ryan, Borysenko & Karnovsky, 1974; Yahara & Edelman, 1975; Schlessinger *et al.*, 1977*b*; Berlin & Oliver, 1978). although inhibition of capping was also described (Aubin, Carlsen & Ling, 1975).

Stabilization of pseudopodial activities

The results described above show that destruction of the microtubular system by antitubulins does not lead to an inhibition of pseudopodial reactions: the cells retain the ability to form and to attach pseudopodia in reactions with the substratum and they are also able to perform phagocytosis and endocytosis. Analysis of the effects of antitubulins on spreading shows, however, that the microtubular system is essential for stabilizing the distribution of sites of formation and attachment of pseudopodia. At the early stage of spreading, the microtubular system stabilizes regular distribution of the sites of formation of pseudopodia along the whole active cell perimenter. At the stage of polarization, the microtubular system stabilizes formation of the cell edge into active and non-active parts. Without microtubules almost all the edge becomes active; that is, it can form pseudopodia. Possibly, the stimulation of endocytosis and capping observed after destruction of the microtubules are also due to an increase of the active part of cell surface: pseudopodia needed for these processes can be formed in all parts of the surface of antitubulin-treated cells.

The mechanisms of microtubule-dependent stabilization are not clear. A radially organized system of microtubules may be essential for interactions between the cell centre and the cell periphery. For instance, the system may be responsible for the radial organization of tensions in the cortical layer and for radial organization of the movements of cellular organelles. A cell with destroyed microtubules retains the ability to develop tensions and to perform intracellular movements, but co-ordination of the processes in different cell parts becomes deficient. Loss of control of pseudopodial reactions may be a result of this randomization of intracellular processes. Other manifestations of the altered co-ordination induced by antitubulins include inhibition of secretory processes of various types (see reviews in Lohmander *et al.*, 1976; Williams & Lee, 1976; Thyberg, Moskalewski & Friberg, 1978) as well as disorganization of other intracellular movements (Freed & Lebowitz, 1970; Kellermayer, Jobst & Szücs, 1978). Microtubules may control intracellular movements and tensions by interacting with microfilaments (Edelman, 1976), or by transducing signals of unknown nature (De Brabander *et al.*, 1977). But we know little about the function of this system and therefore it would be premature to discuss any detailed schemes of microtubule interactions with other cell components.

Although the distribution of active and non-active parts of the cell surface

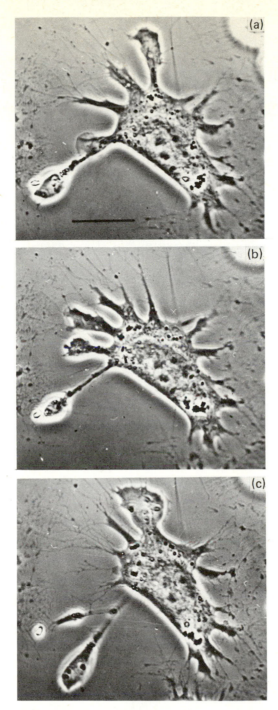

Fig. 3.10. Microtubule-independent stabilization of the cell edge. A series of phase contrast micrographs of transformed L fibroblasts in a medium containing 0.2 μg/ml colcemid. (a) Zero time, (b) 15 min, (c) 40 min. Pseudopodia are being attached and withdrawn in many parts of lateral edge but one area of the edge remains stable. Bar: 20 μm.

depends strongly on the microtubules, this does not mean that cells without microtubules are unable to perform any co-ordination of their surface activities. As described above, antitubulin-treated cells are able to reach and maintain a well-spread radial state; that is, a state in which formation of pseudopodia is restricted to the outer cell edge and pseudopodia are not extended at the upper surface. A detailed microcinematographic study of the activity of these cells has shown (O. Y. Ivanova, J. M. Vasiliev & I. M. Gelfand, unpublished) that certain areas of the cell edge occasionally remain stable for some time: up to 0.5–1.0 h in normal fibroblasts and up to 3.0–6.0 h in poorly attached transformed fibroblasts (fig. 3.10). Each non-active area of the edge in antitubulin-treated cells had a much shorter length than corresponding stable areas in control cells. These data show that the cells without microtubules are able to localize pseudopodial activity at the peripheral edge and also to perform short-range stabilization of certain areas of the cell edge. Long-range stabilization of the edge occurs, however, only in cells with intact microtubules. The mechanism of microbutule-independent stabilization remains to be studied. Possibly it is due to the stretching of the cortical layer produced by attached pseudopodia and perhaps, the new pseudopodia cannot be extended from the stretched part of the surface.

Another microtubule-independent mechanism preventing formation of pseudopodia in certain areas of cell edge, contact paralysis, will be described in the next section.

Formation of cell–cell contacts
Morphology of intercellular junctions

The intercellular junctions formed by epithelial and fibroblastic cells in culture are similar in morphology to those formed by these cells *in vivo*. The main types of these junctions (see descriptions in Farquahar & Palade, 1963; Goodenough & Revel, 1970; Friend & Gilula, 1972; Staehlin, 1974; Hull & Staehelin, 1976; Skerrow, 1978 and others) include desmosomes, tight junctions, zonulae adhaerens and gap junctions. All these structures are local morphological specializations of the contacting membranes. These specializations differ from each other in their organization, function and, probably, in chemical composition.

In tight junctions cell membranes are held in close contact, probably through direct linkages between certain membrane proteins; these proteins form rows of intramembrane particles which are revealed by freeze–fracture microscopy. Gap junctions also probably involve contacts between integral proteins forming cylindrical structures (see below). Direct contact between membrane components is apparently absent in desmosomes and in zonulae adhaerens. Here membrane components appear to be linked via some structures located between the membranes. In particular, desmosomes are characterized by a most complex structure that includes cytoplasmic plaques

– arrays of membrane particles which possibly function as bridging structures transversing the membranes and interspace material located outside the membranes (Skerrow, 1978).

Probably all types of junctions are linked at their cytoplasmic sites with fibrils and, in particular, with actin microfilaments (e.g. zonulae adhaerens; see review in Franke *et al.*, 1978*a*) or with intermediate filaments (desmosomes).

The gap junction is the only type of junction whose molecular structure has been studied in some detail. It is made up of units hexagonally arrayed with a two-dimensional crystal-like regularity in the plane of connected cell membranes; these repeating units, 8–10 nm wide, have been named 'connexons'. Connexons appear to span the bilayer of the plasma membrane and the gap between apposing cells (Caspar *et al.*, 1977). Isolated gap junction material was described to contain one principal protein (connexin) with a molecular weight about 20000 (Goodenough, 1976). According to a recently proposed model, the connexon is a cylindrical assembly of connexin molecules delineating an axial channel with a maximum diameter of about 2 nm (Makowski *et al.*, 1977). The functional role of this channel is to provide a way for the intercellular exchange of ions and small molecules. This intercellular communication takes place mainly or exclusively through gap junctions. The junctions are permeable to small molecules but not to RNA, DNA or proteins (Pitts & Simms, 1977). Cell communication of this type has been observed in practically all normal cultures of epithelial and fibroblastic cells (see review in De Mello, 1977). Cell communication may lead to the mutual exchange of metabolites and regulatory molecules between the cells so that many functions of the interacting cells may be performed in a unified fashion. The exact role of this co-operation remains to be established.

Formation of intercellular junctions

Formation of a specialized junction may require only a short time. For example, gap junctions and tight junctions have been observed to form within a few minutes after the establishment of contact between the surfaces of two cells (Loewenstein, 1967; Hudspeth, 1975). Experiments with cultured epithelia from various tissues showed that the mere contact of two pseudopodial surfaces leads to formation of specialized stable junctions. These junctions were formed when two active edges of substratum-spread homologous epithelial cells touched each other and contact with unspread spherical cells with an inactive upper surface of epithelial sheet did not lead to attachment (Middleton, 1973; Di Pasquale & Bell, 1974, 1975; Elsdale & Bard, 1974; Vasiliev *et al.*, 1975*a*). Cell–cell contact of two active edges of chick fibroblasts also leads to immediate formation of focal junctions with accompanying microfilament bundles (Heaysman & Pegrum, 1973).

Specificity of cell–cell contacts

The cells of different tissue types have different abilities to form specialized contacts. For instance, contact complexes including tight junctions, gap junctions and desmosomes are formed between the lateral surfaces of contacting epithelial cells in culture (Guillouzo *et al.*, 1972; Neupert, 1972; Middleton, 1973; Orci *et al.*, 1973; Pickett *et al.*, 1975). Owing to formation of these complexes, adhesion between epithelial cells is so firm that they are not detached from each other in the course of movement. Fibroblasts, in contrast to epithelial cells, form only unstable cell–cell contacts that are easily broken in the course of movement. Gap junctions and intermediate junctions are predominant morphological forms of specialized contacts in fibroblastic culture (Martinez-Palomo *et al.*, 1969; Pinto da Silva & Gilula, 1972; Cherny, Vasiliev & Gelfand, 1975).

We do not know what factors are responsible for these differences: are they due to different chemical composition of the membranes, or to the different structure of the cortical layer? Differences between the contact-forming abilities of different cells could be quantitative, in which case cells of different types should form contacts with each other. An alternative possibility is that these differences could be qualitative and cells of different types should not form mutual contacts. Experimental results suggest that both situations may exist. Formation of desmosomes between mouse skin epithelial cells and chick corneal epithelium were observed in mixed aggregates (Overton, 1977). Highly permeable junctions were formed between cultured cells belonging to various classes of vertebrates and various tissue types (Furshpan & Potter, 1968; Michalke & Loewenstein, 1971; Pitts & Burk, 1976) but these contacts were not formed between vertebrate and insect cells (Epstein & Gilula, 1977). In contrast to these examples of low selectivity, several other experiments have demonstrated selective formation of contacts between homologous cells. Fentiman, Taylor-Papadimitriou & Stoker, (1976) showed that human mammary epithelial cells easily established intercellular communication and the same was true for human mammary fibroblasts, but communication between epithelial and fibroblastic cells was absent. Similar results were reported for rat liver cells and BHK cells, by Pitts & Burk (1976). These results call for systematic studies of the selectiveness of the formation of cell–cell contacts between various normal cell types.

Cell–cell interaction in three-dimensional aggregates deserves special comment. Three-dimensional aggregates may be formed by suspended cells that have no possibility of becoming attached to the substratum. For a long time these aggregates have been favourite experimental systems for studying mutual 'adhesiveness' of embryonic cells. The most striking phenomenon observed in these aggregates is the ability of cells to sort out and to become attached preferentially to cells of the same tissue type (Holtfreter, 1943, 1944;

Moscona, 1965; Steinberg, 1963, 1964; Roth, 1968). Initial intercellular contacts in the aggregates are made by microvillar processes (Gershman & Rosen, 1978). Sorting of cells of different tissues has also been observed in cultures attached to the substratum (Steinberg & Garrod, 1975). Single suspended cells adhere better to pre-formed aggregates of similar cells than to aggregates of the cells of other tissue types. Species specificity was also observed in experiments with mouse and chick liver cells (Grady & McGuire, 1976). Possibly, the ability for sorting out and 'adhesive selectivity' are related to quantitative or qualitative differences in the formation of cell–cell contacts. However, they may also be due to many other factors; for example, to differences in the ability of cells to form pseudopodia, to develop tension in these pseudopodia, to detach themselves from other cells, etc. (see discussion in Moyer & Steinberg, 1976; Phillips, Steinberg & Lipton, 1977; Harris, 1976*b*). The formation of three-dimensional aggregates can be promoted by 'adhesion factors' released by cells into the medium or obtained from cell extracts. For example, a factor isolated from retinal cells promoted aggregation of these cells but not of the cells of other types. This factor was purified and proved to be a glycoprotein with a molecular weight of 50 000 (Hausman & Moscona, 1976). One can imagine many possible mechanisms of action of these factors on the complex aggregation phenomenon: they may stimulate formation of pseudopodia, act as membrane receptors or as external ligands for these receptors, etc. One particularly intriguing possibility is that specific lectins may be involved in formation of cell–cell contacts. The carbohydrate binding protein of the membranes of the liver cells participates in agglutination of erythrocytes to these cells (Kolb *et al.*, 1978). Aggregation mediated by species-specific lectins reacting with carbohydrate surface receptors have been demonstrated in slime molds (Frazier, 1976). Several lectins have recently been isolated from various animal tissues including muscle, liver and brain (Gartner & Podlesky, 1975; Bowles & Kauss, 1976; Simpson, Thorne & Loh, 1977, 1978; Mir-Lechaire & Barondes, 1978). The role of these lectins in cell–cell adhesion remains to be established.

Contact paralysis

The contact of two cell surfaces may be followed by cessation of pseudopodial activity. This phenomenon is called 'contact paralysis' or 'contact inhibition of pseudopodial activity' (see reviews of Abercrombie, 1970; Harris, 1974). Typical contact paralysis is observed when an active edge of a substratum-spread cell contacts an active or stable edge of another cell (fig. 3.11); in this case establishment of cell–cell contacts inhibits the formation of pseudopodia induced by cell–substratum contact. It is not clear whether the contact of a suspended cell with the surface of another suspended or attached cell induces or inhibits formation of pseudopodia.

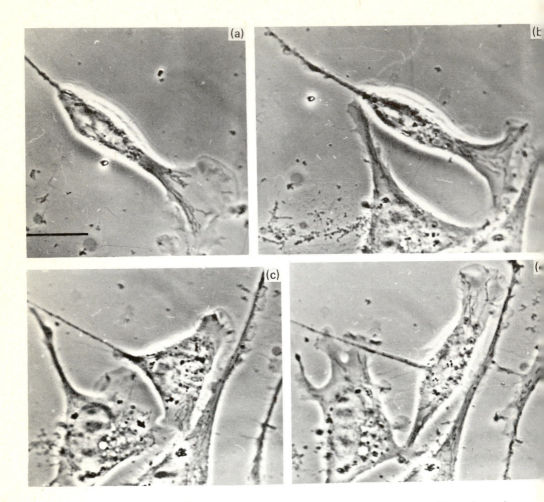

Fig. 3.11. Contact paralysis. A series of phase-contrast micrographs of transformed TLSF fibroblasts. (a) Polarized cell moving on the substratum. The cell is poorly attached and has small anterior lamella (cf. normal mouse fibroblast in fig. 2.1). (b) After 35 min, the anterior active edge of the moving cell has touched the process of another cell. (c) After 55 min, the previously active cell edge facing the surface of the other cell has become paralysed. Two new active edges are formed at the two opposite lateral parts of the previous anterior lamella. The cell body has contracted. (d) After 80 min, the cell has become oriented parallel to the process of the other cell. One of the two previously formed active edges has acquired the leading status. Bar: 20 μm.

Contact paralysis is local: it is observed only in the area of cell–cell contact but not in nearby areas of the cell surface (Trinkaus, Betshaku & Krulikowski, 1971). It is reversible; pseudopodial activity is resumed after the cell–cell contact is broken. Contact paralysis is often accompanied by retraction of the cell edge and pre-established cell–substratum contacts are not broken during this retraction (Abercrombie & Dunn, 1975). This result indicates that retraction is not due to the detachment of the cell surface from the substratum. Cessation of pseudopodial activity at the site of contact is accompanied by inhibition of the ability of this area of cell surface to become cleared of cross-linked receptors (Vasiliev *et al.*, 1976). Contact paralysis is observed only after contact with another cell surface but not after contact with a non-cellular surface.

Contact inhibition of movement has been defined as the restriction of directional cell displacement after contact (Abercrombie, 1970). Usually, contact inhibition of movement is a consequence of contact paralysis (see more detailed discussion of terminology in Abercrombie, 1970; Harris, 1974 and Vasiliev & Gelfand, 1977). Besides contact inhibition of movement, contact inhibition of phagocytosis (Vasiliev *et al.*, 1975a) and of pinocytosis (Vesely & Weiss, 1973) can be regarded as corollaries of contact paralysis.

Mechanisms of contact cell–cell interactions

As described above, extension of pseudopodia is essential for the establishment of specialized cell–cell contacts in cultures. Formation of gap junctions seems to be accompanied by a local accumulation of special membrane components. This accumulation may be similar to patching and/or capping. Probably, formation of other junctions also involves the accumulation of some membrane components but we know little about the nature of the components forming these structures. Extracellular ligands, especially lectins, may promote this local accumulation of receptors in some types of junctions. Unknown extracellular substances present in the gaps between the membranes of intermediate junctions and of desmosomes are possibly special variants of these ligands. Alterations of cortical structures associated with junctional structures may be regarded as manifestations of the 'anchorage' of immobilized membrane receptors. In summary, focal cell–cell contacts, like cell–substratum contacts, are probably formed by pseudopodial attachment reactions. Contact interactions of two cell surfaces are, however, different from cell–substratum interactions in one important aspect: contact with the surface of the non-living substratum induces formation of pseudopodia, while contact with another cellular surface inhibits formation of pseudopodia induced by the substratum.

Nothing is known about the mechanism of this contact paralysis; numerous theories of this phenomenon have been reviewed in the papers of Abercrombie (1970) and Harris (1974). Antitubulin-treated cells retain the ability for contact paralysis (Vasiliev *et al.*, 1970; Domnina *et al.*, 1977). Thus, the mechanism of contact paralysis is different from that of microtubule-dependent stabilization just discussed.

Conclusion

Pseudopodial reactions have been the main topic of this chapter because they play central roles in the various morphogenetic processes observed in cell cultures. Pseudopodial reactions are essentially membrane restoration reactions: they are induced by local alteration of some area of the cell membrane and result in removal of this altered area from the membrane.

These restoration reactions are used by the cells for many different purposes ranging from spreading on the substratum to endocytosis and locomotion. Analysis of these reactions suggests that the surfaces of pseudopodia have a number of special properties including the ability to translocate cross-linked receptors in the plane of the membrane, the ability to form attachments with other surfaces, and the ability to develop characteristic cortical alterations. There are probably many distinct variants of pseudopodial reactions; for instance, different extracellular ligands, different membrane receptors and different cortical structures are probably involved in the formation of diverse types of cell–cell and cell–substratum contacts. The ability of various parts of the cell surface to form pseudopodia is controlled by special mechanisms such as microtubule-dependent stabilization, contact paralysis and, possibly, microtubule-independent stabilization. This list of reactions associated with morphogenesis is, of course, still very far from complete. Nevertheless, it may give some idea about the basic types of cellular processes involved in the alteration of cell shape, adhesion and locomotion. The systematic study of different variants of these processes as well as investigation of their molecular mechanisms remains a task for the future.

4

Morphology and locomotion of normal fibroblastic and epithelial cells in cultures

In the previous chapter we discussed basic morphogenetic reactions of cultured fibroblasts and epithelial cells. Now we will describe the complex morphological reorganizations of single cells and of multicellular systems in culture that are due to different combinations of these basic reactions.

Fibroblasts

We will describe first the morphology and behaviour of normal and minimally transformed fibroblasts in the suspended or in the substratum-attached state, and the transitions between these states. The subsequent sections will be devoted to cell–cell interactions on flat substrata and to cell orientation induced by heterogeneities of the substratum (contact guidance). In many parts of our description we will use normal mouse embryo fibroblasts as prototype cells because our personal experience is based mostly on work with these cells. Other recent reviews on the morphology and locomotion of fibroblasts may be found in Trinkaus (1976), Abercrombie, Dunn & Heath (1976, 1977) and Vasiliev & Gelfand (1977).

Spherical cells

Suspended cells unattached to any substratum have an approximately spherical shape. Scanning electron microscopy usually reveals the presence of numerous micro-extensions on the surface of spherical cells (fig. 4.1). These micro-extensions belong to three main types: blebs (about 1 μm in diameter), microvilli 0.1–0.2 μm in diameter and up to 1.0–1.5 μm long) and flat ruffles of various sizes. Systematic comparisons of the surface topography of various suspended normal and minimally transformed cells have not yet been made.

The reported results show that this morphology can vary considerably. For instance, in the experiments of Ukena & Karnovsky (1977) most suspended cells from the cultures of many lines of normal and transformed fibroblasts were found to be covered with microvilli. Erickson & Trinkaus (1976) have found that freshly trypsinized, rounded cells of the BHK line are usually covered with blebs, while the cells of the same line rounded for mitosis have

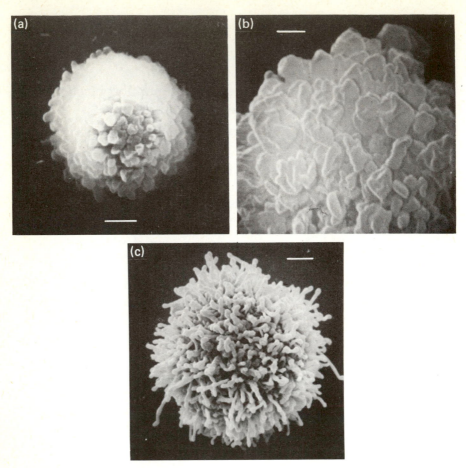

Fig. 4.1. Surface topography of spherical normal mouse fibroblasts as seen by scanning electron microscopy. (a, b) Surface covered with blebs; (c) microvillous surface. Bars: (a), 2 μm; (b), 1 μm; (c), 2 μm.

predominantly microvillous surfaces. In our unpublished experiments, the surface topography of mouse fibroblasts suspended by treatment with trypsin or EDTA proved to be heterogeneous: about 60–80% of the cells were covered with blebs, 20–30% with microvilli, 5–10% had ruffles, and there were also a few cells with mixed topography. This distribution was similar in suspensions obtained using trypsin or EDTA.

It is not clear which factors are responsible for the variations of surface topography in suspensions. Cell cycle-dependent alterations of the surface topography of neoplastic cells in synchronized suspension cultures of certain neoplastic cells were described by Enlander, Tobey & Scott, (1975) and by

Knutton, Sumner & Pasternak (1975). We do not know whether the surface topography of suspended normal fibroblasts is also affected by the phase of the mitotic cycle at which these cells are removed from the substratum.

Microfilament bundles were not seen in the cytoplasm of spherical cells and microtubules were rarely seen by electron microscopy (Goldman *et al.*, 1973). This gave reason to suggest that microtubules are depolymerized during cell transition into the rounded state. However, our immunomorphological studies of detergent-extracted mouse fibroblasts have shown that these cells contain numerous bent microtubules radiating from a central point. In this connection it would be interesting to investigate in detail the effects of antitubulins on suspended cells. Possibly, from these experiments we would learn something about the function of the microtubular system in suspended cells. Studies of the morphology and physiology of suspended cells are in their infancy. These studies are of particular interest with regard to the ability of normal and transformed cells to proliferate differently in the suspended state (chapter 8).

Morphology and movements of individual fibroblasts spread on flat substrata

Time course of spreading. The contact of a suspended cell with various types of solid surfaces induces it to spread. In the previous chapter we outlined the crucial events occurring in the course of spreading: initial attachment of cell, extension and attachment of pseudopodia leading to the formation of lamellar cytoplasm, radial spreading of the ring of lamellar cytoplasm and, finally, polarization of the cell (fig. 3.3). In the course of spreading on glass in a serum-containing medium, the first attached pseudopodia of fibroblasts are seen 5–10 min after the seeding of the cells; in most cells formation of the ring of lamellar cytoplasm is complete at about 30–40 min. Radial spreading of the ring of lamellar cytoplasm accompanied by flattening of central part of the cell body is observed during the following 2–3 h. Towards the end of this stage, the extension of pseudopodia and ruffling at the peripheral cell edge are gradually diminished until the disc-shaped cell comes to an almost complete standstill. At 3–4 h after seeding, the stage of polarization begins, manifested by reactivation of pseudopodial activity at the cell edge. In contrast to the stage of radial spreading, this activity now becomes unevenly distributed along the edge, so that several directions of preferential spreading can be seen. As a result, the cell contour that has been circular at the radial stage, gradually acquires a stellate shape. At the beginning of polarization these 'stars' usually have numerous 'beams' (up to eight to ten) but eventually many of them disappear and the number of directions of spreading is decreased to two to four; discrete out-growths of lamellar cytoplasm. The active parts of the cell edges are located at the periphery of these outgrowths. These alterations occurring at 6–10 h complete the process of cell spreading.

Later the cells remain polarized but continuously change their shape and their position by preferential spreading of certain lamellar areas.

The total projected area of a spread, polarized mouse fibroblast on the plane of the substratum reaches about 2000–3000 μm^2: about two-thirds of this area is occupied by lamellar cytoplasm and the remaining third by endoplasm. Lamellar cytoplasm and endoplasm are easily distinguished in the living cell: only endoplasm contains accumulations of refractile vesicles visible by phase contrast microscopy. Electron microscopy shows that vesicular organelles accumulated in the endoplasm are of a diverse nature: cysternae of rough endoplasmic reticulum, filled with intracysternal material, lysosomes, fat droplets, etc. The lamellar cytoplasm contains free ribosomes, flattened cysternae of endoplasmic reticulum and occasionally other organelles. The total surface area of a spherical cell covered with blebs or microvilli and the same cell maximally spread on the substratum are similar (Erickson & Trinkaus, 1976; A.D. Bershadsky, personal communication). In other words, cell spreading and polarization are not accompanied by a net increase of the plasma membrane area but only by a smoothening of the surface protrusions of suspended cells.

The well-spread cell may change the degree of its spreading and surface topography spontaneously, depending on the stage of mitotic cycle. The most obvious change of this type is cell rounding during mitosis. Mitotic cells are often similar to suspended cells and remain attached to the substratum by only a few retraction fibres. After mitosis the daughter cells regain a flattened morphology. In artificially synchronized cultures of several permanent cell lines, it was found that the cells in the G$_1$ phase had numerous microvilli and blebs whereas cells entering the S phase became more smooth and better spread (Porter, Prescott & Frye, 1973a; Hale, Winkelhake & Weber, 1975; Lundgren & Roos, 1976). Somewhat different results were obtained by Wetzel, Jones & Sanford (1977a) who labelled asynchronous cultures of CHO–KI line by [^{3}H]thymidine and examined the same individual cells first by autoradiography and then, by scanning electron microscopy. In these experiments no correlation was found between the morphology of interphase cells and the phase of cycle. Probably, the rate of flattening after mitosis may be different depending on cell type and on the treatment of the culture: in some cases the cells may reach a maximally flattened state soon after mitosis while in other cases this state may be achieved only at the beginning of S phase. It is also possible that cycle-associated changes of shape are better expressed in cultures of poorly spread transformed cells than in the cultures of well-spread cells.

Fibrillar systems in spread cells. Cell spreading is accompanied by marked reorganization of the microfilaments, filaments and microtubules. Microfilament bundles are absent in the suspended cell that has not yet formed

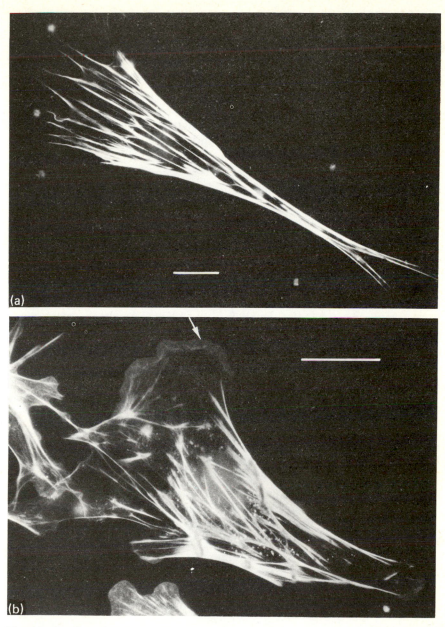

Fig. 4.2. Bundles of actin microfilaments in polarized normal mouse fibroblasts using the indirect immunofluorescence method with the application of anti-actin antibodies. (a) Fibroblasts fixed after extraction with Triton X-100 (Bershadsky *et al.*, 1978*b*). (b) Fibroblast fixed without previous extraction. Notice the diffuse accumulation of actin in the area near active anterior edge (arrow). Bars: 20 μm.

filopodia; however, the attached filopodia invariably contain these bundles (Bragina *et al.*, 1976). The radially spread cell already contains a complex system of bundles. In radially spread mouse fibroblasts at 0.5–2 h after seeding there are two main groups of these bundles: radial and tangential. The circular bundle near the boundary of the lamellar cytoplasm and endoplasm is often the largest of the tangential bundles. According to Lazarides (1976c) 40% of rat fibroblasts at 3–8 h after the beginning of spreading contain a regular net of microfilament bundles approximately perpendicular to each other. Similar nets and 'stars' can be present after 2–4 h of spreading in many mouse fibroblasts.

Polarization of a radially spread cell is accompanied by the gradual disappearance of bundles parallel to the active edge and of the 'nets' of the bundles. The polarized cell contains bundles that are mostly parallel to the stable edges and approximately perpendicular to the active ones (fig. 4.2); these bundles are present both in the outer and in the inner cortical layers. Prominent bundles are usually seen near the stable cell edges at the boundary between the upper and lower cortical layers. Each bundle is about 0.1–0.3 μm in diameter. The bundles in the lamellar cytoplasm are usually located 1–3 μm from each other. A meshwork of diagonally oriented microfilaments is seen between the bundles (Small & Celis, 1978b). As already mentioned, the distal end of each microfilament bundle is associated with a focal contact of the lower cell surface (Heath & Dunn, 1978). Wide areas of close contacts are seen between the focal contacts; both types of contact are located preferentially at the lower surface of the lamellar cytoplasm (Izzard & Lochner, 1976).

Just after the beginning of attachment, the cells of BHK-21 and other cell lines were found to contain intermediate filaments localized in a birefringent sphere near the nucleus. In the course of spreading this sphere gradually became smaller and eventually disappeared. Simultaneously, filaments appeared in the peripheral parts of the spreading cells (Goldman & Follet, 1970; Goldman *et al.*, 1973). The organization of intermediate filaments at the stages of radial spreading and polarization has not yet been studied in detail.

As first shown by Osborn & Weber (1976a, b) in 3T3 cells, microtubules grow from the centre into the peripheral lamellar parts of the cells during spreading. Similar results were obtained in our experiments with normal mouse fibroblasts. In these experiments (fig. 3.8) microtubules were not present at the early stages of spreading in attached filopodia and lamellae. The more-spread cells, however, usually contained microtubules in their ring of lamellar cytoplasm. These microtubules extended radially from the perinuclear zone towards the cell edges; at various distances from the centre their direction was changed from radial to parallel to the edge. Often distal parts of the microtubules formed circular bundles. Polarization was accompanied by the disappearance of the circular bundles of microtubules. Most microtubules in polarized cells were approximately parallel to the stable edges and perpendicular to the active edges. Microtubules radiating from the centre often formed bundles; one or several bundles extended into each zone of lamellar cytoplasm and radiating microtubules ended near the active edges.

The role of microtubules in the organization of pseudopodial activity during spreading and in the maintenance of the polarized state of fibroblasts was discussed in the previous chapter.

Polarization of the spread fibroblast is accompanied by alterations of the shape and orientation of intracellular organelles. The most obvious of these changes is the elongation and orientation of the nucleus (Weiss & Garber, 1952; Margolis *et al.*, 1975). The projection of the nucleus on the plane of the flat substratum usually has an elliptical shape. In fan-shaped cells, the long axis of the ellipse is located on a line connecting the centres of two active edges. Elongation and orientation of the nuclei are possibly caused by the anisotropic tension in the cortical layer that compresses the internal organelles. Also, the possibility of the direct attachment of fibrillar structures to the nuclear membrane cannot be excluded. Mitochondria and probably other organelles may be aligned along the microtubules radiating from the central parts of the cell (Heggeness, Simon & Singer, 1978).

Locomotion of polarized cells

Extension, attachment and contraction of lamellipodia take place at the active edges of polarized fibroblasts. A detailed quantitative study of pseudopodial activity at the leading active edge of chick embryo fibroblasts was carried out by Abercrombie *et al.* (1970*a*). They found that any point of the leading edge undergoes repetitive extensions and withdrawals. The net forward movement of the edge results from more time being spent moving forwards than backwards. Ruffles are continually formed at the active edges and they move centripetally along the upper surface of the lamellar cytoplasm (Abercrombie *et al.*, 1970*b*). Because of the repetitive formation and attachment of pseudopodia, each active zone tends to expand. This extension involves centrifugal displacement of the active edge and its lateral widening. Extension seems to proceed in a chain-like manner and new pseudopodia are formed near the sites of attachment of previously formed pseudopodia. Extension is limited by the tension developed by attached pseudopodia and also by the microtubule-dependent stabilization described in previous chapter. Various active zones of the same polarized cell may have different initial sizes as well as a different efficiency of pseudopodial attachment. Owing to the chain-like character of attachment, these differences tend to be increased.

As a result, one of the active zones usually acquires the leading status and becomes considerably larger than the others. The tension from the leading zone is not equilibrated by the tensions from other active zones. From time to time, the tension from the leading zone detaches lamellar zones located near the non-leading edges from the substratum; then the cell body contracts and moves in the direction of the leading edge. Thus, competition between several active edges leads to translocation of the cell on the substratum. Translocation is most efficient in fan-like cells that have only two active zones: a leading one, and a very small one located at the end of the tail process. Contraction of the tail is an active process; it may be induced by ATP in glycerinated cells (Goldman *et al.*, 1976*a*). Retraction of the tail is often followed by activation of the extension of lamellipodia at the opposite leading edge of the cell (Chen, 1978).

From time to time the moving cell changes direction. This may be done in two ways. (*a*) The leading status of one active zone may be lost and acquired by another area. For instance, when the leading edge of the fan-like cell meets an obstacle the cell may reverse the direction of its movement by expanding the size of its rear active zone and reducing the size of the previous zone. (*b*) The leading zone may begin to spread assymetrically. This may happen, for instance, when only one side of the leading edge meets an obstacle.

Owing to the competitive pseudopodial activities of various active edges and different parts of the same edge, the cell is able to 'compare' the efficiency of attachment in different areas of the substratum and move towards the area where this efficiency is highest. It is not clear what factors determine the directions of movement of a single cell located on a homogeneous substratum. Possibly, even in this case, the direction of movement may be affected by hidden irregularities of the substratum.

Until recently, statistical investigation of the characteristics of cell loco-motion (e.g. of the average rate of cell movements, the frequencies and directions of the turns, etc.) was a laborious task that required long-term cinematography of many cells followed by analysis of the films. Now this task becomes much easier due to ingenious technique of Albrecht-Buehler (1977). He proposed covering the glass surface with the particles of colloidal gold. In the course of their locomotion the cells remove the particles from the substratum leading to particle-free tracks. These 'phagokinetic tracks' repre-sent a record of the movements of the cell on the substratum. The method of Albrecht-Buehler opens up new possibilities for the detailed assessment of different parameters of cell locomotion.

Effects of environmental factors on spreading and locomotion

The properties of the substratum essential for spreading (rigidity, polarity, and the presence of certain substratum-absorbed proteins) have been discussed in the previous chapter.

Mechanical factors may affect cell–substratum attachment. For instance, it can be promoted by centrifugation (Milam *et al.*, 1973). In contrast, attachment of cell can be prevented by a medium flow of more than 100 μm/s (Doroszewsky, Skierski & Przadka, 1977). Of course, low temperature inhibits attachment (Juliano & Gagalang, 1977; Nath & Srere, 1977).

Many biologically active substances added to the medium may affect the shape of the spread fibroblasts. For instance, one of the growth factors inducing proliferation of fibroblasts, epidermal growth factor, was found to stimulate ruffling activity of glial cells (Brunk, Schellens & Westermark, 1976). Another growth factor, fibroblast growth factor, added to the medium of BALB/3T3 cells made them less flattened on the substratum (Gospodarovicz & Moran, 1974*b*). Insulin induced formation of microvilli on the upper cell surface of 3T3 cells (Evans *et al.*, 1974). Similar alterations of the topography on the upper surface of normal mouse fibroblasts were induced by incubation of these cells with small lipid vesicles (liposomes) added to the culture medium (Margolis & Bergelson, 1979). Only liposomes made from lipids that were in a fluid state at 37 °C induced formation of microvilli; liposomes made from lipids that are in a crystalline state at that temperature were ineffective. It is not clear what the mechanism is for the action of growth factors and liposomes on the cell shape. Their possible effects on cell locomotion also deserve investigation.

The presence of calcium is essential for spreading; this is most easily demonstrated by the detachment of fibroblasts by agents which chelate calcium (see below). Cell spreading is prevented in media with low calcium concentration and the rate of cell movement is also decreased (Gail, Boone & Thompson, 1973). An analysis of the effects of calcium and other divalent ions on different processes involved in spreading and locomotion has not yet been made.

Detachment of cells from the substratum

Several groups of agents may cause the rounding of spread fibroblasts. This rounding eventually leads to the detachment of the cells from the substratum and their transition into suspended cells with a spherical shape. The most widely used of these detaching agents are proteases (trypsin, pronase, papain, etc.) and calcium-chelating agents (EDTA or EGTA). Other agents that induce the partial or complete rounding of cultured cells are urea (Weston & Hendricks, 1972) and local anaesthetics (Poste, Papahadjopoulos & Nicolson, 1975). Alexandrov & Wolfenson (1956) showed that fibroblasts of rat connective tissue spread on collagen fibres may undergo reversible rounding after various treatments, e.g. with 5–10% ethanol, 2% calcium chloride, or adrenalin. It is a common knowledge that cultured cells damaged by certain chemicals or infection may undergo rounding but this process has not been studied in detail. Morphological changes accompanying rounding have been studied in several types of fibroblastic cells and have proved to be

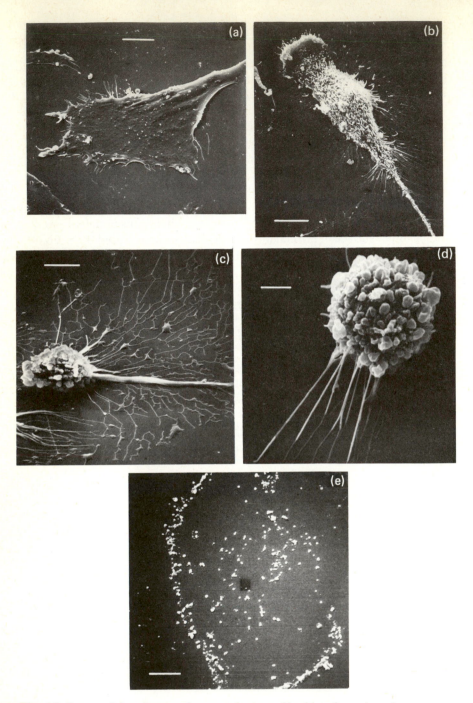

Fig. 4.3. Stages of detachment of a normal mouse fibroblast from the substratum as seen by scanning electron microscopy. Detachment was induced by 1 M urea. (a) Untreated cell. (b) Cell with retracted lamella. (c) Cell with thick and thin cytoplasmic processes. (d) Rounded cell attached by a few retraction fibres. (e) 'Traces' of cell left on the substratum after detachment. Bars: (a, b), 10 μm; (c), 5 μm; (d), 2 μm; (e), 5 μm.

similar in all cases (Dalen & Todd, 1971; Rosen & Culp, 1977; Vogel, 1978; Furcht & Wendelschaffer-Crabb, 1978; Lyubimov, 1978*a*).

As an example we will discribe the detachment of normal mouse fibroblasts induced by trypsin, EDTA or high concentrations (1 M) of urea. Treatment of the spread cells with each of these agents induced a similar sequence of changes (fig. 4.3) which could be divided into several stages.

(1) *Retraction of lamellar cytoplasm.* At this stage, the area of lamellar cytoplasm decreased and the thickness of remaining areas increased. The peripheral edges of retracted lamellar cytoplasm remained attached to the substratum by cylindrical processes (about 0.2 μm in diameter) usually designated 'retraction fibres'.

(2) *Formation of thick cytoplasmic processes.* At this stage the lamellar cytoplasm disappeared almost completely. The central part of the cell body was attached to the substratum by several thick cylindrical processes (about 1–2 μm in diameter) which radiated from the central part of cell body; these processes were attached to the substratum only at their ends where small lamellar areas or groups of retraction fibres were formed.

(3) *The formation of a spherical cell with retraction fibres.* At this stage the thick cytoplasmic processes disappeared and the almost spherical cell body was surrounded by numerous retraction fibres. The proximal parts of these fibres, unattached to the substratum, were more straight and thick than the distal attached parts of the same fibres. In the course of progressive rounding (stages a–c in fig. 4.3) numerous blebs, more rarely, microvilli and ruffles appeared at the surface of the cell body.

(4) *Detachment of the rounded cell from the substratum.* After detachment, pieces of attached cytoplasm could often be seen remaining on the substratum. Most probably, these pieces were the peripheral ends of the retraction fibres that had been broken in the course of detachment. Sometimes larger lamellar areas remained attached to the substratum. These 'traces' of detached cells have also been designated 'footpads', 'microexudates' or as 'substratum-attached materials'. The amount and the character of this material may vary depending on the cell type and the procedure of detachment. Recently, Badley *et al.* (1978) using mechanical streaming of the medium as a detaching agent isolated substratum-attached material that, by interference–reflection examination, corresponded to focal contacts. Chemical analysis of these and other samples of substratum-attached materials revealed the presence of a fibronectin-like protein, the components of microfilament bundles (actin, myosin, α-actinin) and various proteins associated with glycosaminoglycans (Culp, 1976; Badley *et al.*, 1978; Culp & Bensusan, 1978).

The treatment of spread cells with proteases soon leads to the disappearance of microfilament bundles, and this alteration preceeds advanced alterations of cell shape (Pollack & Rifkin, 1975). Alterations of other fibrillar components have not been specially studied.

The mechanisms of action of detaching agents remain unknown. The shape of the spread cell is the result of an equilibrium between the forces contracting extended parts of the cytoplasm and of cell–substratum adhesion which counteracts this contraction. This equilibrium can be upset either by an increase in the contracting tension or by a decrease of adhesion. Increased tension may be a result of alterations of the cell membrane leading to secondary changes in the cortical layer (e.g. of increased permeability for calcium) or of the primary alterations of cortical structures. Decreased adhesion may be a result of the alteration of extracellular ligands participating in adhesion, of cell membrane receptors or of anchoring cortical structures. Trypsin not only acts extracellulary but also penetrates into the cell (Hodges, Livingston & Franks, 1973). Thus, detaching agents may have many potential extracellular and intracellular targets. We do not know which cellular component is primarily affected in each case: the targets of various agents may be different. In particular, fibronectins are very sensitive to proteases (see reviews in Yamada & Olden, 1978; Vaheri & Mosher, 1978) and are likely to be their primary targets. Even slightly diminished adhesion would shift the equilibrium in favour of contraction so that remaining attachment sites would be destroyed by the pull of contracting cortical components or, more probably, these attachment sites with adjoining pieces of cytoplasm would be broken from the cell by contraction and form the 'footpads' remaining on the substratum. External mechanical stresses, e.g. turbulence of the medium, may also break retraction fibres and detach the cell. A glycerinated cell model mechanically detached from the substratum in a medium without ATP retains its flat shape; the addition of ATP causes the rounding of these detached cells. A cytochalasin-treated cell retains its arborized shape after detachment (Vasiliev *et al.*, 1975*c*). Thus, the cell rounding accompanying detachment is a result of an active ATP-requiring contraction of cytochalasin-sensitive structures. It is not clear whether microfilament bundles are essential for this contraction. As described above, these bundles possibly disappear earlier than when protease-treated cells become rounded. The cells without microfilament bundles (transformed fibroblasts) may undergo contraction as well as normal fibroblasts with bundles. Possibly, the main structures responsible for the rounding are not the bundles but some other cortical components (matrix microfilaments?).

Cell–cell interactions in cultures of fibroblasts

Sparse cultures. Cell–cell collision occurs when one moving fibroblast touches another moving or non-moving cell. One can distinguish two groups of collisions: those between an active leading edge of one cell and an active edge of another cell (head–head collisions) and those between an active leading edge of one cell and an inactive stable edge of another cell (head–side collisions). Collisions of both groups may have three types of outcome.

(1) *Halting.* This is the cessation of forward translocation of one of the active edges in head–head collisions. Usually, the halting is accompanied by contact paralysis of one or both contacting active edges.

(2) *Underlapping.* The movement of an active edge between the surface of another cell and the substratum.

(3) *Overlapping.* The forward movement of an active edge over the upper surface of another cell.

Theoretically, a fourth type of head–head collision is possible: combined *underlapping–overlapping* where one active edge continues to move forward over the surface of a second cell while the active edge of this second cell continues to move forward on the substratum. However this type of collision outcome has not been observed in any culture.

Analysis of time-lapse films in cultures of mouse fibroblasts (Guelstein *et al.*, 1973) has shown that most head–head collisions lead to halting; about 50% head–side collisions ended in halting and another 50% in underlapping. Overlappings were rarely (in less than 10%) observed after head–head and head–side collisions. The degree of underlapping and of overlapping was usually small: the leading edge usually stopped after moving 2–3 μm forward over or under the surface of the other cell.

Analysis of the collisions in 3T3 cell cultures gave essentially similar results (Bell, 1977). A high efficiency of contact inhibition of movement was also observed in cultures of fibroblasts of other species (Abercrombie, 1961). Head–head collisions may lead to the formation of specialized contacts (Heaysman & Pegrum, 1973). There are no direct observations proving that head–side collisions may also result in formation of specialized contact structures, although this seems probable. Numerous cell–cell collisions may lead to two statistical results: (*a*) cells in sparse culture acquire a monolayered distribution; (*b*) collisions alter the directions of cell locomotion and eventually lead to mutual parallel orientation of cells in the monolayer. When two neighbour cells are approximately parallel to each other, they do not change the direction of their movement after collision but continue to slide along each other. To measure the mutual orientation of a cell on a flat substratum it is convenient to use an index based on the determination of the direction of the long axis of the projection of nuclei on the plane of the substratum. These measurements confirm that the degree of mutual orientation in groups containing the same number of cells increases with time (Margolis *et al.*, 1975). The size of cell groups having the same orientation also increases with time (Elsdale & Bard, 1972).

Dense cultures. As cell density on the substratum increases, cultures of normal mouse and human fibroblasts become multilayered (Elsdale & Bard, 1972; Cherny *et al.*, 1975). There may be up to 6–10 cell layers in dense cultures of mouse fibroblasts. These cells remain flattened and have well developed lamellar cytoplasm (fig. 4.4). The mean area of their projection on the plane of the substratum is not significantly different from that in monolayered sparse cultures (Cherny *et al.*, 1975). Partial multilayering was also observed in 3T3 cell cultures (Bell, 1977). Collagen fibres, glycosamino-glycan-containing materials and fibrous aggregates of fibronectin are revealed between the cells in dense cultures of various types. Partial and even complete multilayering of cells in dense cultures may remain unnoticed by light microscopy as central cell parts and nuclei of cells in adjacent layers are rarely located immediately over each other; usually the central part of one cell contacts peripheral lamellar parts of the cells located in adjacent layers (Cherny *et al.*, 1975; Bell, 1977).

The mechanisms of formation of multilayered sheets in dense fibroblastic cultures are not quite clear. As we have seen, contact inhibition of movement effectively prevents multilayering of cells in sparse cultures. It seems improbable that the efficiency of contact inhibition decreases in dense cultures: our cinematographic observations indicate that overlapping remains rare in dense areas of culture. There are several possible ways of formation of multilayers without loss of contact inhibition.

(1) The cells may underlap other cells after head–side collisions and this may lead to detachment of the upper cells from the substratum and their subsequent spreading on the surface of other cells. These detachments have been observed in mixed cultures of normal and transformed cells (see chapter 5). It is not clear whether detachment and subsequent spreading take place in dense normal cultures. In this connection it may be significant that pre-labelled suspended mouse fibroblasts are easily attached to the upper surfaces of dense fibroblastic cultures (Domnina *et al.* 1972; Vasiliev *et al.*, 1975a).

(2) Poorly attached mitotic cells may be detached from the substratum and then daughter cells may spread over the surface of other cells.

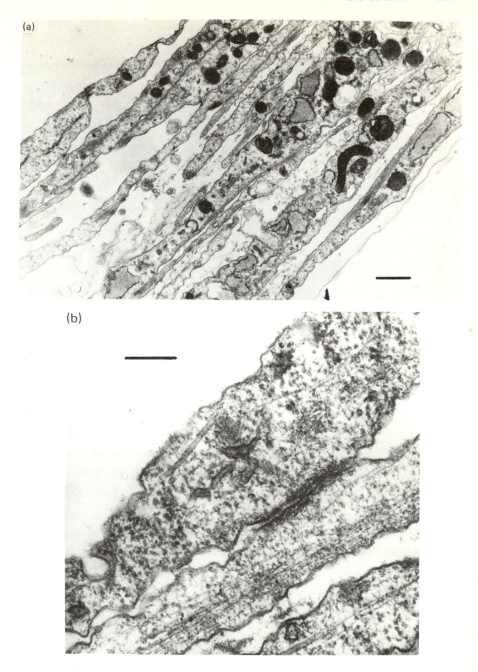

Fig. 4.4. Multilayered dense cultures of normal mouse fibroblasts. Transmission electron microscopy of sections perpendicular to the plane of the substratum. Cell–cell contacts between cells in adjacent layers (b). Bars: (a), 1 μm; (b), 2 μm.

(3) Cells may spread in dense culture not directly over the surface of other cells but over the fibrillar structures present in these cultures such as collagen fibres and fibronectin aggregates. The cells spread on the fibrils may form specialized contacts with cells in adjacent layers through holes in the fibrillar nets. These contacts were in fact observed in dense cultures (Cherny *et al.*, 1975). The mechanisms of multilayering listed above are not mutually exclusive; possibly, they all participate in the formation of multilayered dense cultures; the relative importance of each particular mechanism remains to be established.

In addition to contact inhibition there may be an additional factor increasing mutual orientation of cells in dense cultures: the surfaces of underlying cells and of extracellular fibrils may serve as orienting substrata for upper cells. Labelled fibroblasts seeded on the surface of homotypic cells (Stoker, 1964) or on the surface of glutaraldehyde-fixed cultures of these cells (V. I. Samoilov, personal communication) orient themselves in parallel with the underlying cells. Mechanical removal of a part of a dense culture leads to migration of the cells from the edge of the wounded culture on the free substratum (Ephrussi, 1935; Fischer, 1946; Vasiliev, Gelfand & Erofeeva, 1966; Vasiliev *et al.*, 1969). Migration of fibroblasts is accompanied by breaking cell–cell contacts: each cell migrates into the wound individually. Directional migration is probably a statistical result of numerous cell–cell collisions. A similar situation is observed during cell migration from the explanted tissue fragments (Abercrombie & Heaysman, 1953, 1954).

Contact guidance and the choice of substratum

We have described the morphology and locomotory behaviour of fibroblasts on flat, homogeneous, adhesive substrata. If the surface of the substratum is not flat and/or contains areas with different adhesive properties the cell may choose a definite place and orientation on this substratum. For instance, in the experiments of Carter (1967*a*) and Harris (1973*b*), when metals were evaporated on certain areas of a cellulose acetate surface the cells migrated preferentially on the metal-coated areas. This migration probably was a result of the different efficiency of attachment of the active cell edges to different areas of the substratum (Harris, 1973*a*). The same differences may also lead to cell orientation along the boundary between the more- or less-preferred substrata. For instance, normal fibroblasts can orient themselves along the boundary between a glass surface and that of a non-adhesive fluid lipid film (Ivanova & Margolis, 1973). This orientation is an example of 'contact guidance'; that is, of the cell orientation by substrate structures. The term 'contact guidance' was introduced by P. Weiss (see review in Weiss, 1958). Cells can respond not only to the chemical characteristics of the surface substratum but also to its geometrical shape. In particular, cells are found to respond to two types of geometrical patterns of the substratum: they orient themselves along the long axes of certain cylindrical surfaces and they avoid crossing those areas of the substratum where two surfaces intersect each other (grooves or ridges). Orientation along cylindrical substrata was observed in many experiments (Curtis & Varde, 1964; Rovensky, Slavnaja & Vasiliev, 1971). A detailed study of this phenomenon was made by Dunn & Heath (1976) who showed that chick fibroblasts can orient themselves on cylindrical fibres of glass with a diameter smaller than 100 μm.

In our laboratory, normal rodent and human fibroblasts were seeded on plastic substrata containing cylindrical areas (radii from 61 to 333 μm) separated from each other by grooves. During the first hours of spreading the cells migrated from these grooves and avoided the bottom of the grooves in the course of subsequent locomotion; the cells oriented themselves along the directions of the grooves (Rovensky *et al.*, 1971; Samoilov *et al.*, 1975, 1978; Slavnaja & Rovensky, 1975). Dunn & Heath (1976) studied cell behaviour on the surface of silica prisms and found that chick fibroblasts are usually deflected from crossing the ridge and take up a new course more parallel to it. Cell orientation along the ridge became obvious with a ridge angle of 8° or higher.

What is the mechanism of cell reactions to the geometrical shape of the substratum? Dunn & Heath (1976) suggested that linear bundles of micro-filaments in the cytoplasm of the fibroblast cannot assemble or operate in a bent condition. Due to this restriction the cell on a non-flat substratum would have a tendency to achieve eventually a position in which most bundles are linear and oriented in a similar direction. In fact, as predicted by this hypothesis, when chick fibroblasts were fixed while crossing the ridge of a silica prism their microfilament bundles on different sides of the ridge were found to be discontinuous and oriented in different directions. (Dunn & Heath, 1976). It should be noted that when the cell is located at the bottom of the large groove and oriented along the direction of this groove its microfilament bundles will probably be oriented in the same direction and will therefore be linear. Nevertheless this position of the cell will not be stable – as mentioned above, fibroblasts migrate from the areas at the bottom of the grooves. Possibly, linearity of the longitudinal microfilament bundles is an essential but not sufficient condition for the stability of cell shape: some structures of the lamella perpendicular to cell axis also have to be linear. In other words, as suggested by Vasiliev & Gelfand (1977), the well-spread cell may have a tendency to preserve its flat shape; that is, the shape in which its active edges and its central part are located in the same plane. Positions in which different parts of lamellar cytoplasm are not located in the same plane would not be stable.

Cells migrate from grooves in the presence of colcemid (Rovensky *et al.*,

1971). Thus, the 'tendency to preserve flatness' does not depend on the presence of microtubules. Microtubules are, however, essential for the stabilization of an elongated shape and of cell orientation induced by contact structures: the colcemid-treated cells are not elongated and do not have any definite orientation on the grooved substrata (Rovensky *et al.*, 1971) and at the boundaries between the different substrata (O. Y. Ivanova & L. B. Margolis, unpublished).

An aligned hydrated collagen gel is another example of a substratum that can orient cells. Drying of these hydrated gels led to their flattening and diminished their ability to orient the fibroblasts (Dunn & Ebendal, 1978). These results suggest that the orienting effect of nature fibrillar matrices is due to their non-flat geometrical shape.

Epithelia
Morphology

The main distinctive feature of epithelia in culture is their ability to form coherent sheets. Many types of cultured epithelia form monolayer sheets including kidney epithelium, intestinal epithelium, hepatic epithelium, and pigmented retina epithelium of chick embryos. Endothelia also form monolayers. Certain types of epithelia, especially epidermis, may form multilayered sheets with a basal layer of relatively undifferentiated cells and upper keratinized layers (Rheinwald & Green, 1975*a, b;* Sun & Green, 1977; Indo & Wilson, 1977). This structure is very similar to that formed by epidermis *in vivo*. Keratinized layers of epidermis in culture may even form 'whirls' somewhat similar to dermatoglyphs (Green & Thomas, 1978). Each type of epithelium has its morphological and locomotory peculiarities and the following description is based mainly on the results of experiments with a few well studied types such as mouse kidney epithelium, mouse and rat hepatocytes and the pigmented retina epithelium of chick embryos.

Little is known about the spherical state and spreading of single epithelial cells because these cells are difficult to separate from each other. When seeded into the culture, single cells are readily joined with each other to form coherent cellular islands and sheets. Middleton (1976, 1977) has shown that pigmented retina epithelial cells lacking contacts with other cells spread poorly on the substratum. However, when these isolated cells form stable lateral contacts with other cells their spreading is promoted. These observations of 'contact-induced' spreading stress once again that an isolated state in culture is hardly normal for an epithelial cell. It is somewhat easier to observe spreading and locomotion of isolated epithelial cells from certain continuous epithelial lines such as the MPTR line of mouse kidney epithelium used in our experiments. Although the cells of this line have been transformed with SV-40 virus they continue to form coherent sheets. Our unpublished observations show that single cells of this line spread on the substratum, although much more slowly than normal fibroblasts. These cells become polarized but do not acquire a very elongated shape. Their polarization is reversed by colcemid: in a medium containing this drug the cells display pseudopodial activity around the whole cellular edge.

Early stages of spreading of rat hepatocytes were studied by Miettinen, Virtanen & Linder (1978). They found that the cells were attached to the substratum by lamellipodia and that spreading was relatively slow: microfilament bundles within the attached pseudopodia were formed after 24 h.

Epithelial cells are probably more selective in their requirements for substrata than fibroblasts, although systematic comparisons of these requirements have not been made. Different varieties of collagen surfaces have been found to be relatively good substrata for hepatic epithelia (Ehrmann & Gey, 1956; Sattler *et al.*, 1978). The role of fibronectins or of other substratum-attached proteins in the spreading of epithelia remains to be studied.

The upper surface of monolayered sheets of normal kidney epithelium is usually smooth except for a few microvilli; characteristic rows of these microvilli are seen over the belts of cell–cell contacts (fig. 4.5). These belts go around all the lateral surfaces of the central cells of the sheets. Electron microscopy reveals in these areas several types of contact structures such as desmosomes, tight junctions and highly permeable gap junctions. Adult rat hepatocytes in culture form complex sets of specialized junctions identical with bile canaliculi formed between hepatocytes *in vivo* (Wanson *et al.*, 1977; Sattler *et al.*, 1978). Probably, many different variants of contact complexes can be revealed at the lateral cell surfaces in cultures of various epithelia. Formation of lateral cell–cell contacts between hepatocytes does not require serum but requires calcium (Rubin, Kjellen & Öbrink, 1977).

Epithelial sheets *in vitro*, like their prototypes *in vivo*, have dorso-ventral polarity: the upper and lower surfaces of these sheets have different transport properties and intracellular structures are distributed regularly with regard to these surfaces. In particular, renal epithelial monolayers transport water through the sheet from the apical to the basal surface and maintain transepithelial electrical potential (Misfeldt, Hamamoto & Pitelka, 1976; Cereijido *et al.*, 1978). Pisam & Ripoche (1976) found that concanavalin A receptors are localized only at the apical surfaces of frog urinary bladder epithelium cells. After EDTA-induced dissociation of epithelial sheets into separate cells, concanavalin A receptors migrated from the apical surfaces to all other parts of the cell surface. These experiments show that lateral cell–cell contacts are essential for the maintenance of differences between apical and basal surfaces. The possible role of the microtubular system in the maintenance of dorso-ventral polarity remains to be established. Polarity of cell differentiation is obvious in the multilayered sheets formed by keratinocytes. Only the cells in the upper layers undergo differentiation, while basal

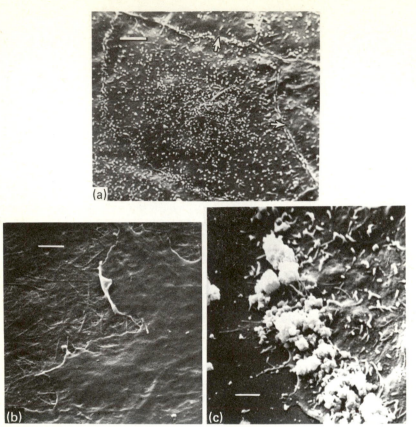

Fig. 4.5. Surface topography of normal mouse kidney epithelium grown on a collagen substratum by scanning electron microscopy. (a) Surface of central cells; area of cell–cell contact (arrow). (b) Lamellipodia at the lateral cell edge. (c) Carmine particles attached to the surface of the active edge of epithelial sheet. Carmine was added to the medium 2 h before fixation. Bars: 2 μm.

substratum-contacting cells continue to proliferate. The loss of cell–substratum attachment induces differentiation in these cells: keratinocytes suspended in methylcellulose without the contact with any substratum invariably underwent keratinization (Green, 1977; Rice & Green, 1978). Due to contact paralysis, formation of pseudopodia is not observed at the lateral surfaces of central cells

---

Fig. 4.6. Microtubules in the epithelial cells of normal mouse kidney as revealed by an indirect immunofluorescence method with the application of anti-tubulin antibodies. (a) Part of the epithelial sheet; (b) The same field of view as in (a) seen in phase contrast. (c) Microtubule-organizing centre associated with a cilium. The culture was incubated with colcemid (0.5 μg/ml) for 6 h and then for 30 min without colcemid. Colcemid destroyed all the microtubules but not the cilium-associated centre. Growth of microtubules begins from the centre in a colcemid-free medium. Bars: (a, b), 20 μm; (c), 10 μm.

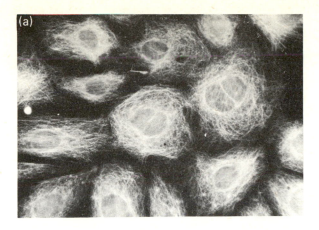

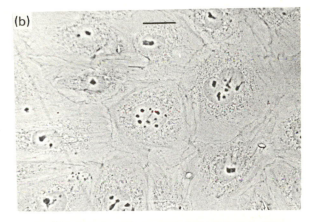

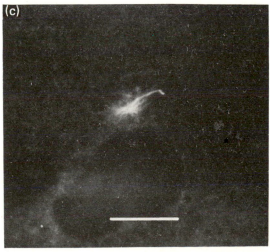

linked on all sides to other cells. In contrast, pseudopodia (lamellipodia and filopodia) are continuously extended and retracted at the free edges of marginal cells of the sheet. The quantitative characteristics of displacements of active edges in epithelial cultures were found to be similar to those observed at the active edges of fibroblasts (Di Pasquale, 1975a). The areas of lamellar cytoplasm are located near these free edges.

All the marginal and central cells of the sheets of kidney epithelium contain a well-developed system of microtubules (fig. 4.6); these microtubules radiate from the central cilium-like structures and fill the whole cytoplasm except the lamellar areas near the free edges of the marginal cells (Bershadsky *et al.*, 1978a). As described above, microtubules in polarized fibroblasts are usually oriented parallel to the stable cell edges. In contrast, the orientation of microtubules in epithelial cells is not correlated with the activity of the edges: microtubules parallel to the edges are predominant near certain stable edges while in other cases those located near a stable edge are perpendicular to that edge. Destruction of microtubules by colcemid does not lead to alterations of distribution of pseudopodial activity in epithelial cultures: this activity remains restricted to the free edges of marginal cells (Vasiliev *et al.*, 1975a; Domnina *et al.*, 1977; Di Pasquale, 1975b). It is not clear whether colcemid treatment is able to induce any minor alterations in the morphology of pseudopodia at active edges. Thus, differentiation of stable and active edges in epithelial cells of the sheet, in contrast to that in individually moving fibroblasts, is not dependent on the microtubular system. The distribution of microfilaments and intermediate filaments in epithelial sheets has not yet been studied in detail.

As mentioned in the previous chapter, the properties of the surface of pseudopodial and lamellar cytoplasm at the free edges of sheets are different from those of other parts of the upper cell surface of the sheet. One of these properties is adhesiveness. Inert cell particles adhere to the active parts of the cell surface and are phagocytosed preferentially by marginal cells having these active parts (the so-called contact inhibition of phagocytosis). Cell–cell contacts between homologous epithelial cells are formed also by pseudopodia at the free cell edges; the upper surface of the central cells is non-adhesive for these cells (see chapter 3). Most cell–substratum contacts of epithelial sheets are located at the lower surface of the marginal and sub-marginal cells; the lower surface of the central cells has few, if any, contacts with the substratum (Di Pasquale, 1975a). The morphology of the cell–substratum contacts of epithelial cultures has not yet been studied in detail.

The selective ability to become cleared of cross-linked surface receptors is another characteristic property of the active surface of marginal cells of epithelial sheets (see chapter 3). The marginal zones of endothelial cultures perform receptor-mediated endocytosis of serum very low density lipoprotein much more actively than the central zones (Vlodavsky *et al.*, 1978b). Thus the upper surface of the central parts of epithelial sheets, in contrast to marginal areas, is relatively non-responsive to various types of external ligands such as solvated macromolecules, particles, and the surface of the substratum or of other cells. We do not know anything about the reactivity of the basal surfaces.

Locomotion

When part of the monolayer of mouse kidney epithelium has been removed mechanically, the marginal cells begin to extend and to attach pseudopodia; as a result, these cells start to move into the wound (J. M. Vasiliev & I. M. Gelfand, unpublished). The cell–cell contacts are rarely broken in the course of these movements, so that translocation of marginal cells is followed by the translocation of adjacent rows of central cells. Mutual distribution of cells is not much changed in the course of these translocations and usually the cells retain the same neighbours. As a result of translocation the cells located near the edge of the wound become more stretched on the substratum. The mean area of the cells on the substratum gradually decreases from the edge of the wound to the central part of the monolayer. These observations suggest that the central cells are passively dragged by the moving marginal cells, although it is not yet clear that this dragging is the only mechanism of movement of the central cells. Cinematographic observations show that cells within the epithelial islands of human mammary epithelium rately change their neighbours in the course of displacements (Stoker, Piggott & Riddle, 1978). Thus, cell–cell contacts in monolayers of certain epithelia are stable and do not permit significant individual movements of the cells within the sheet. Somewhat different results were obtained by Steinberg and collaborators (Steinberg, 1973; Garrod & Steinberg, 1975) in experiments with monolayers of chick embryonic liver parenchyma cells and of mouse 3T3 line. As discussed in chapter 2, the tissue origin of 3T3 cells is not clear. Both these cell types were found to display considerable directional cell movements and re-shuffling of relative cell positions with frequent exchange of neighbours. This movement was not correlated with the formation of visible gaps between the cells. Possibly, the cultures studied in thise experiments had less firm cell–cell contacts than other epithelia. This question needs further investigation.

In summary, the main distinctive feature of morphogenetic reactions of epithelia is the ability to form stable cell–cell contacts; pseudopodial cell–cell contact between two epithelia cells leads to the formation of much more complex, widespread and firm junctional structures than similar contact between two fibroblasts. The stability of junctional structures leads to permanent contact paralysis of movements of contacting cell edges. As a result, the upper surface of the epithelial sheet becomes non-adhesive and relatively unresponsive to particles and macromolecules. In fibroblastic cultures, cell–cell contacts are less stable and the cells move individually. The microtubular system is essential for the maintenance of the direction of translocation of fibroblasts. In contrast, in epithelial cultures, cell–cell contacts are not broken in the course of locomotion. These cells contain a well-developed microtubular system but it is not essential for directional locomotion. Here the directional character of cell translocation is simply a result of the contact paralysis inhibiting pseudopodial activity at the lateral edges of the marginal cells.

Prevention of polarization of isolated epithelial cells by colcemid suggests that in these cells, in contrast to cells forming sheets, microtubular systems may be essential for directional movements. It is possible that individually moving fibroblasts and contact-locked epithelial cells represent two extreme cellular types and cell types with intermediate morphogenic features will be found after more detailed investigations. These cells may, for instance, form monolayered sheets like epithelium, but cell–cell contacts in these monolayers may not be stable, so that individual movements of a cell within monolayers would be possible, the upper cell surface of the monolayer would retain certain pseudopodial activity and adhesiveness, etc. Further studies are needed to prove the existence of cell types with these 'intermediate' properties.

Two homotypic epithelial cells easily form lateral cell–cell contact at their active edges. We do not know yet how two heterotypic epithelia interact with each other. In-vivo epithelia establish stable contacts with each other at their boundaries, e.g. at the boundary between morphologically different epithelia lining various parts of the digestive tract, urogenital organs, etc.

Mixed cultures of epithelia and fibroblasts

Interactions of fibroblasts and mammary epithelial cells in mixed cultures were studied by Stoker *et al.* (1978). Fibroblasts touching the edge of epithelial islands formed close contacts with the marginal cells but specialized junctions at the boundary were not revealed. On reaching the epithelial edge a fibroblast always changed its direction of movement: it started to move either backwards or laterally along the edge of the epithelium. Invasion of the epithelium by fibroblasts was never observed.

Essentially similar results were obtained in our unpublished experiments with mixed cultures of the MPTR line of kidney epithelium and mouse primary embryo fibroblasts. Contact between the active edges of epithelial and fibroblastic cells usually led to contact paralysis of the fibroblastic edge; activity of the epithelial edge also seemed inhibited but additional observations are needed to establish whether contact paralysis at the boundary of fibroblasts and epithelial cells is reciprocal. Formation of stable contacts between the epithelium and fibroblasts was not observed. Usually a fibroblast developed activity in some other part of its edge and moved away from the epithelium or oriented itself along the edge of the epithelium. Dense mixed cultures consisted of epithelial islands and streams of fibroblasts oriented approximately in parallel to these islands.

These results indicate that fibroblasts and epithelial cells in a two-dimensional culture are not merged together but remain as different cell types. This sorting is due to the effective contact inhibition of fibroblasts by epithelium combined with the inability of these two cell types to form stable cell–cell contacts.

Conclusion

Interactions of cultured cells with each other and with the substrata can lead to the formation of a number of rather complex multicellular structures. These complex behavioural acts can be regarded as the net result of a few basic morphogenetic reactions performed many times by numerous cells. Using various combinations of these reactions even a single fibroblast is able to solve problems of considerable complexity: the cell can compare the properties of various parts of the substratum and migrate on the 'most favourable' areas of the substratum, orient itself relative to other cells, etc.

The main groups of morphogenetic processes performed by different cell types are probably similar; however, cells of each tissue type have certain distinctive features. For instance, pseudopodial attachment reactions of fibroblasts and of epithelial cells lead to the formation of cell–cell contacts with various properties. The different stability of these contacts may be responsible for many differences between the morphological structures formed by epithelial and fibroblastic cells. Various subtypes of epithelia are obviously different from each other in their ability to form different types of specialized contact structures and in other properties, but they have certain common features; in particular, the relative stability of their cell–cell contacts. The cells of other tissue types also have their own distinctive features of morphogenetic reactions which remain to be studied.

The morphogenetic processes in cell cultures obtained from isolated cells attached to artificial substrata are, of course, considerably simplified in comparison with those which occur *in vivo*. Nevertheless, the main features of these processes are probably similar in the organism and in culture. The main argument supporting this conclusion is the similarity of multicellular structures formed by the tissue cells *in vivo* and *in vitro*. Epithelial sheets in culture are similar in pattern to those formed by the same cells in the organism. The general morphology of fibroblasts spread *in vivo* on the network of extracellular fibres in subcutaneous connective tissue is very similar to that of polarized fibroblasts spread on artificial substrata in culture. The lamellar cytoplasm of cultures fibroblasts corresponds to the transparent peripheral part of the cytoplasm of fibroblasts spread *in vivo;* histologists formerly designated this part as 'ectoplasm' (Jasswoin, 1928, 1930). As stressed by P. Weiss (1933, 1961), the contact guidance of fibroblasts by fibrin and collagen fibres and by other structures may occur not only in culture but also in the organism. Wound healing in epithelial and fibroblastic cultures is rather similar morphologically to wound healing processes *in vivo*. The micro-environment in the organism is of course much more diverse and complex than in culture. We will return to the discussion of this question in chaper 13.

5

Morphology and locomotion of transformed cells in culture

Introduction

In this chapter we will describe one group of changes characteristic of neoplastic transformations in culture, namely, morphological transformations of fibroblastic and epithelial cells. As we will see, manifestations of morphological transformations in different cell lines may vary considerably. A general description of these multistage and manifold alterations would be justified if they had some common trends. We will try to show that such a common feature can, in fact, be found and may be a deficiency of cell–substratum and cell–cell attachments. Each morphological transformation affects attachment processes and the degree of their deficiencies increases with each consecutive transformation.

Different cell lines may have different degrees of morphological transformation. Of course, each particular transformed line, besides these deficiences, may also have additional, secondary alterations. As described in chapter 2, one may distinguish between cells with minimal and strong morphological transformations. In studies of morphological transformation strongly transformed cultures are usually compared with their progenitors: either with primary non-transformed cultures or with minimally transformed permanent cell lines. Cells with a temperature-sensitive transformed phenotype examined at permissive and non-permissive temperatures are also used in comparative studies; these are the cell lines transformed with viruses having temperature-sensitive mutations of the oncogene or specially selected temperature-sensitive variants of chemically transformed cells.

Fibroblasts
Spherical cells

Alterations of the surface topography associated with transformation in spherical cells have been investigated by several authors. Willingham & Pastan (1975a) reported that transformed murine cells (L 929 line) had a microvillous cell surface while microvilli were absent on the surface of weakly transformed cells. Pastan & Willingham (1978) regard the presence of long microvilli at the surface as a characteristic feature of the transformed phenotype. In our unpublished experiments, most transformed cells of the L line had a microvillous relief in suspension, while normal mouse fibroblasts in suspension contained less than 50% of cells with microvilli; other cells had blebs or ruffles (see chapter 3). Ukena & Karnovsky (1977) did not find any differences between the topography of suspended cells from normal cultures of mouse fibroblasts, weakly transformed 3T3 lines or a number of strongly transformed lines: cells with microvillous relief were predominant in all types of supension. Collard & Temmink (1976) found that suspended 3T3 cells had a microvillous surface while their strongly transformed counterpart, SV 3T3 cells, had a smooth surface.

This diversity of results may be due to some unidentified differences in the experimental procedures used for the preparation of cell suspensions. One hopes that the development of more standard and less damaging methods will allow a systematic basis to be revealed for differences in surface relief or other morphological features of suspended non-transformed and transformed cells.

Cell spreading on flat substrata

After contact with the appropriate substratum, transformed cells, like normal ones, undergo a transition from the spherical state the polarized, spread state. However, the spreading of strongly transformed cells is always deficient as can be seen from the final morphology of the cells, which is different in many respects from the morphology of parent-normal or weakly transformed cells. The morphological alterations most characteristic of transformed cells in the spread state include: (*a*) decreased area and altered shape of the lamellar cytoplasm; (*b*) decreased number of focal cell–substratum contacts; (*c*) altered topography of the upper surface; (*d*) deficient formation of microfilament bundles. We will now discuss each of these alterations in more detail.

Deficiency of the lamellar cytoplasm. The morphological symptoms of a deficiency of lamellar cytoplasm include a decrease of its areas, a more jagged contour and an increase of thickness. The projected area of the lamellar cytoplasm of L cells on the plane of the substratum was found to be about three times less than that of normal mouse fibroblasts while the areas of endoplasm were not very different (Domnina *et al.*, 1972). Several other lines of transformed cells also showed a decrease of their area of lamellar cytoplasm. (As stated in the previous chapter, 'lamelloplasm' in these experiments was defined as part of the cell devoid of vesicular organelles visible by phase-contrast microscopy.)

Fox, Dvorak & Sanford (1976) and Fox *et al.* (1977) examined seven paired examples of cultured fibroblastic cells obtained from mouse and rat embryo. Each pair consisted of cells of the same line taken from two different stages of spontaneous transformation: before and after the development of oncogenic properties (see chapter 2). Projected cytoplasmic and nuclear areas as well as nuclear and cytoplasmic dry mass were measured. It was found that both the projected area and the dry mass of the lamellar cytoplasm were consistently decreased in oncogenic lines compared with their non-oncogenic counterparts. These results confirm that a decreased area of lamellar cytoplasm can be regarded as an important symptom of morphological transformation. Reduced spreading of spontaneous transformants on the substratum was also observed in other experiments of Sanford's group (Sanford *et al.*, 1970; Sanford, Handelman & Jones, 1977).

Another change of the lamellar cytoplasm which often accompanies transformation is the alteration of its external contour: this contour becomes more jagged and more irregular than in normal cells (Vasiliev & Gelfand, 1973; Ambrose & Easty, 1976; Ambrose, 1976). For instance, instead of a single anterior lamella, polarized cells may have several elongated bands of lamellar cytoplasm divided by deep indentations of their exterior edges. Active parts of the edges are located at the ends of discrete bands of the lamellar cytoplasm. Probably, this change is accompanied by a decrease of the total length of the active cell edges; however, quantitative comparisons of these lengths in normal and transformed cells have not yet been made.

An additional morphological alteration of the lamellar cytoplasm is the increase in its thickness which is sometimes observed in strongly transformed cells. The thickness of lamellar cytoplasm may also be more variable, in transformed cells.

It would be important to compare the shape of pseudopodia and quantitative parameters of their protusion and extension in normal and transformed cells. In certain transformed cells, lamellipodia at the active edge become more narrow or are replaced by cylindrical outgrowths of the filopodial type. Inter-relations between alterations of the area of lamellar cytoplasm and its shape are not clear; certain mouse lines from plastic sarcomas have more jagged contours than normal embryo mouse fibroblasts but their projected areas are not different from those of normal cells (O. Y. Ivanova, unpublished).

The alterations of the lamellar cytoplasm listed above may be expressed to various degrees in different transformed cultures and in different cells of the same culture (fig. 5.1). Several degrees of alteration in the lamellar cytoplasm can be distinguished (Vasiliev & Gelfand, 1973).

(1) *Cells with an almost complete absence of lamellar cytoplasm.* These cells may have an almost spherical, hemispherical or elongated, cigar-like shape; the latter may have lamellar areas at the poles.

(2) *Cells with reduced areas of lamellar cytoplasm.* These areas are often located at the ends of long cytoplasmic processes; the lateral edges of these

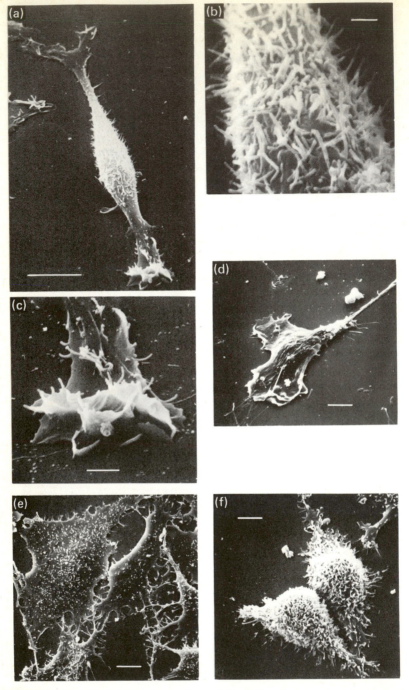

Fig. 5.1. Surface topography of transformed fibroblasts by scanning electron micro-scopy. (a–c) A transformed fibroblast of the mouse L line: (a), general view; (b), microvilli on the upper surface; (c), ruffles on the small lamella. (d) Mouse fibroblast of M 22 line transformed with SV-40 virus. (e) Minimally transformed rat kidney cell of the NRK line. (f) Cells of the NRK line additionally transformed with Kirsten mouse sarcoma virus. Bars: (a), 10 μm; (b), 1 μm; (c), 2 μm; (d), 5 μm; (e), 10 μm; (f), 5 μm.

processes are unattached to the substratum. These lamellar areas often have the morphological irregularities already described (p. 133).

(3) *Quasi-normal cells*. These cells have an almost normal total area of lamellar cytoplasm, although some of them may have morphological changes, especially jaggedness of their contours.

Of course, there are no sharp boundaries between these classes. Micro-cinematographic studies show that with time the same cell can pass from one class to another. Cells of different lines have various proportions of different classes. Strongly transformed lines usually contain cells of the first and second classes but the proportion of quasi-normal cells is high in weakly transformed cultures. When chick fibroblasts infected with a temperature-sensitive mutant of Rous sarcoma virus are transferred from a non-permissive to a permissive temperature, a progressive decrease of cell spreading is observed until the cells acquire a rounded shape (Wang & Goldberg, 1976). These descriptions are based on the visual examination of various cultures. It would obviously be important to continue quantitative studies of the alterations of cell spreading brought about by transformation.

The surface of the lamellar cytoplasm of transformed cells has certain properties that are similar to those of normal lamellar cytoplasm: it may become cleared of cross-linked surface receptors and it may translocate attached particles centripetally. As the fraction of the cell surface occupied by lamellar cytoplasm decreases in transformed cells the part of the surface cleared of cross-linked receptors also decreases (Vasiliev *et al.*, 1976). We do not know whether the rates of clearance of cross-linked receptors in the lamellar cytoplasm of normal and transformed cells are similar. The rates of centripetal movement of particles on the surfaces of these cells also require comparative studies.

Morphology of the lower cell surface. It is natural to expect that deficient formation of the lamellar cytoplasm should be accompanied by an alteration of the number and morphology of focal cell–substratum contacts. In fact, examination of the lower surface of strongly transformed mouse L cells (Bragina, 1975) revealed a striking decrease of the number of these contacts. Focal contacts were present only in the immediate vicinity of active edges and there were no focal or close contacts behind these edges, so that the lower surface had an arc-like shape. The contacts were not connected with the ends of microfilament bundles. The lower surface of L cells often contained numerous microvilli that were never seen at the lower surface of substratum-spread normal cells (fig. 5.2). It is interesting that microvilli were present only at the lower surface of the anterior part of L cells but not at the posterior part. Possibly, formation of these microvilli can be regarded an an abortive manifestation of pseudopodial activity of the anterior part of the cell surface.

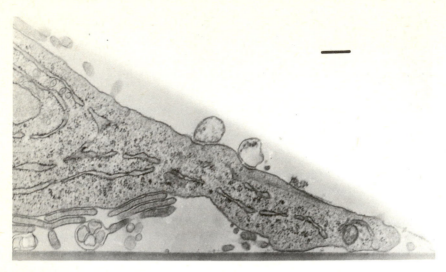

Fig. 5.2. The anterior edge of a transformed L fibroblast: section perpendicular to the plane of the substratum. The lower cell surface makes contact with the substratum only at a few points. Numerous microvilli are present at the lower surface. Bar: 0.5 μm.

As mentioned by Abercrombie *et al*. (1977) interference–reflection microscopy of several types of sarcoma cell revealed the total absence of focal contacts; only close contacts with the substratum were seen.

Morphology of the upper surface. Substratum-spread normal fibroblasts usually have a smooth upper surface both in the zone of the endoplasm and in the zone of the lamellar cytoplasm. This smoothness may disappear in strongly transformed cells: their upper surface becomes covered by numerous protrusions such as ruffles, microvilli and blebs. Cinematographic observations show that these protrusions move continuously (Vesely, 1972). These alterations of the topography of the upper cell surface have been described in many types of cell lines transformed by different agents (Easty & Mercer, 1960; Boyde, Weiss & Vesely, 1972; Hodges & Muir, 1972; Vesely & Boyde, 1973; Perecko, Berezesky & Grimley, 1973; Porter & Fonte, 1973; Porter, Todaro & Fonte, 1973*b*; Borek & Fenoglio, 1976; Gonda *et al*., 1976; Malick & Langenbach, 1976; Glaser *et al*., 1977; Rossowski *et al*., 1977). Various types of extension may be predominant in different cell lines. Microvilli are numerous at the upper surface of many cell types; there are however, certain transformed cells (Pontén, 1975; Ambros, Chen & Buchanan, 1975) in which complex ruffles are predominant. A number of lines retain a predominantly smooth surface. For instance, in the experiments of Wetzel *et al*. (1977*b*), the spontaneous mouse transformants obtained by Sanford and collaborators were shown by scanning electron microscopy to have predominantly smooth surfaces. The same was true for certain chemically transformed rat cells (Cloyd & Bigner, 1977) and for the M-22 line of mouse fibroblasts transformed by SV-40 virus, observed in our experiments (Fig. 5.1). Thus, formation of protrusions at the upper surface is frequently observed in strongly transformed lines but cannot be regarded as a universal symptom of morphological transformation. Probably, formation of protrusions at the upper surface and the deficient spreading of transformed cells are somehow inter-related. Theoretically, at least two types of inter-relationship are possible. (*a*) Formation of extensions may be a secondary result of the decreased spreading of transformed cells. We know that the surface of normal cells becomes smooth only after spreading on the substratum. (*b*) Formation of extensions and deficient spreading could be two independent consequences of some common cause; for instance, a deficiency in the organization of the cortical layer. At present, it is difficult to choose between these two possibilities.

Microfilaments and intermediate filaments. Decrease in the number and size of microfilament bundles is characteristic for many transformed lines. This change may be revealed by electron microscopic observations and by immunomorphological studies using anti-actin antibodies. It has been observed in various virus-transformed and spontaneously transformed cell lines (Ambrose *et al.*, 1970; McNutt, Culp & Black, 1971, 1973; Domnina *et al.*, 1972; Pollack, Osborn & Weber, 1975; Goldman, Jerna & Schloss, 1976*b*; Tucker *et al.*, 1978). When cells transformed by temperature-sensitive mutants of oncogenic viruses are shifted from a non-permissive to a permissive temperature a reversible disorganization of actin bundles is observed (Pollack *et al.*, 1975; Altenburg, Somers & Steiner, 1976*a*; Edelman & Yahara, 1976; Vollet *et al.*, 1977). Disorganization of microfilament bundles is not an all-or-none phenomenon: certain transformed cells may still contain bundles although their number and the thickness of each bundle decrease in comparison with homologous non-transformed cells. In particular, Tucker *et al.* (1978) studied a series of spontaneous transformants of mice and found that all the oncogenic lines had thinner and more sparse bundles than all the paired, non-oncogenic lines. Probably, the degree of disorganization of the microfilamental system in different individual cells and in different lines is correlated with the degree of deficiency of spreading, but rigorous proof of this correlation has not yet been obtained. The causal relationships between deficient spreading and disorganization of microfilaments are not clear. Possibly, disorganization of microfilaments is a secondary result of decreased spreading; as mentioned in previous chapter, microfilament bundles disappear at the early stages of detachment of normal cells from the substratum. Alternatively, the disorganization of the bundles in transformed cells may be the cause of their inability to maintain a well-spread shape.

No characteristic changes of intermediate filaments accompanying trans-

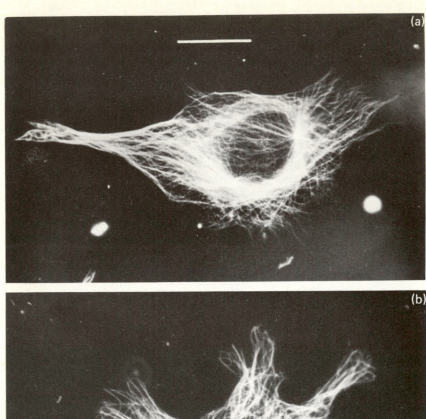

Fig. 5.3. Microtubules in transformed mouse fibroblasts extracted with Triton X-100 and viewed by indirect immunofluorescence. (a) L line. (b) M-22 line transformed with SV-40 virus. Bars: 20 μm.

formation have yet been described. According to Hynes & Destree (1978) when minimally transformed NIL 8 cells were treated with colchicine, intermediate filaments formed a circum-nuclear ring while in strongly transformed NIL-8-HSV sub-lines these filaments formed not a ring but a coil, after colchicine treatment. The significance of this altered reaction to microtubule disruption is not clear.

Microtubules. Electron microscopic and immunomorphological investigations employing anti-tubulin antibodies (Osborn & Weber, 1977; Tucker *et al.*, 1978; De Mey *et al.*, 1978; J. M. Vasiliev & I. M. Gelfand, unpublished) have revealed a system of microtubules radiating from perinuclear centres in transformed fibroblasts. The pattern of the microtubule system in transformed cells is very similar to that in normal cells (fig. 5.3). Application of anti-tubulin antibodies to unextracted transformed cells often does not reveal the microtubular system (Brinkley, Fuller & Highfield, 1975; McClain, D'Eustachio & Edelman, 1977). This difficulty is due to the increased thickness of the cytoplasm of the poorly spread transformed cells. In these cases extraction of the cells with detergent before application of antibodies (Osborn & Weber, 1977; Bershadsky *et al.*, 1978*b*) facilitates a clear demonstration of the microtubular system.

The microtubular system is not only present in transformed cells but it is functionally active. At least one normal function of the microtubules can be demonstrated in morphologically transformed cells: these structures are essential for the stabilization of the normal, polarized state of the cell surface (Svitkina, 1977).

Incubation of strongly transformed L and M-22 cells with antitubulins changes their shape: the long, lateral, stable parts of the edge that are a characteristic of non-treated cells disappear and the cells start to sprout cytoplasmic processes at various parts of the edge. These alterations are similar to those observed after incubation of normal fibroblasts with antitubulines. Transformed cells with destroyed microtubules remain poorly attached to the substratum. In contrast to normal colcemid-treated cells which form a band of lamellar cytoplasm with lamellipodia at the outer edge, transformed cells are attached mainly by long cylindrical or fusiform processes and the cell areas between these processes remain unattached (figs. 3.10 and 5.4).

In summary, microtubules are essential for the maintenance of a polarized state and directional locomotion in both normal and transformed cells. The destruction of microtubules does not abolish the difference in attachment between these cells.

Locomotion of transformed cells. Statistical analysis of the tracks of transformed L fibroblasts recorded by cinematography have shown that locomotion

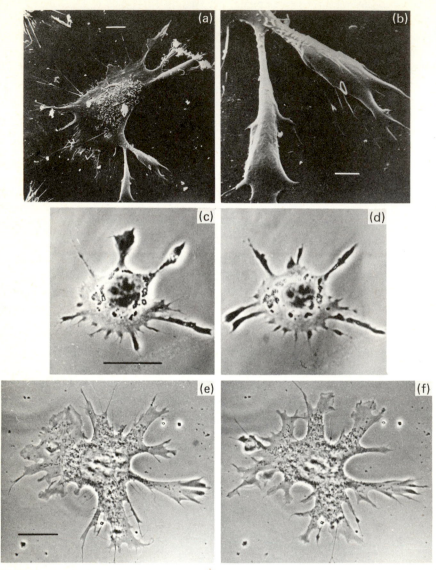

Fig. 5.4. Colcemid-treated transformed cells forming poorly attached processes along the whole perimeter. (a, b) An L cell by scanning electron microscopy. (c, d) Two phase contrast micrographs of the same L cell taken with an interval of 30 min. (e, f) A cell of the M-22 line transformed with SV-40 virus; two phase contrast micrographs of the same cell taken with an interval of 15 min. Bars: (a), 5 μm; (b), 2 μm; (c–f), 20 μm.

is more random than that of normal fibroblasts. L cells make more turns and are less consistent in the direction of their movements (Andrianov *et al.*, 1971). Also, an analysis of the movement of non-neoplastic and neoplastic spontaneous transformants made by Sanford *et al.* (1977*b*) revealed a randomization of locomotion. The non-neoplastic cells tended to maintain the same direction of locomotion in sequential 2.5-h periods whereas neoplastic cells did not. As suggested by Andrianov *et al.* (1971) and by Sanford *et al.* (1977*b*), the more random locomotion of transformed cells may be a secondary consequence of their deficient attachment to the substratum: the anterior active edge of these cells retracts more easily and loses its leading position more often than large well-attached, leading lamella of normal cells. As a result, transformed cells change the direction of their locomotion more frequently.

Early stages of spreading. As we have seen earlier, spreading of strongly transformed cells leads to an abnormal final result; to formation of cells with a deficient lamellar cytoplasm. It is natural to expect that the morphology of transformed cells at an early stage of spreading would be different from that of normal cells at a similar stage. A special study of these early stages has been made with only one transformed line: L cells (Vasiliev & Gelfand, 1973; Bragina, 1975). The main differences between these cells and normal mouse fibroblasts at the early stages of spreading were as follows. (*a*) At the stage of initial cell–substratum contact (before the formation of pseudopodia) L cells had numerous microvilli and folds at the lower surface, while the lower surface of the normal cells was smooth (Fig. 5.5). (*b*) At the radial stage of spreading, filopodia formed around the body of an L fibroblast were not quickly replaced by lamellar cytoplasm; filopodia might persist for several hours. (*c*) Many L cells did not form the ring of lamellar cytoplasm but underwent polarization directly from the filopodial stage by forming small lamellae which spread in several directions so that the cell acquired a stellate or fusiform shape. (*d*) At the early stages of spreading, as at the final ones, filopodia and lamellae of the L cells did not contain microfilament bundles.

It would be interesting to study early stages of spreading of other transformed cells and, in particular, to find out whether the alterations of initial contact and the shape of pseudopodia can be observed in other lines.

Cell rounding and detachment from the substratum. Strongly transformed cells are poorly spread on the substratum. Naturally, it is easier to detach them from the substratum by the action of such agents as proteases or EDTA and this occurs in a shorter time than with non-transformed cells in similar conditions. The detachment of transformed cells, like that of normal cells is preceeded by cell rounding but the intermediate stages of the rounding process may be somewhat different. L cells treated with trypsin or EDTA are

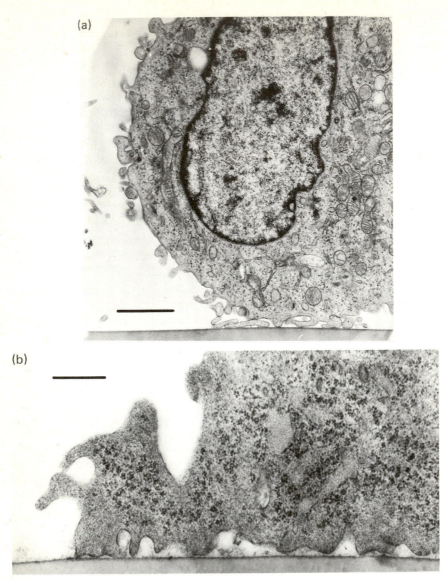

Fig. 5.5. Early stages of spreading of a transformed L cell. Electron micrographs of sections perpendicular to the plane of the substratum. (a) Initial contact with the substratum at 10 min after seeding showing microvilli at the lower surface. (b) The lateral edge of a spreading cell at 30 min after seeding. Focal cell–substratum attachment sites and microfilament bundles are absent. Bars: (a), 2 μm; (b), 0.5 μm.

transformed directly into spheres attached by retraction fibres but the intermediate state the cell being attached by thick cytoplasmic processes (see chapter 4) which is characteristic for normal mouse fibroblasts is absent. Certain strongly transformed lines are so poorly attached to the substratum that they may acquire a spherical shape, not only after treatment with standard detaching agents but also after the action of certain agents that do not significantly alter the shape of normal cells. We (Bershadsky *et al.*, 1979) selected from the L line of mouse fibroblasts several sub-lines that were spread on the substratum even more poorly than parent cells. It was found that most cells of these sub-lines (about 60%) reversibly acquired a spherical shape when they were transferred from 37 °C to room temperature (18 °C). Under the same conditions only 5–7% of the parent 'wild-type' L cells became rounded. The rounded cells remained attached to the substratum by numerous retraction fibres. However, when 'wild-type' L cells had been pretreated with low concentrations of trypsin or EDTA, they acquired the ability to become rounded after transfer to 18 °C. Thus, moderate cooling could produce rounding of both the parent L line and of the more poorly attached sub-lines; however, only in experiments with these sub-lines was the cooling alone sufficient for this alteration of shape. It is not clear how moderate cooling induces cell rounding in poorly attached cells. One possible explanation is that the cooling inhibits extension of pseudopodia at the active edge and thus shifts the equilibrium between the formation of new contacts and the detachment of the old ones; this shift may lead to partial detachment if the average life-time of the contact structure is short. It is very probable that, besides cooling, there are many other factors that may cause rounding of poorly attached transformed cells. For example, Paranjpe & Boone (1975) described the spontaneous periodic rounding of transformed cells attached to a poorly adhesive substratum.

In the course of mitosis cells are partially detached from the substratum. This alteration of shape may be more evident in transformed cells. In fact, Fox *et al.* (1976) found that the percentage of cells in metaphase that acquired a rounded shape was significantly higher in neoplastic cell lines of mouse fibroblasts than in non-neoplastic paired lines.

Cell–cell interactions

In this section we will describe the interactions of substratum-spread fibroblasts with each other. We will begin by analysing paired collisions of transformed cells and then we will describe the morphological patterns formed by transformed cells in dense culture as well as functional and morphological alterations of cell–cell contacts in these cultures.

Cell–cell collisions. As described in chapter 4, the active edge of a moving cell may collide with the active edge of another cell (head–head collisions) or with its stable lateral edge (head–side collisions). These collisions may lead to three types of outcome: halting, underlapping, or overlapping. The analysis of individual collisions in cultures of mouse fibroblasts transformed by mouse sarcoma virus and in a cell line derived from mouse sarcoma (Guelstein *et al.*, 1973) showed that most head–head collisions resulted in haltings, while head–side collisions ended either in haltings (see fig. 3.11) or in underlappings. Partial overlappings in cultures of transformed cells were as rare as in cultures of normal fibroblasts. An increased degree of underlapping was the main feature which distinguished collisions in transformed cultures from those in normal cultures. In a transformed culture, the whole body of one cell often passed under the unattached lower surface of another cell. Essentially similar results were obtained by Bell (1977) who analysed collisions in cultures of 3T3 cells and of Py3T3 cells: 75% and 60%, respectively, of head–head collisions resulted in haltings; all other collisions resulted in underlappings, while overlappings were practically absent. Underlappings were approximately twice frequent after head–side collisions of strongly transformed Py3T3 cells as compared with weakly transformed 3T3 cells (86% and 46% respectively). The increased degree of underlapping in transformed cultures is obviously a result of the poor attachment of large parts of the cell bodies to the substratum. More exactly, it is probably a result of an increased ratio of the length of unattached lateral parts of cell edge to that of attached active parts.

The results presented above show that contact inhibition of locomotion may be as efficient in transformed cultures as in normal ones: overlappings remain rare and head–head collisions usually result in haltings. The degree of underlapping is not relevant to the question of the efficiency of contact inhibition: the active edge of an underlapping cell continues to move on the substratum and may not touch the surface of another cell at all. Gail & Boone (1971*b*) have observed that cell motility decreases dramatically with increasing density in 3T3 cultures but does not decrease in SV 3T3 cultures. Possibly, this difference is also a result of increased underlapping in transformed cultures. Usually, contact inhibition of locomotion is a result of contact paralysis of pseudopodial activity at the active edges, accompanied by the retraction of contacting edges. Is contact paralysis effective in transformed cultures? Our observations of these cultures suggest that head–head collisions are often accompanied by paralysis and retraction of the contacting edges (fig. 3.11). The degree of retraction is often increased in transformed cultures compared with normal ones. This increase is probably a result of the deficient attachment of the active edges to the substratum and of deficient formation of stable cell–cell contacts by these edges. In experiments with cultures of rat fibroblasts transformed by Rous sarcoma virus, Vesely & Weiss (1973) observed the apparent absence of contact paralysis: the ruffling of active cell edges continued after cell–cell contact. Obviously, it is possible that the ability to perform contact paralysis may be different in various transformed cultures. It should also be taken into consideration that even when contact paralysis remains efficient, its demonstration may be difficult in cultures with poorly attached cells and retractions of the contacting edges may lead to rapid activation of the paralysed zones. The pseudopodia of these cells may become narrower so that paralysis in the area of their contact with another cell may remain unnoticed. In other words, because of deficient attachment, contact paralysis in transformed cells may be less long-lasting and may involve smaller areas of the edges, so that its demonstration would require more detailed cinematographic investigation.

To summarize, investigation of collisions in several types of transformed cultures revealed efficient contact inhibition of locomotion and overlapping was practically absent. Possibly, the haltings observed after cell–cell collisions in transformed cultures are not always accompanied by contact paralysis but this question needs further investigation. Increased underlappings after head–side collisions are a characteristic feature of many transformed cultures.

Morphology of dense cultures. Morphology of different dense transformed fibroblastic cultures is very diverse. Each transformed culture has its own characteristic cell distribution pattern that depends on the type of transforming agent and of the target culture. The individual variability of transformed clones as well as the progression of changes in the course of cultivation may contribute to this morphological diversity (fig. 5.6). Among the morphological variants of strongly transformed cultures the following ones may be mentioned.

(1) *Monolayered or multilayered cultures consisting of poorly spread elongated fibroblasts oriented in parallel.* Our preparations of normal mouse fibroblasts transformed by mouse sarcoma virus belonged to this morphological type.

(2) *Multilayered cultures consisting of poorly spread fusiform or stellate cells which criss-cross each other without any visible orientation.* Many cultures transformed by various chemical agents have this morphology.

(3) *Multilayered cultures consisting predominantly of rounded cells.* Certain variants of Rous sarcoma virus produce this type of chick fibroblast (see chapter 2).

(4) *Monolayered cultures consisting of tightly packed hemispherical cells.* In our preparations, dense cultures of the L line had this morphology.

Measurements of the average projected areas of cells on the substratum in dense cultures of several lines of transformed mouse fibroblasts showed that

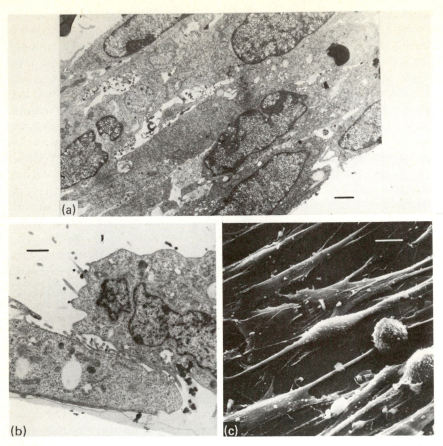

Fig. 5.6. Dense cultures of transformed cells. (a, b) Electron micrographs of sections perpendicular to the plane of the substratum. (a) The M-22 line of mouse fibroblasts transformed with SV-40 virus; a multilayered culture of poorly spread cells. (b) The L line, a monolayer of poorly spread cells. (c) Scanning electron micrograph of a primary culture of mouse fibroblasts transformed by mouse sarcoma virus showing the mutual orientation of poorly spread cells. Bars: (a, b), 1 μm; (c), 10 μm.

these areas were considerably smaller than those in dense cultures of normal mouse fibroblasts (Cherny *et al.*, 1975). For instance, normal mouse fibroblasts had average areas of about 1800–1900 μm^2 in both sparse and dense cultures. Probably, decreased spreading is a general feature of the different morphological variants of dense transformed cultures. We do not know the cellular peculiarities which are responsible for the differences between multicellular structures formed by various lines of transformed cells in dense cultures. Possibly, one of the variables is the degree of deficiency of spreading. For instance, if cells have a decreased size of their active edge, but the lamellar cytoplasm near this edge is relatively well attached to the substratum, then these cells will effectively underlap each other and this may lead to the formation of multilayered criss-cross patterns. The cells in which even the anterior lamella is badly attached, will be unable to underlap other cells; these cells may form monolayered cultures.

As we have seen, contrary to widespread belief, multilayering is not a distinctive property of transformed fibroblastic cultures: the dense cultures of normal fibroblasts may be highly multilayered while certain transformed cultures may form monolayers.

Cell–cell contacts. The morphological examination of sections of dense, strongly transformed cultures of various types suggests that specialized cell–cell contacts are formed more rarely in these cultures than in their weakly transformed or normal counterparts (Cornell, 1969; Martinez-Palomo *et al.*, 1969; Cherny *et al.*, 1975). Certain types of specialized contacts may disappear completely. For instance, dense cultures of the M-22 line of mouse fibroblasts transformed by SV-40 virus had the same types of specialized contacts (gap junctions and intermediate junctions) as dense cultures of normal fibroblasts but these junctions were seen much more rarely in the transformed cells. Dense cultures of the L line of transformed fibroblasts had no specialized junctions (Cherny *et al.*, 1975).

Certain lines of strongly transformed cells may have deficient metabolic co-operation and deficient cell–cell coupling. These alterations are due to the defective formation of gap junctions responsible for intercellular communication involving direct molecular transfer (see chapter 3). For instance, certain sub-lines of the mouse L line are apparently unable to perform cell–cell communication (Boitsova, Vasiliev & Potapova, 1975). This loss of cell–cell communication is not a general feature of all strongly transformed lines; many such lines have been found to form functioning, highly permeable contacts (see review in Loewenstein, 1979). It is probable that the complete loss of cell–cell communication ability is associated only with very advanced stages of the deficiency of cell–cell contact formation. If permeable contacts are formed, even in reduced numbers, cell–cell transfer of ions and molecules would take place and could be detected experimentally.

To measure quantitative alterations of contact-mediated communication, Corsaro & Migeon (1977) developed an assay system based on the co-cultivation of 'HPRT-positive' cells,* which formed a toxic metabolite (6-thioguanilic acid) in media containing 2-amino-6-mercaptopurine, and 'HPRT-negative' cells which were unable to form this metabolite. In mixed cultures, toxic nucleotide was transferred through contacts from HPRT-positive to HPRT-negative cells and the proliferation of recipient cell types was diminished as the number of HPRT-positive cells in the mixture

* HPRT, hypoxanthine phosphoribosyltransferase.

increased. With this assay it was shown that a number of human cell lines transformed by SV-40 virus and cell lines derived from human tumours have a decreased ability for contact-mediated communication compared with normal human fibroblasts. Wright, Goldfarb & Subak-Sharpe (1976), and Wright *et al.* (1976) used the contact-mediated transfer of a toxic metabolite for the selection of variant sub-lines with an additional reduction of the ability for metabolic co-operation over 'usual' transformed lines. They selected a number of the co-operation-defective sub-lines from line BHK C13 transformed with polyoma virus. The deficiency of co-operation was accompanied by morphological alterations which suggested a decreased spreading on the substratum. These results confirm that the complete loss of ability for intercellular transfer of molecules is associated with very severe deficiencies of cell–cell and cell–substratum attachment.

Deficiency of contact guidance

In this section we will discuss reactions of transformed fibroblasts to substrata that are able to induce contact guidance in normal fibroblasts. The results of experiments with normal cells were described in chapter 4.

A series of detailed comparative studies of the behaviour of normal and transformed cells on substrata with large grooves and cylindrical protrusions were performed by Slavnaya & Rovensky (1977) and by Samoilov *et al.* (1978). Nine cell lines transformed by different viruses and other agents were examined: the control, normal cells included cultures of normal embryo fibroblasts of several different species. Behaviour of all the transformed lines examined were found to differ in several aspects from that of their normal counterparts.

(1) *Cell migration from the grooves was decreased.* In contrast to corresponding normal cells, a considerable proportion of the cells in transformed cultures of mouse, rat, chicken and human fibroblasts was retained near the bottom of the grooves even after 1–2 days of culture. Both normal and transformed hamster fibroblasts migrated poorly from the grooves.

(2) *The degree of cell orientation on grooved substrata was considerably decreased.*

(3) *The elongation of nuclei on grooved substrata compared with a flat substratum was absent* in all the experiments with transformed lines, although it was observed in the experiments with all the examined types of normal fibroblasts.

These alterations of the behaviour of transformed cells on grooved substrata can be regarded as corollaries to their deficient spreading. As discussed in the previous chapter, well spread normal cells have a tendency to attain a flat shape and this is probably associated with the rigidity of the microfilament bundles. This may explain the migration of normal cells from the grooves as well as their orientation on cylindrical substrata. Transformed cells, possibly, have a decreased preference for a flat shape: their disorganized microfilament system may show less resistance to the bending of the cell body. On the other hand, transformed cells have decreased lamellar areas therefore they will more easily reach the state in which all these areas are located on the same plane. These factors may lead to a less regular distribution of the cells on the substrata, decreased preference being shown to certain areas of the substrata and a decreased orientation. It seems probable that the ability for contact guidance may be decreased but not completely lost in many transformed lines. Possibly, certain types of substrata would be able to orient even strongly transformed cells. In particular, these may be substrata with a considerable curvature or with a small distance between orienting structures which may even affect the orientation of cells with narrow active edges. In fact, orientation of L cells was observed on the surface of bundles of polymer threads about 10–20 μm in diameter (Slavnaja *et al.*, 1974) as well as on 10 μm wide strips of glass surface between two non-adhesive lipid films (Vasiliev & Gelfand, 1977). Cylindrical substrata of varying diameter similar to those used by Dunn & Heath (1976) may be particularly suitable for future quantitative studies of contact guidance in different normal and transformed lines.

Aggregation of normal and transformed cells on poorly adhesive substrata and in suspension

Normal and transformed cells can adhere to each other and form aggregates on certain substrata and in suspension. Two different sets of experimental conditions for aggregation have been described: in one set, transformed cells form aggregates less readily than non-transformed ones while in the second set this relation is reversed; that is, transformed cells form aggregates at a greater rate and form larger aggregates than their non-transformed analogues.

Results of the first type were observed when cells were seeded on poorly adhesive substrata such as the surface of millipore filters or that of a glutaraldehyde-fixed monolayer of normal fibroblasts (Friedenstein, Rapoport & Luria, 1967; Ambrose & Ellison, 1968; Bershadsky & Guelstein, 1973, 1976; Bershadsky & Lustig, 1975). Normal embryo fibroblasts of different species were poorly spread on these substrata; at 20–24 h after seeding these cells started to assemble into three-dimensional aggregates regularly distributed over the surface of the substratum and aggregation continued during subsequent days. Certain cells in the peripheral parts of the aggregates remained partially attached to the substratum, so that the aggregates were not detached from it. Many transformed lines seeded on the same substrata did not form aggregates: even after several days of cultivation they remained randomly distributed on the substratum. Other lines formed aggregates more slowly than non-transformed cells and these aggregates reached a smaller final size. In the experiments of Bershadsky & Lustig (1976) only one of the nine transformed lines examined (HEK line of hamster fibroblasts) formed aggregates of the same size as those formed by normal fibroblasts.

The relationship between aggregation of normal and transformed cells was reversed in experiments in which the cells remained in suspension (Steuer & Ting, 1976; Wright *et al.*, 1977). In these conditions, transformed cells demonstrated a greater ability to aggregate. For instance, in the experiments of Wright *et al.* (1977) the initial rate of aggregation was expressed as a percentage decrease of single cells after incubation of a suspension for 15 min. The figures varied from 7 to 18% for poorly aggregating cells and from 35 to 55% for rapidly aggregating cells. The list of poorly aggregating cells included primary fibroblasts of different species and three of the four clones of 3T3 cells. The rapidly aggregating cells included six virus-transformed lines, the line of B-16 melanoma and one clone of 3T3 cells.

In the experiments of the first and second types, the conditions for aggregation were significantly different. In the first, a cell had the choice between spreading on the poorly adhesive substratum or the surface of another cell; in the second, a cell had either to attach itself to the surface of other cells or remain in suspension. We do not know why normal and transformed cells have different responses to the choice between these two alternatives. In this connection it would be important to learn more about the processes accompanying cell aggregation. For instance, we do not know whether cell–cell contact in suspension can induce the extension of pseudopodia, whether the aggregation involves attachment of these pseudopodia or that of the microvilli formed before the contact, etc.

The inhibition of the aggregation of embryonic cells produced by added transformed cells (Maslow & Mayhew, 1975; Maslow, Mayhew & Minowada, 1976; Maslow & Weiss, 1978) is another poorly understood phenomenon.

Lectin-mediated cell agglutination

Concanavalin A and many other lectins added to suspended cells can induce agglutination. This agglutination can be regarded as a special type of cell aggregation in suspension but lectin-mediated agglutination can be observed in more dilute suspensions and after a shorter incubation time than the spontaneous aggregation discussed above. Relative rates of concanavalin A agglutination and of spontaneous aggregation of different cells are not always correlated (Wright *et al.*, 1977). In experiments with many transformed lines it was found that their agglutination requires considerably lower concentrations of lectins than that of their non-transformed parent cells (see reviews in Burger, 1973; Nicolson, 1974, 1976*b*). For instance, when BHK cells transformed with temperature-sensitive polyoma virus were grown at a permissive temperature (32 °C) and then suspended and incubated with wheat germ agglutinin at 23 °C for 5 min, 40–50 μg/ml of this agglutinin induced aggregation of 75% of the cells. The same cells grown at a non-permissive temperature needed 160–180 μg/ml of agglutinin to produce the same degree of aggregation (Eckhart, Dulbecco & Burger, 1971).

The agglutinability of non-transformed cells can be increased by treatment with low concentrations of trypsin or other proteases (Burger, 1971; Starling *et al.*, 1977*a*, *b*). Diverse agents such as 5-bromodeoxyuridine (Biquard, 1974), 7,12-dimethylbenzo(a)anthracene (Vasiliev, 1976) and low ionic strength solutions (Killion & Kollmorgen, 1975) also increase agglutinability. The agglutinability of normal cells temporarily increases during mitosis (Fox, Sheppard & Burger, 1971; Smets & DeLey, 1974).

The increased agglutinability of transformed cells has been observed in experiments with different lectins specific for various carbohydrate groups.

Increased agglutinability is characteristic for many lines of transformed fibroblasts and also for certain lines of transformed epithelial cells. There are, however, several exceptions where the agglutinability of transformed lines is as low as that of normal cells. For instance, Marciani & Okazaki (1976) found that chick fibroblasts infected with a thermo-sensitive mutant of Rous virus (RSV-BH-Ta) showed a negative correlation between agglutinability and transformation. At the permissive temperature for transformation, the cells were not agglutinable with concanavalin A, whereas at the non-permissive temperature they showed high agglutinability.

The different reactivity of normal and transformed cells with lectins can be seen not only in the agglutination reaction but also in the lectin-dependent haemadsorption reaction. When pre-fixed cultures of substratum-attached cells are incubated with lectin and then with a suspension of erythrocytes, attachment of erythrocytes to the surface of the cultured cell is observed. Much lower concentrations of lectins are needed to obtain haemadsorption on the surface of a transformed culture compared with that of a normal culture (Gelfand, 1973).

The reason for increased agglutination and haemadsorption properties of transformed cells remains a mystery. Apparently, these properties are not associated with altered density of lectin receptors on the surface of transformed cells because the surfaces of normal and transformed labelled cells can be shown to bind similar concentrations of radioactive lectins. Also, efforts to find a correlation between the ability of lectins to induce patching of corresponding receptors on the surfaces of different cells and their agglutinability have not yielded convincing results (see review in Nicolson, 1976*b*). Willingham & Pastan, 1975*a*) suggested that high agglutinability is associated with a particular topographical relief of the cell surface; namely, with the presence of numerous microvilli. It is not clear whether the presence of microvilli is a distinctive character of transformed cells (see above). In the experiments of Oppenheimer *et al.* (1977), treatment of mouse sarcoma 180 cells with cytochalasin B increased their agglutinability and caused formation of broad ruffles instead of microvilli on the surface of untreated cells. Papain-treated hepatocytes had a high concanavalin A agglutinability but a smooth surface while control cells with low agglutinability had numerous microvilli (Starling *et al.*, 1977*b*). In our unpublished experiments performed in collaboration with A. D. Bershadsky & Y. A. Rovensky, treatment of normal mouse fibroblasts with trypsin increased the concanavalin A agglutinability but did not change the ratio of cells with different reliefs present in the suspension. We still hope that further improvement of the methods of morphological examination of the suspended cells will reveal that some consistent alterations of their surface topography are associated with changes of agglutinability.

Epithelia

Manifestations of morphological transformation in epithelial cultures are much less well known than those in fibroblastic cultures. Strongly transformed epithelial cells lose their ability to form coherent epithelial sheets. The cells form 'piled-up' colonies (Leibovitz *et al.*, 1976; Owens *et al.*, 1976; Ishiwata *et al.*, 1977; Rhim *et al.*, 1977; Rosenthal *et al.*, 1977). Our preparations of mouse hepatoma 22a cultures consisted of individual polymorphous cells (Fig. 5.7) that either did not form lateral, stable cell–cell contacts or formed small numbers of these contacts so that larger or smaller groups of mutually linked cells were seen but coherent large sheets were not formed even in dense cultures (Zakharova, 1976). The cells in these cultures were often polarized, and moved individually into the wound. These cells often acquired a fibroblast-like shape.

The upper surface of normal epithelial sheets is non-adhesive due to the stable inactivation of pseudopodial activity by cell–cell contacts. As the formation of contact becomes deficient in transformed cultures, the upper surface remains adhesive for other cells and particles (Vasiliev *et al.*, 1975*a*). Formation of multilayered areas is often observed in cultures of hepatoma 22a as in other strongly transformed epithelial lines. By analogy with transformed fibroblastic cultures, it seems probable that in these epithelial cultures poorly attached cells often underlap each other leading to the formation of multilayered structures. Strongly transformed epithelial cells, like transformed fibroblastic cells, may have numerous irregular microvilli and other extensions on their upper surface; this surface is less smooth than that of non-transformed cells (Kahan *et al.*, 1977; Karasaki, Simard & de

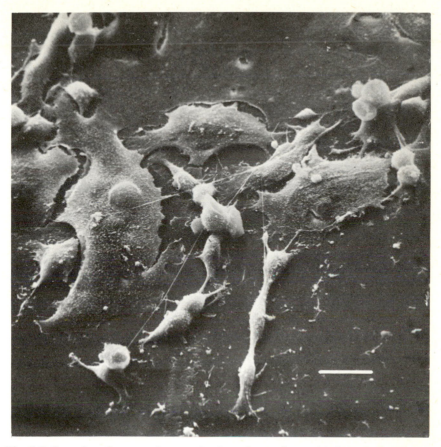

Fig. 5.7. Scanning electron micrograph of a culture of an epithelial cell line derived from mouse hepatoma 22. The cells do not form a coherent sheet and many are poorly spread. Bar: 20 μm.

Lamirand, 1977). The ability to form highly permeable cell–cell contacts is completely lost by certain lines of rat hepatoma cells; in other lines formation of these contacts takes considerably more time than in cultures of the normal parent cells (Azarnia & Loewenstein, 1976, 1977; Loewenstein, 1979). Other epithelial lines may lose their selectivity in the establishment of cell–cell communication. This is probably the case with transformed mammary epithelial cells: normal human mammary epithelial cells exchange metabolites through their contacts with homologous cells but not with mammary fibroblasts. In contrast, neoplastic epithelium may non-selectively establish metabolic co-operation with epithelium and with fibroblasts (Fentiman & Taylor-Papadimitrou, 1977). Possibly, in the course of progression of epithelial transformation, cell–cell communication first becomes non-selective and then is lost completely.

An additional manifestation of strong transformation in epidermal cultures is abnormal keratinization. Instead of lamellar sheets of keratinized cells covering the upper surfaces of normal cultures, irregular foci of keratinization are seen in transformed cultures (Indo, 1977).

In summary, a deficient ability to form stable cell–cell contacts, possibly, associated with deficient spreading on the substratum, is probably the main characteristic of strongly transformed epithelial cultures.

A number of epithelial cultures that have oncogenic properties do not demonstrate any striking morphological alterations (Weinstein *et al.*, 1975; Pickett *et al.*, 1975). In particular, these cells retain an apparently normal ability to form monolayered coherent epithelial sheets with a non-adhesive upper surface. A detailed investigation of certain minimally transformed lines of this type may, however, reveal certain alterations of cell–substratum attachment. For instance, in our laboratory, the morphology of an MPTR line of kidney epithelial cells transformed with SV-40 virus was examined (Zakharova, 1976). These cells, like normal mouse kidney epithelium, formed coherent sheets but the lamellar cytoplasm on their free edges had a decreased size; not only lamellopodia but also filopodia were often seen at their edges. The density of microvilli on the upper surface was higher than in normal kidney cultures. It is not clear whether these changes are common for many minimally transformed cultures. Montesano *et al.* (1977) examined the cytological characteristics of a number of oncogenic and non-oncogenic lines of hepatic epithelial cells; they found that the acquisition of oncogenicity is correlated with cytological changes. A correct diagnosis of oncogenic *versus* non-oncogenic lines was made by several independent observers who examined coded stained cultures of these lines. Analysis of morphological differences between these cultures suggested than an increased basophilia of the cytoplasm and an increased nucleus: cytoplasm ratio are the most valuable symptoms for characterizing oncogenic lines. Possibly, these alterations are consequences of decreased cell spreading.

These data show that further analysis of 'quasi-normal' epithelial cultures may be fruitful and may reveal certain minimal alterations of cell–cell and cell–substratum attachment or some other characteristic morphological changes.

Mixed cultures of transformed and normal cells

Investigation of cell behaviour in mixed cultures is interesting in at least two respects. First, it may increase our knowledge of the morphogenetic properties of transformed cells, and second, it may serve as a very simplified model of the processes taking place in the course of the invasion of normal tissues by neoplastic cells *in vivo*.

Invasive behaviour between populations of normal and transformed fibroblasts

When populations of normal fibroblasts and of transformed fibroblasts migrating in the opposite directions meet each other, a mutual intermixing of both populations takes place in the area of encounter. In other words, normal and transformed populations invade each other. This phenomenon of mutual invasion was described and studied by Abercrombie and his collaborators (Abercrombie & Heaysman, 1954, 1976; Abercrombie, Heaysman & Karthauser, 1957) and recently also by E. M. Stephenson & N. G. Stephenson (1978). In each of these experiments two explanted fragments of various tissues were placed on the substratum at some distance from each other. The cells migrated in all directions from each explant and after some time the cells of one explant met those from the other migrating in the opposite direction. The cultures were fixed at different times before or after this junction of confronted explants and the distance the cells had migrated from each explant was measured in stained preparations. When explants of chick embryo fibroblasts were confronted with those of neonatal mouse fibroblasts, migration of both types stopped almost completely after junction of the out-wanderings. In contrast, in experiments in which explants of normal chick fibroblasts were confronted with mouse sarcomas the cells of both types continued to migrate towards each other after junction. Neoplastic cells in the junction zone were often located on the upper surface of the fibroblasts. Different sarcoma strains confronted with chick fibroblasts demonstrated different degrees of invasion.

Cell–cell collisions between normal and transformed fibroblasts

As just described, migration of two confronted populations of normal fibroblasts is stopped when they meet each other, while populations of normal and neoplastic cells invade each other. This different distribution is obviously the 'statistical' result of different outcomes of many individual cell–cell collisions taking place in the zone between the two confronted explants. What are the distinctive features of cell–cell collisions between normal and neoplastic cells which distinguish them from collisions between two normal fibroblasts? Inhibition of the migration of normal cells towards each other after they contact, is probably due to their effective contact inhibition: head–head collisions between the two normal fibroblasts usually lead to haltings (see p. 117). One possible reason for the mutual invasion of normal and neoplastic cells could be a failure of contact inhibition, leading to active overlapping of normal cells by transformed fibroblasts. Individual collisions between the chick fibroblasts and sarcoma cells used in the mutual invasion experiments of Abercrombie's group have not been analysed. However, analysis of collisions between other types of normal and transformed fibroblasts have not revealed active overlapping; head–head collisions of cells moving in the opposite directions usually lead to the halting of both cells. This was observed in our experiments with normal mouse fibroblasts and transformed cells of the L line (Domnina *et al.*, 1972) and of the S-40 line (Guelstein *et al.*, 1973). In the experiments of Bell (1977) overlapping of 3T3 cells by strongly transformed Py3T3 cells was also absent after contact; in contrast, they often underlapped the 3T3 cells. It is not quite clear whether the halting of normal and transformed cells after head–head collisions is accompanied by contact paralysis of both active edges.

Heaysman (1970) and Vesely & Weiss (1973) described non-reciprocal contact paralysis between normal fibroblasts and sarcoma cells: only the leading edges of the normal cell were inhibited but those of the neoplastic cells were unaffected. In our experiments (Domnina *et al.*, 1972; Guelstein *et al.*, 1973 and unpublished) leading edges of both normal and neoplastic cells were apparently paralysed after collision.

The difficulties of diagnosing contact paralysis in transformed cells have already been discussed. We wish only to stress that the failure of contact paralysis (if any) in transformed cells would not be sufficient for invasion without the active overlapping of normal cells.

On the other hand, mutual invasion may be a result of processes other than active overlapping. When groups of normal fibroblasts meet neoplastic fibroblasts, their small anterior lamellas may be easily accomodated in the spaces between individually migrating normal cells. As a result, after the junction of the populations of normal and neoplastic cells, directions of cell migration will be changed more rarely than after junction of two normal populations. As we have observed in mixed cultures of normal mouse fibroblasts and L cells (Domnina *et al.*, 1972) a transformed fibroblast surrounded by better-attached normal cells retracts its edges and is eventually detached from the substratum. Its place on the substratum is immediately occupied by normal cells, while the detached transformed cell has to attach itself to the upper surface of the normal fibroblasts. In this way, transformed fibroblasts may become located on the upper surface of normal cells without actively migrating on this surface. Thus, the deficient spreading of transformed cells may be sufficient to explain their invasive properties in mixed cultures.

Behaviour of transformed cells located on the upper surfaces of normal cultures of fibroblasts

As described above, transformed cells may become located on the upper surfaces of normal cells in the zone of intermixing of the two populations. Alternatively, transformed cells may be seeded on the upper surface of dense cultures of normal fibroblasts where they attach themselves as easily as to glass (Domnina *et al.*, 1972). The differences in morphology between transformed

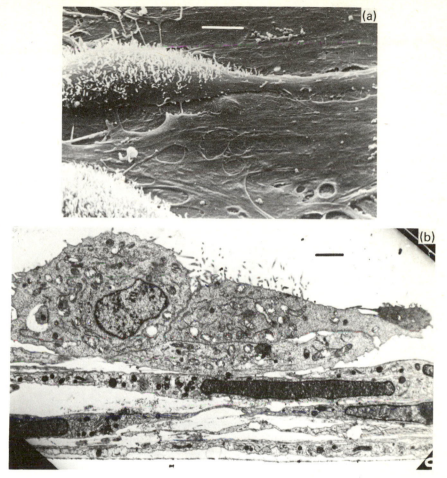

Fig. 5.8. Poorly spread transformed L cells on the upper surface of dense culture of normal mouse fibroblasts. (a) Scanning electron micrograph; (b) electron micrograph of a section perpendicular to the plane of the substratum. Bars: (a), 5 μm; (b), 2 μm.

and normal cells are usually retained in this situation despite their mutual contact: transformed cells are much less spread (fig. 5.8) and may also have an uneven surface relief. At the same time, their contact with underlying normal cells may have certain influences: (*a*) transformed cells may become oriented parallel to underlying normal cells (Stoker, 1964; Abercrombie & Heaysman, 1976); (*b*) specialized cell–cell contacts can be formed between the normal and transformed cells. In particular, normal mouse fibroblasts and L cells in mixed culture were found to form permeable intercellular contacts, although L cells in isolated cultures did not form these contacts (Boitsova &

Potapova, 1974). An exchange of metabolites through permeable contacts probably takes place between normal and neoplastic cells. In particular, toxic metabolites of carcinogenic hydrocarbons formed only by normal cells exert a local toxic effect on neoplastic cells in mixed cultures (Mittelman, Sharovskaja & Vasiliev, 1972). Co-operation between neoplastic epithelial cells and normal fibroblasts was discussed above (p. 153) as an example of 'non-selective communication'. The physiological effects of this co-operation remain unknown.

Mixed cultures of normal epithelium and transformed fibroblasts

The upper surface of epithelial sheets is as non-adhesive for transformed cells as for normal fibroblasts: when mouse transformed fibroblasts of the L line were seeded on the upper surface of islands of kidney epithelial cells, the density of attached cells per unit area of this surface was 10–16 times lower than that on the adjacent glass surface (Vasiliev *et al.*, 1975a). Our unpublished observations show that when a transformed L cell moving on glass meets the active lateral edge of an epithelial island, it immediately retracts its anterior lamella and changes its direction of movement. thus, contact inhibition of movement is effective between these two cell types. Barski & Belehradek (1965) showed that transformed mouse fibroblasts are stopped at the boundary of a sheet of vascular endothelium.

Mixed cultures of normal fibroblasts and transformed epithelium

The only description of interactions between these two cell types is that of Stoker *et al.* (1978). In their experiments, moving fibroblasts actively invaded the zone occupied by human mammary malignant epithelium. In contrast, the sheets of non-malignant mammary epithelium were never invaded by normal fibroblasts. The invasion of malignant epithelium by fibroblasts was probably due to the defective formation of epithelial sheets: fibroblasts had enough space to move between epithelial cells which were unattached or poorly attached to each other.

Conclusion: diagnosis of morphological transformation

In this chapter we have described a number of changes characteristic of strongly transformed fibroblasts as compared to normal fibroblasts. The list of these changes includes the following: (*a*) decreased spreading on the substratum in sparse cultures manifested by decreased size and abnormal morphology of the lamellar cytoplasm, by the scarcity of microfilament bundles and by the irregular relief of the upper surface; (*b*) decreased spreading in dense cultures associated with the formation of various abnormal multicellular patterns; (*c*) decreased formation of specialized cell–cell contacts; (*d*) deficient contact guidance: decreased orientation and elongation on the substrata with regular oriented relief as well as decreased migration from large grooves on the substratum; (*e*) invasive behaviour in mixed cultures with normal fibroblasts; (*f*) decreased formation of aggregates on poorly adhesive substrata; (*g*) enhanced formation of aggregates in suspension; (*h*) increased ability for lectin-mediated agglutination of haemadsorption.

Symptoms of morphological transformation in epithelial cells have been studied in less detail; but these symptoms probably include deficient formation of monolayered epithelial sheets and decreased attachment to the substratum. Each particular strongly transformed line of fibroblasts may demonstrate many, but not necessarily all these symptoms. Of course, it is not always necessary to test for the presence of all these symptoms in order to conclude from morphological evidence that a culture has been transformed. Diagnosis of strong transformation can often be made after light microscopic examination of the morphology of the culture. However, when one needs to confirm or refute the diagnosis of minimal transformation, one has to carry out various tests. Numerous tests are also essential if one wishes to compare the degrees of transformation of two lines. Quantitative tests such as the measurement of the area of lamellar cytoplasm, estimation of contact guidance and aggregation may be of particular value in these situations. Systematic studies of manifestations of the minimal morphological transformations in various fibroblastic cultures, and especially in epithelial cultures remain a task for the future.

Most of the transformation symptoms listed above are manifestations of decreased cell–substratum and cell–cell attachment; therefore, these symptoms can be revealed only in substratum-attached cells. We do not know at present any reliable morphological differences between normal and strongly transformed cells that are revealed in the suspended state. The only exceptions are the different abilities of transformed cells for lectin-mediated agglutination and for spontaneous aggregation, but the cellular basis of these phenomena remains obscure.

6

Alterations of morphogenetic reactions responsible for morphological transformation

Altered pseudopodial attachment reactions of transformed cells

In the previous chapter, we showed that most manifestations of morphological transformation are associated with decreased cell–substratum attachment and cell–cell attachment. The experimental results suggest that pseudopodial attachment reactions responsible for spreading and for the formation of cell–cell contacts are altered in transformed cells. The distribution of pseudopodia on the cell surface is controlled in normal cells by special morphogenetic processes: by microtubule-dependent stabilization and by contact paralysis. Immuno-morphological studies show that transformed cells have a microtubular system. Results of experiments with microtubule-destroying drugs show that the microtubular system in transformed cells retains the ability to stabilize the surface: this system remains essential for specialization of the cell edge into stable and active zones. At least some lines of transformed fibroblasts retain the ability for contact paralysis of their surfaces; possibly, some other cell lines lose this ability. In any case, possible failures of contact paralysis in transformed cultures will not lead to considerable alterations of the culture morphology, as they are not accompanied by active cell–cell overlapping.

In summary, alterations of pseudopodial attachment can be regarded as the main alteration of morphogenetic processes responsible for various manifestations of morphological transformation. Are other pseudopodial reactions altered in transformed cells? Strongly transformed cells may retain the ability for phagocytosis (Moizhess, 1969) but no detailed comparative studies of phagocytosis by normal and transformed cells have been reported.

Reactions of normal and transformed cells induced by soluble ligands have been studied in different cells growing in various conditions (see review by Nicolson, 1976b). The results obtained are rather variable. In a number of experiments patching and/or capping of concanavalin A receptors induced by the ligand were enhanced in transformed cells compared with their normal counterparts (Nicolson, 1971; Bretton, Wicker & Bernhard, 1972; Rosenblith et al., 1973; Garrido et al., 1974). In our experiments (Vasiliev et al., 1976; Domnina et al., 1977) patched receptors were cleared from the lamellar cytoplasm of substratum-attached, strongly transformed L cells. As these cells had smaller areas of lamellar cytoplasm than normal mouse fibroblasts, the relative area of cell surface was correspondingly smaller. In other experiments ligand-induced redistribution of receptors was not correlated with pseudo-podial activity, although this might be important for interpretation of the results. For instance, microvilli and other extensions are present on the upper surface of many transformed spread cells but not on the surface of normal cells. The membranes of microvilli, like those of other extensions, may be selectively cleared of cross-linked concanvalin A receptors. Possibly, the increased patching observed by a number of authors on the surface of transformed cells was associated with the presence of these extensions. Alterations of cell–cell contacts in transformed cultures may also affect pseudopodial activity and therefore may change receptor movements. We are not familiar with any data about the rate of ligand-induced endocytosis in normal and transformed cells. Thus, it is not clear whether only pseudopodial attachment reactions induced by solid surfaces and other cells, or all types of pseudopodial reactions, are altered in transformed cells.

The pseudopodial attachment reaction involves the following.

(1) *Formation of pseduopodia induced by cell contact with another surface*. This induction probably involves the transmission of some signal from the membrane into the adjacent cortical layer.

(2) *Attachment of pseudopodia to the other surface*. This attachment probably involves interaction of some membrane components with molecules of other surface; the interaction can be either direct or mediated by some intermediate ligand molecules.

(3) *Alterations of the cortical layer*. These possibly included the anchoring of membrane receptors by cortical structures.

This scheme suggests that many possible molecular sites of alteration may be responsible for the decreased attachment of transformed cells: alterations of extracellular ligands, of membrane receptors, of cortical molecules and of the systems transmitting the signals from the surface into the cell interior. In theory, alterations of the fluidity of the lipid layer may also affect the redistribution of receptors and thus change the course of pseudopodial reactions. At present we do not know what particular change is really responsible for altered attachment. Below we will discuss briefly several types of chemical alterations of different surface components that have been actively studied during the recent years.

Alterations of fibronectin and other matrix components

As described in chapter 3, fibronectins are glycoproteins forming extracellular aggregates and fibrils at the cell surface. Fibronectins seem to be important for cell spreading on the substratum; possibly, they serve as intermediate ligands between some unknown membrane receptors and the surface of the substrata. The amount of cell-associated fibronectin is often considerably decreased in the course of neoplastic evolution in culture. In cultures of primary fibroblasts from various species this protein constitutes 1–3% of the total cellular protein. Minimally transformed permanent cell lines contain reduced concentrations of fibronectin (0.8–0.15%) (Yamada, Yamada & Pastan, 1977). Strongly transformed chick fibroblasts contain 5–7 times less fibronectin than normal fibroblasts (Olden & Yamada, 1977). Transformation of human fibroblasts and glial cells is accompanied by a more than ten-fold decrease of fibronectin (Vaheri *et al.*, 1976). A decrease of fibronectin was demonstrated in numerous lines derived from fibroblasts of various species, transformed by various agents (see reviews in Vaheri & Mosher, 1978; Yamada & Olden, 1978). Lines in which fibronectin was not decreased have also been described. Summarized data from many papers given in the review of Yamada & Olden (1978) include 77 lines in which fibronectin was decreased and 12 lines in which a decrease was not observed. Material in our laboratory (Lyubimov, 1978*b*) includes three strongly transformed lines with decreased fibronectin and one line (the M-22 line of mouse fibroblasts transformed with SV-40 virus) in which fibronectin apparently was not decreased (fig. 6.1). The degree of decrease in fibronectin also varies from line to line. In certain transformed lines the cells retain some amount of surface-associated fibronectin that is preferentially localized in the areas of cell–cell contacts (Chen, Teng & Buchanan, 1976) or is diffusely distributed on the surface (Furcht, Mosher & Wendelschafer-Crabb, 1978).

The decrease of surface-associated fibronectin in transformed cultures can be due to several factors: decreased synthesis of this protein, decreased ability of transformed cells to retain aggregates of fibronectin at their surfaces, increased proteolytic degradation of fibronectin. An immuno-morphological study of mixed cultures of normal and transformed fibroblasts of different types showed that the close proximity between these cells had no effect on the distribution of fibronectin: the fibronectin-positive cells remained as positive and the fibronectin-negative cells remained as negative (Chen *et al.*, 1977*b*). In other experiments transformed cells in mixed culture were observed to increase the turnover of radioactively labelled fibronectin associated with normal cells (Mahdavi & Hynes, 1978). Close proximity between normal and transformed cells was required to produce this effect. Further experiments are needed to resolve this apparent discrepancy between the results of various groups. In any case, the secretion of soluble proteases is unlikely to be the

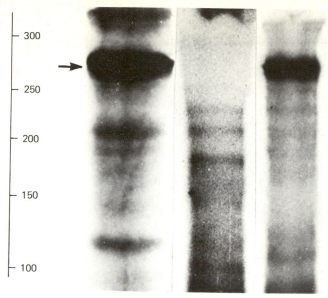

Fig. 6.1. Surface proteins of normal and transformed cells. *Left*, mouse embryo fibroblasts; *middle*, L cells; *right*, M-22 fibroblasts transformed with SV-40 virus. Note the normal content of fibronectin (arrow) in M-22 cells and complete loss in L cells. Washed cells were iodinated with lactoperoxidase, cell proteins analysed by sodium dodecyl sulphate polyacrylamide gel electrophoresis (5.6% acrylamide), and autoradiograms prepared by exposure of dried gels to Kodrirex (Kodak) X-ray film. Equal amounts of protein were applied to each gel. Low molecular weight proteins are not shown. Scale indicates weights in daltons ($\times 10^{-3}$).

major cause of the disappearance of fibronectin, because long range effects of transformed cells on normal cells are not observed in mixed culture.

The nature of cellular changes reponsible for the decreased amount of fibronectin in transformed cultures remains an open question. Diminished synthesis of fibronectin was observed in transformed chick fibroblasts (Olden & Yamada, 1977). A decreased amount of messenger RNA (mRNA) coding this protein was observed in transformed cells (Adams *et al.*, 1977). A decreased capacity of certain transformants to retain fibronectin of the cell surface has been also reported (Vaheri & Ruoslahti, 1975; Vaheri *et al.*, 1976).

Is the decreased spreading of transformed cells a result of decreased surface-associated fibronectin? Experiments which reconstitute this protein on the surface of transformed cells are in favour of this suggestion. Purified cellular fibronectin added to various transformed cultures improved their spreading, increased the number of microfilament bundles in the cells, restored parallel alignment of the cells, decreased the number of microvilli and ruffles on the upper cell surfaces (Yamada, Yamada & Pastan, 1976; Ali *et al.*, 1977; Hynes *et al.*, 1977), and improved the ability for directional migration (Ali & Hynes, 1978a). In summary, addition of fibronectin to these cultures diminished several manifestations of morphological transformation; probably, this diminishment was a result of improved cell–substratum attachment. Several reservations should be made at this point. (*a*) Besides fibronectin, a number of other physiological and non-physiological agents are able to restore a more normal morphology in transformed cultures (see below). It would be important to confirm that fibronectin affects cell attachment directly and does not act more indirectly, e.g. by changing the ability of cells to extend pseudopodia or by altering some other steps of pseudopodial reactions. (*b*) The cause of abnormal morphology in 'exceptional' lines, in which the amount of fibronectin apparently remains at the almost normal level, remains to be found. (*c*) It is not clear to what degree external fibronectin can restore normal morphology. Application of quantitative tests such as measurements of the area of lamellar cytoplasm, of the degree of nuclear orientation and elongation may be useful to clear up this point. (*d*) The role of fibronectin, if any, is not clear in the establishment of cell–cell contacts between fibroblastic cells as well as in the establishment of cell–substratum and cell–cell contacts between epithelial cells. Accordingly, it is not clear whether alterations of these forms of contact caused by transformation may be due to decreased fibronectin.

Despite these reservations, the results obtained during recent years give reason to suggest that a decrease of fibronectin is one of the factors responsible for the altered morphology of transformed cells.

In addition to a decrease of fibronectin, a number of other changes of the components of the extracellular matrix were observed in different experimental systems.

(1) Inhibition of the ability to synthesize collagen (Levinson, Bhatnagar & Liu, 1975), and a decreased amount of collagen mRNA (Adams *et al.*, 1977) were found in chick fibroblasts transformed with Rous sarcoma virus. Decreased post-translation modification of type I procollagen and possibly of type III procollagen secreted by human fibroblasts was observed after transformation with SV-40 virus (Sundarraj & Church, 1978).

(2) The total production of glycosaminoglycans was increased in malignant gliomas compared with normal glial cells and the composition of the glycosaminoglycans was altered (Glimelius *et al.*, 1978). Cells transformed with SV-40 virus were reported to secrete a glycosaminoglycan, heparan sulfate, with altered properties (Underhill & Keller, 1976). As collagens and glycosaminoglycans may be important for cell–substratum attachment these changes may have some relationship to morphological transformation.

Alterations of diverse surface proteins

Several alterations of other surface proteins that have been recently reported, will be listed here.

(1) Deletion of a glycoprotein with a subunit weight of 170 kd was observed in clones of polyoma-transformed hamster cells (Carter & Hakomori, 1978).

(2) A new non-virion protein of 80–100 kd was found in the plasma membranes of chick and hamster fibroblasts transformed by Rous sarcoma virus or by polyoma virus (Isaka *et al.*, 1975; Tarone & Comoglio, 1977; Comoglio, Tarone & Bertini, 1978). Possibly, this protein is identical with the 90 kd protein which is induced in normal and transformed fibroblasts by glucose deprivation (Shiu, Pouyssegur & Pastan, 1977; Pouysségur, Shiu & Pastan, 1978).

(3) A glycoprotein of 100 kd was found to have an altered carbohydrate composition in a variety of tumourigenic cells in contrast to non-tumourigenic ones (Bramwell & Harris, 1978). More specifically, in malignant cells this protein bound more concanavalin A (mannose and glucose specific lectin) and less wheat germ agglutinin (*N*-acetylglucosamine specific lectin) than the same protein in non-transformed cells. The relationship between this protein and those mentioned in the point (2) remain obscure.

(4) Alterations of fucosyl glycopeptides were observed in several cell lines transformed by various agents: the glycopeptides from transformed cells had increased amounts of sialic acid residues and possibly other alterations of their carbohydrate chains (Buck, Glick & Warren, 1971; Ogata, Muramatsu & Kobata, 1976; van Beek, Emmelot & Homburg, 1977; Smets *et al.*, 1978; Warren, Buck & Tuszynki, 1978).

(5) New surface-located antigenic specificities were found immunologically in chemically transformed and virally transformed cells (Smith & Landy, 1975; Schmidt-Ullrich, Wallach & Davis, 1976). The chemical nature of these components and their functional significance are not clear.

As seen from the data listed above, many surface proteins may be changed in the course of transformation. At present we do not know what the normal functions of these proteins are and which alterations of the behaviour of transformed cells may be associated with their changes.

Alterations of lipids

Characteristic alterations in the composition of a class of carbohydrate-containing lipids called gangliosides were observed in several lines of mouse fibroblasts transformed by various viruses and chemical carcinogens (see review in Hakomori, 1975; Fishman & Brady, 1976). Oligosaccharide chains of gangliosides are made by sequential addition of monosaccharide residues catalysed by specific glycosyltransferases. Plasma membranes of normal cells contain a series of gangliosides corresponding to sequential steps in the synthesis of oligosaccharide chains. Transformed cells have been found to contain decreased amounts of the more complex gangliosides. Loss of these gangliosides is correlated with a diminished activity of specific glycosyltransferases catalysing their synthesis. These alterations may have considerable functional significance, as specific gangliosides are receptors of cholera toxin and possibly other ligands. In particular, levels of ganglioside G_{M1}, the natural receptor for cholera toxin, are markedly decreased in a number of transformed lines. For instance, 3T3 cells transformed with MSV bind only 2% of the cholera toxin bound by parent cells (O'Keefe & Cuatrecasas, 1978). Langenbach, Malick & Kennedy (1977) showed that long-term cultivation of one mouse fibroblastic line resulted in its spontaneous morphological transformation and a simultaneous change in the synthesis of gangliosides. Altered synthesis of gangliosides is not, however, a common feature of all transformed cells: several virus-transformed lines were found to have an apparently normal pattern of gangliosides and normal activity of glycosyltransferases. These results suggest that, at least, in certain cases, alterations of gangliosides may have some relation to morphological transformation, and therefore deserve further study.

Examination of membrane fluidity is another experimental approach that has been recently used by lipid-oriented investigators studying transformation. Obviously, any alterations of fluidity could have considerable effects on cell behaviour. Unfortunately, the results of fluidity measurements in the plasma membranes of normal and transformed fibroblasts, thus far have not revealed any characteristic changes. Decreased fluidities of the membranes of transformed fibroblasts (Inbar, Yuli & Raz, 1977) have been reported, but several other authors (Gaffney, 1975; Micklem *et al.*, 1976; Hatten *et al.*, 1978) did not find any differences between the fluidities of normal and virus-transformed plasma membranes.

Secretion of proteases

Poorly spread transformed cells are morphologically similar to protease-treated normal cells at the early stages of their detachment from the substratum. This similarity makes it natural to suggest that the abnormal morphology of transformed cells is due to the effect of proteases secreted into the medium. Years ago, Fischer (1946) described the ability of cultured tumour cells to lyse the fibrin clot that had served as the substratum. Fibrinolytic properties of transformed cells have been investigated in detail during recent years. It has been shown that many types of transformed cells secrete into the medium enzymes which, after interaction with a component of normal serum, result in dissolution of fibrin. Some enzymes secreted by transformed cells have been identified as plasminogen activators: proteases which by limited proteolysis transform non-active serum plasminogen into active plasmin. Plasmin is the protease directly involved into the process of fibrinolysis. Elevation of the level of plasminogen activator has been observed in mouse and chick cultures transformed by viruses and chemical carcinogens (Ossowsky *et al.*, 1973; Unkeless *et al.*, 1973, 1974; Pollack *et al.*, 1974; Quigley, 1976; Jones *et al.*, 1975; Barret *et al.*, 1977). Several types of plasminogen activators can be secreted by one cell culture (Danø & Reich, 1978). Normalization of the morphology of transformed cells grown in a medium containing serum with low plasminogen content (Ossowski *et al.*, 1973) or with inhibited proteolytic activity (Weber, 1975) has been described. However, a number of facts show that secretion of plasminogen activator cannot be regarded as a general and specific feature of transformed cells. Certain normal cells have been found to secrete high levels of plasminogen activators, while some transformed lines produced low levels of these enzymes (Mott *et al.*, 1974; Chibber *et al.*, 1975; Pearlstein *et al.*, 1976). Certain strongly transformed lines obtained from Balb/3T3 cells formed no more plasminogen activator than parent minimally transformed line (Rifkin & Pollack, 1977). Elegant experiments were done by Wolf & Goldberg (1976) who isolated a number of clones from cultures of chick fibroblasts transformed by the Rous sarcoma virus. Of these clones, 23% had a low activity of plasminogen activator but did not differ significantly from clones possessing high levels of activator in their morphology or in any other transformed characters. In experiments with a special line of melanoma cells which expresses a transformed phenotype only in the presence of 5-bromodeoxyuridine it has been found that this chemical inhibits secretion of a plasminogen activator; in other words, it has been shown that in this line secretion of plasminogen activator is not associated with a transformed morphology (Rosenthal, Zucker & Davidson, 1978). Thus, elevated secretion of plasminogen activator frequently but not always accompanies transformation.

Besides plasminogen activator, certain transformed cells are able to secrete other proteases (Poste, 1975; Chen & Buchanan, 1975*b*; Wu *et al.*, 1975; Schultz, Wu & Yunis, 1975); certain cell-associated proteases were reported to be elevated in transformed cells (Bosmann *et al.*, 1974; Spataro, Morgan & Bosmann, 1976). The role of these proteases has not yet been studied in detail. Experiments with mixed cultures (see chapter 5 and p. 162) show that transformed and normal cells may grow in close proximity without the morphology of either cell type changing. These experiments argue against the role of any secreted proteases in the alterations of morphology accompanying transformation. Possibly, transformed cells carry on their surface non-secreted proteases with a strictly local action (Hatcher *et al.*, 1976, 1977); but the nature and exact localization of these enzymes as well as their role, if any, in morphological transformation are not yet clear.

Cyclic nucleotides and ions

Alterations of the synthesis of cyclic nucleotides (cyclic AMP, cyclic GMP) induced by external ligands serve as membrane-generated signals changing the course of many intracellular processes. It is very probable that these nucleotides have considerable effect on morphogenetic reactions but their concrete role in these reactions remains to be established. It has been suggested that the altered morphology of transformed cells may be associated with a decreased level of cyclic AMP. In fact, low cyclic AMP concentrations and a decreased activity of adenylate cyclase synthesizing this nucleotide were found in many transformed cell lines (see review in Otten, Johnson & Pastan, 1971; Gidwitz, Weber & Storm, 1976; Willingham, 1976; Pastan & Willingham, 1978). Cyclic nucleotide-dependent protein kinase activity was present in the membranes of normal chick fibroblasts but none could be measured in preparations of cells transformed with Rous sarcoma virus (Branton & Landry-Magnan, 1978). However, 3T3 cells transformed with SV-40 virus contained cytoplasmic cyclic AMP-dependent kinase that was not found in parent 3T3 cells (Gharret, Malkinson & Sheppard, 1976). Changes in the hormone sensitivity of adenylate cyclase and, in particular, reduction or loss of sensitivity to prostaglandin E_1 were also observed in transformed fibroblasts (Ayad & Foster, 1977). Large concentrations of dibutyryl cyclic AMP added to the medium may reverse many effects of morphological transformation but it is not clear whether this effect is due to specific replacement of the deficiency of endogenous cyclic AMP (see below).

Alterations of the production or function of cyclic GMP as well as alterations to the membrane permeability to ions, especially Ca^{2+}, may also serve as 'signal changes' affecting morphogenetic processes. At present, there are no consistent data about the characteristic alterations of ionic permeability or of cyclic GMP-associated processes in transformed cells. Low intracellular concentrations of cyclic GMP in virus-transformed NRK cells have been reported (Nesbitt *et al.*, 1976).

Cortical components

Almost nothing is known about the chemical alterations of components of the various fibrillar structures in transformed cells. Total synthesis of actin was not significantly diminished in 3T3 cells transformed with SV-40 virus (Fine & Taylor, 1976) while synthesis was increased in chick fibroblasts transformed by Rous sarcoma virus (Wickus *et al.*, 1975). Total actin content was similar in minimally transformed normal rat kidney (NRK) cells and in the viral transformant of NRK cells (Rubin *et al.*, 1978). The distribution of actin and myosin is different between normal and transformed cells. The amounts of these proteins associated with the membrane fraction were decreased (Wickus *et al.*, 1975). The fraction of actin which could be sedimented 100000 *g* (the 'particulate' fraction) decreased by about two-fold in 3T3 cells transformed with SV-40 virus compared with parent 3T3 cells (Fine & Taylor, 1976). These chemical alterations possibly correspond to the morphologically observed deficiency of microfilament bundles associated with contact sites in transformed cells. McClain, Maness & Edelman (1978) injected cytoplasmic extracts of 3T3 cells transformed with Rous sarcoma virus into normal cells and observed a dissolution of microfilament bundles within 30 min after injection. These results suggest that microfilament bundles may be direct or indirect targets of the product of the viral *src* gene.

Further studies on the synthesis and distribution of actin and other cortical proteins in cells before and after transformation are very important. Of special interest would be comparative studies of cortex components in suspended cells which do not have any contacts with other surfaces. These studies may reveal changes that are not secondary consequences of the different attachment abilities of normal and transformed cells.

Phenotypic reversion of morphological transformation

The morphology of transformed cells described in chapter 5 is observed when the cells are cultivated in 'standard' culture media. However, the expression of morphological transformation can be considerably diminished by certain alterations of the humoral culture media. Analysis of these phenotypic reversions can be valuable for understanding the mechanisms of morphological transformation. We have mentioned already (see p. 163) one example of this reversion: the partial normalization of cellular morphology produced by large concentrations of exogeneous fibronectin added to the culture medium. Here we will discuss several other examples of these reversions, especially, those induced by serum deprivation and by dibutyryl cyclic AMP.

Serum dependence of the transformed phenotype

In experiments with LSF and TLSF sub-lines of L cells adapted to growth in serum-free media, Bershadsky *et al.* (1976) studied the effect of serum deprivation on the expression of the transformed phenotype. It was found that the presence of serum is essential for the expression of many phenotypic traits. Cells growing in a serum-free medium were more flattened than those growing in the usual medium containing 10% bovine serum and they also had larger areas of lamellar cytoplasm and a smoother upper surface (fig. 6.2). Concanavalin A-mediated agglutination of cells grown in serum-free media required greater concentrations of lectin than agglutination of cells grown with serum. Several other traits of the transformed phenotype, including the ability to grow in semi-solid media and the inability to metabolize benzo(a)pyrene, also disappeared in serum-free media. All these alterations were completely reversible: they disappeared soon after cell transfer into a medium with serum.

The serum dependence of the transformed phenotype was confirmed by Tomei & Bertram (1978) who used another cell system: the 10T 1/2 line of mouse fibroblasts additionally transformed by 3-methylcholanthrene. The cells grown in a serum-supplemented media had a variable morphology with considerable 'piling up' and poor attachment. The same cells adapted to a serum-free medium acquired a more uniform morphology; they were better spread and formed monolayers. Reversion of these cultures to the transformed phenotype could be caused by the addition of whole serum (2%) or serum albumin (0.1%).

The nature of the serum components essential for the expression of the transformed phenotype is unknown. Plasminogen may be one of these components. However, as discussed above a number of facts suggests that the activation of plasminogen is not always associated with transformation. In our unpublished experiments, insulin decreased the spreading of LSF cells cultivated in a serum-free medium. In the experiments of Corwin, Humphrey & Shloss (1977) 3T3 cells transformed by Kirsten sarcoma virus, when grown in a medium containing de-lipidized fetal calf serum, regained many of the properties of the parent cells; they occupied a larger area on the substratum, were more flattened and more adherent to the substratum. Cholesterol and linoleic acid added in combination to the de-lipidized serum prevented the change of cell phenotype induced by this serum. This result suggests that certain lipid components of the serum may affect cell morphology. Possibly, the effect of serum on the expression of morphological transformation may be a result of the combined action of several different molecules.

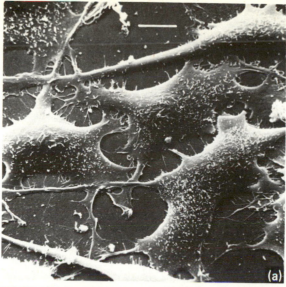

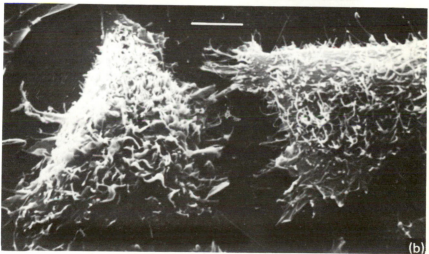

Fig. 6.2. Scanning electron micrographs of transformed mouse fibroblasts of the LSF line grown in a medium without serum (a) and with 10% bovine serum (b). Bars: (a), 10 μm; (b), 5 μm.

Effect of dibutyryl cyclic AMP and other agents

Various substances that are not standard components of culture media may induce phenotypic reversions in transformed cells when added to these media. N^6,O^2-dibutyryl cyclic adenosine 3′,5′-monophosphate (Bt$_2$cAMP) is the best studied of these substances (see review in Willingham, 1976; Pastan &

Willingham, 1978). When added at 10^{-4} M to a medium containing a permanent line of Chinese hamster ovary cells (CHO), alone or in combination with testosterone, this agent produced considerable alterations of cellular morphology: poorly spread polygonal ('epithelial-like') cells were transformed into better attached and better spread 'fibroblast-like' cells with long cytoplasmic processes (Hsie & Puck, 1971; Schröder & Hsie, 1973). Elongation of cytoplasmic processes associated with the increased formation of microtubules was also induced by Bt$_2$cAMP in hamster melanoma cells (Di Pasquale *et al.*, 1976) and in mouse neuroblastoma cells (Prasad & Hsie, 1971). Transformed mouse fibroblasts of several lines also altered their shape after addition of Bt$_2$cAMP: they developed long cytoplasmic processes containing microtubules and microfilament bundles. Minimally transformed 3T3 cells become more flattened on the substratum after Bt$_2$cAMP treatment (Johnson, Friedman & Pastan, 1971; Johnson, Morgan & Pastan, 1972; Willingham & Pastan, 1975*a*, *b*). Thus Bt$_2$cAMP decreases various manifestations of morphological transformation in many cell lines; all these effects seem to be completely reversible. Similar effects were obtained in CHO cells treated by certain other analogues of cyclic AMP (N^6-monobutyryl-cAMP and 8-bromo-cAMP); cyclic AMP itself was also effective in the presence of an inhibitor that prevented phosphodiesterase-catalysed degradation of this nucleotide (Storrie, Puck & Wenger, 1978; reviews in Pastan & Johnson, 1974; and in Willingham, 1976).

These 'reverse transformation' effects of cyclic AMP and its analogues are not characteristic for all transformed lines. An interesting example of a line with exceptional properties was described by Somers, Rachmeler & Christensen (1975) and Somers, Weberg & Steiner (1977). This was the normal rat kidney (NRK) line transformed with a cold-sensitive mutant of mouse sarcoma virus. At the permissive temperature (39 °C) these cells had a transformed phenotype; 8-bromo-cAMP treatment at that temperature did not result in any significant morphological changes. At a non-permissive temperature (33 °C) the control cells had non-transformed morphology and after treatment with 8-bromo-cAMP they became less flattened, more refractile and acquired a polygonal or rounded shape. Thus, in this line, in contrast to the others, the cyclic AMP analogue apparently induced the expression of transformation at a non-permissive temperature. This result suggests that expression of morphological transformation is not always negatively correlated with the intracellular level of cyclic AMP.

A number of agents, besides cyclic AMP analogues, are reported to decrease the expression of morphological transformation in certain lines. One of these factors is butyrate which caused flattening and/or extension of cytoplasmic processes in several lines (Wright, 1973; Ginsburg *et al.*, 1973; Ghosh *et al.*, 1975; Simmons *et al.*, 1975; Prasad & Sinha, 1976; Henneberry & Fishman, 1976; Storrie *et al.*, 1978); this flattening was accompanied by the development of microfilament bundles in the cytoplasm (Altenburg, Via & Steiner, 1976b). The list of other agents reported to produce certain symptoms of reverse transformation includes dextran sulphate (Goto *et al.*, 1973), dimethylsulphoxide (Kish *et al.*, 1973), a medium containing galactose instead of glucose (Gahmberg & Hakomori, 1973), and retinoids (Adamo *et al.*, 1979). The effects of these treatments were not analysed in detail.

Phenotypic transformation

Normal and minimally transformed cells may temporarily acquire a morphology which mimicks that of strongly transformed cells. These transformation-like effects can be produced by at least two groups of agents: detaching agents and growth factors. As mentioned above, normal and minimally transformed cells treated with moderate doses of proteases look similar to strongly transformed cells: they become less spread, lose their microfilament bundles, their surface becomes less smooth; they are more agglutinable with lectins. Most of these effects can also be produced by other detaching agents, e.g. urea (Weston & Hendricks, 1972). The morphology and behaviour of cells grown for long periods in media containing low doses of detaching agents have not been studied in detail.

Growth factors form another group of agents that can cause transformation-like changes (see chapter 10 for a general description of their properties). It seems that these factors affect considerably the morphogenetic reactions of the cells. We mentioned in chapter 4 the effect of epidermal growth factor on the surface activity of normal cells; this factor may also alter fibronectin synthesis in normal cells (Chen *et al.*, 1977a) and induce the production of plasminogen activator (Lee & Weinstein, 1978). Induction of microvilli by insulin was described in chapter 4. Cells treated with large doses of fibroblast growth factor acquired a morphology similar to that of transformed cells (Gospodarowicz & Moran, 1974b). A similar result was obtained in experiments with NRK cells treated with a growth factor isolated from sarcoma cells: the cells incubated with this factor were more refractile than untreated cells and they grew in a criss-cross pattern instead of an orderly monolayer. These changes were completely reversible (De Larco & Todaro, 1978a).

Certain phorbol esters, especially 12-*O*-tetradecanoyl-phorbol-13-acetate, may also induce alterations mimicking symptoms of transformation. In particular, they induce plasminogen activator (Wigler & Weinstein, 1976) and decrease the amount of cell-associated fibronectin (Blumberg, Driedger & Rossow, 1976). These substances act *in vivo* as promoters of skin carcinogenesis.

The phenomenology of the effects of all these agents has not been studied in detail. It would be important to find out whether they are able to induce all the symptoms of morphological transformation that were listed in chapter 5.

Conclusion

Morphological transformation can be regarded as a consequence of the decreased ability of transformed cells to perform pseudopodial attachment reactions. The molecular basis of these altered attachment reactions is an open question. Among the known chemical changes, a decrease of fibronectin is most frequently associated with morphological transformation. Several other alterations (altered glycolipid pattern, secretion of plasminogen activator, etc.) were observed in certain transformed lines but their association with morphological transformation is less consistent.

It is probable that not a single alteration but manifold molecular alterations of the membrane and of the cortex accompany morphological transformation; the exact list of these alterations may vary from one line to another. These alterations might result from some genetic defects which are characteristic of transformed cells and lead to their inability to synthesize fibronectin or other specific molecules involved in attachment reactions. However, morphological transformation in many lines can be temporarily reversed by a variety of physiological and non-physiological factors. These 'phenotypic reversions' suggest that transformed cells have no innate defect in the synthesis of any component essential for attachment; the ability to perform normal attachment reactions seems not to be lost but is only 'hidden' in transformed cells growing in usual culture conditions. Probably, morphological transformation is due not to genetic alterations of some components responsible only for attachment but to alterations of some more general functions regulating cell interactions with the environment.

The ability of normal cells to perform morphogenetic reactions can be modified by various external agents and, in particular, by certain growth factors. These external agents seem to be able to regulate the cell's ability to perform pseudopodial reactions in such a way that in certain conditions the cell may acquire a transformed phenotype. This regulation may be achieved by co-ordinated changes of many cellular components involved in attachment. However, genetic transformation may alter 'the reaction norm' of the cell to external factors regulating morphogenetic reactions. As a result, the cell will have a transformed phenotype in standard culture conditions. Thus, it is possible that in transformed cultures alterations of components such as fibronectins which participate in attachment are secondary consequences of some changes in the central mechanisms regulating morphogenetic reactions and other cellular reactions. One corollary to this hypothesis is that external factors may decrease the expression of the transformed phenotype in two different ways: (*a*) by altering general cell reactivity, i.e. the cell's ability to perform morphogenetic reactions; (*b*) by altering the expression of one of the secondary symptoms of altered attachment.

Possibly, serum deprivation and cyclic AMP analogues are the 'reversing' factors in the first type of action while externally added fibronectin may act as the 'symptomatic therapy' of the second type. To test these suggestions, detailed comparative studies of the various cases of phenotypic reversions are essential. We will return to the problem of 'central' and 'peripheral' changes in transformed cells in chapter 14.

Part III

Regulation of growth in normal and transformed cultures

7

Introduction: growth-regulating factors of environment

The growth of normal cells in culture depends on many factors present in their environment. The main groups of environmental factors controlling growth are the following.

(1) *The substratum.* Normal fibroblasts and epithelia actively proliferate only if they are attached to a solid substratum.

(2) *Local cell density.* Normal cells actively proliferate only if the local population density does not exceed a certain level.

(3) *Growth factors.* The presence of certain specific proteins (growth factors) in the medium is essential for active proliferation. Usually, serum added to the culture medium is the source of these growth factors.

(4) *Ions.* Normal cells proliferate only if ionic concentrations in the medium do not exceed certain limits. Ca^{2+}, Mg^{2+} and H^+ seem to be the most important of these ions.

(5) *Nutrients.* Normal cells proliferate only when their medium contains a set of essential nutrients.

In the course of neoplastic transformation cell growth may become less dependent on any of these factors. At present we have no data proving that the growth-controlling effects of any environmental factors are a secondary result of alterations of other factors, e.g. that density effects can be explained by decreased spreading of the cells on the substratum in dense culture or by altered availability of serum growth factors in the culture. In various transformed cultures, the dependence of growth on different factors may be changed to a varying degree; therefore we distinguish between anchorage transformants, serum transformants and density transformants. The same is true for revertants selected from transformed cultures (see chapter 2). Therefore, at the first step of analysis the effects of each particular environmental factor should be discussed separately from the effects of other factors.

Alterations in the composition of the environment may induce a normal

179

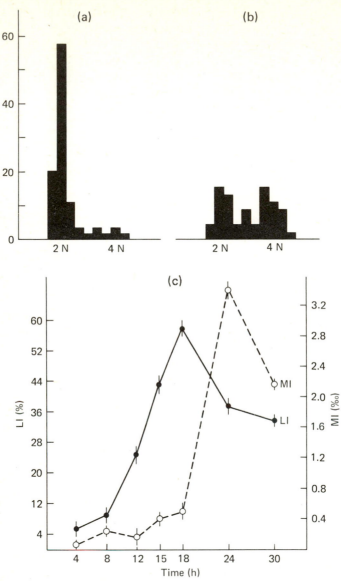

Fig. 7.1. Transition of a culture of mouse fibroblasts from the resting state to the growing state. (a, b) Histograms showing the DNA content of dense cultures of normal mouse fibroblasts at 24 hours after wounding, determined by micro-spectrophoto-metry: (a), the cells remain in the dense part of the wounded cultures; (b), the cells of the same culture which have migrated into the wound. Abscissae: DNA content per nucleus expressed in ploidy units (N). Ordinates: percentage of cells in particular, arbitrarily defined classes. (c) Stimulation of proliferation after wounding. Abscissa: time after wounding in hours. Ordinates: *left*, labelling index (LI) after pulse labelling with [³H]thymidine; *right*, mitotic index (MI).

culture to assume one of the two extreme states: the state of active proliferation (growing state) or a state in which proliferation is low or absent but the cells remain alive (the resting state). Actively growing asynchronous cultures contain a considerable proportion of cells which enter hourly the phase of DNA synthesis (S phase) or mitosis; the number of cells in these cultures grows with time. Accordingly, these cultures have high labelling indices after pulse labelling with [³H]thymidine; they also have high mitotic indices. Spectrophotometric measurements show that the DNA content of individual cells in these cultures varies between the values corresponding to the post-mitotic content of DNA to doubled pre-mitotic content. Histograms of the distribution of DNA content in various cells show no sharp peak (fig. 7.1).

Resting cultures are characterized by a very low proportion of cell entering the S phase or mitosis each hour. Accordingly, indices after pulse labelling with [³H]thymidine, and mitotic indices are very low. Histograms showing the distribution of cells with various DNA contents show a clear modal peak: the cells are accumulated in one particular part of the cycle. This peak corresponds to the post-mitotic value of DNA; that is, the cells are accumulated at a stage preceeding DNA synthesis. An accumulation of cells between the end of the S phase and the beginning of mitosis (G_2 phase) has also been described in certain cultures (see chapter 12).

Of course, the intensity of proliferation in culture may vary continuously so that there can be many intermediate conditions between the resting and growing states. When we talk about 'growing' or 'resting' cultures, we are describing population states which are generalizations of the states of individual cells. We do not yet know exactly which alterations of the kinetic parameters of individual cells correspond to the transition of cultures from the resting to the growing state or *vice versa*. These may be alterations of the rate of passage through the G_1 phase of the cell cycle, or alterations of the probability of cell entry into the so-called G_0 phase, etc. These problems will be discussed in chapter 12.

In the next three chapters we will describe the effects of the substratum, cell population density, and components of the fluid medium (growth factors, ions and nutrients) on the growth of normal and transformed cultures. In the subsequent two chapters we will discuss possible inter-relationships between the reactions to different growth-controlling factors as well as possible mechanisms of transitions from the resting to the growing state.

8

Anchorage dependence and its loss

Assessment of proliferation in suspended cells

Cells that have been able to proliferate in the substratum-spread state may reversibly stop proliferating when they are removed from the substratum and kept in suspension. This phenomenon of 'anchorage dependence' was described by Macpherson & Montagnier (1964) and by Stoker *et al.* (1968). To assess their ability to proliferate it is necessary to keep these cells in suspension for a sufficiently long time. This may be done either by growing the cell suspension in fluid media or by suspending the cells in semi-solid gels. The cultivation of cell suspensions in fluid media (suspension culture) is very convenient for mass production of strongly transformed cells which have lost anchorage dependence for growth. However, this method does not permit a comparison of the ability of individual cells in the population to multiply in suspension. In other words it is not convenient for quantitative clonal assessment of anchorage dependence. Cultivation of suspended cells in gel-like media is, in contrast, very suitable for this assessment: one can count the percentage of cells that have formed colonies in the gel. Several types of substances are added to the medium in order to obtain gels: agar, agarose or methylcellulose. The chemical composition of agar may vary and certain components, in particular sulphated polysaccharides, may inhibit the growth of suspended cells but the agarose fraction of agar is devoid of these polysaccharides (Montagnier, 1971). Methylcellulose (methocell) seems to be the least damaging of the available substances used for the preparation of gels. Detailed comparative studies of the behaviour of suspended cells in fluid cultures and in semi-solid media have not yet been performed. Obviously, conditions are different in these two types of cultures. For instance, many suspension fluid techniques require continous stirring of the medium but this mechanical movement is absent in semi-solid media. It would be very interesting to learn more about the differences in survival rate, morphology, growth rate and other cell parameters in fluid and semi-solid suspension cultures.

The cloning efficiency of cells in semi-solid media is usually regarded as a measure of the degree of loss of anchorage dependence: strictly dependent cells do not form colonies in these conditions. However, a low percentage of colonies in suspension can be due not only to anchorage dependence but also to the general low cloning efficiency of the culture. In the later case, the cells will form a low percentage of clones, not only in suspension but also in the substratum-attached state. Therefore, the ratio of cloning efficiency in suspension to that on the substratum can be regarded as the best index of anchorage dependence. The actual number of colonies counted per dish, of course, depends on the definition of 'colony' used by an investigator. These definitions vary: in some works only colonies with eight or more cells have been counted; in others, only colonies larger than 0.1 mm in diameter, etc.

The cloning efficiency in suspension may depend on a number of factors. In particular, it may be affected by the density of seeded cells (Montagnier, 1971), so that ideally one should compare efficiencies at several initial densities. Cloning efficiency in suspension can be increased by the addition of 'filler' cells which are unable to multiply in these conditions (Lernhardt *et al.*, 1978). Possibly, the effect of these cells is similar to the effect of 'feeder' layers in increasing the cloning efficiencies of substratum-attached cells (see chapter 10).

The amount of serum and growth factors in the medium may also affect the cloning efficiency in suspension (see chapter 11). The addition of collagen to the medium may increase the number of colonies (Sanders & Smith, 1970). At present we cannot answer many questions that may be important for the evaluation of the data obtained in experiments with cells grown in semi-solid media. For instance, we do not know the time of survival of the cells which do not form colonies in these conditions. Stoker *et al.* (1968) have shown that anchorage-dependent BHK cells may survive for at least seven days in suspension but the maximal survival time of this line is not known. Data about the survival of other cell types are lacking. It would be important to find out to what degree differences in cloning efficiencies on the substratum and in suspension are due to decreased survival of suspended cells. Another unsolved problem is that of the effect of cell–cell contacts between suspended cells on their cloning efficiencies: we do not know whether two-cell aggregates and single cells have different probabilities of forming colonies. Besides its theoretical interest, this question might be important for the evaluation of results of experiments in which a low percentage of colonies has been obtained. It would be good to know whether this percentage could be affected by the small percentage of non-single cells that could be present in the initial suspension.

Anchorage dependent and independent cells

The ratio of the cloning efficiency in suspension to that on the substratum is near to zero ($< 10^{-7}$) in lines with a strict anchorage dependence but it may increase to values of about 0.1–1.0 in strongly transformed lines. Strictly dependent cells include the primary cultures of various fibroblasts and epithelial cells as well as several continuous lines such as certain sub-lines of BHK and 3T3 cells. Many strongly transformed lines of fibroblasts with a decreased anchorage dependence have been obtained from primary cultures and from minimally transformed lines. This change is observed after transformation induced by various viruses or by chemical carcinogens as well as after spontaneous transformation. Epithelial cells may also acquire anchorage independence in the course of transformation (see chapter 2). Loss of anchorage dependence is not an all-or-none phenomenon. In the course of transformation different cell lines may acquire different degrees of independence. For instance, in the experiments of Tucker *et al.* (1977) anchorage dependence of several spontaneously transformed oncogenic lines of mouse fibroblasts were compared. Several lines had plating efficiency* of zero in agarose while in other lines it varied from 0.8 to 1553. In the experiments of Freedman & Shin (1977) described in the chapter 2, incubation of human fibroblasts with a chemical carcinogen was followed by an increase of plating efficiency in a semi-solid medium from 10^{-8} to 10^{-5}. Thus, the appearance of even a small fraction of anchorage-independent cells in a population can be an important symptom of transformation. It would be important to study in detail distinctive characteristics of the cells forming this fraction. Are these cells genetically stable mutants, anchorage-independent 'stem cells' giving dependent progeny, or those in some special reversible phenotypic state? The kinetics of growth of colonies of anchorage-independent cells in a semi-solid medium ('spheroids') were studied by Folkman & Hochberg (1973). All spheroids enlarged for 5–23 weeks until a critical diameter was reached beyond which there was no further expansion ('dormant phase'). For the V-79 line of hamster lung cells the critical diameter was 4.0 mm ± 0.8 mm reached at 175 days, for B-16 (mouse melanoma cells) it was 2.4 mm ± 0.4 mm reached at 100 days. The number of viable cells per one spheroid usually reached a maximum before the spheroid diameter stopped increasing; after that, further enlargement of a spheroid was accompanied by an expanding volume of cell necrosis. Autoradiography with [³H]thymidine showed that proliferating labelled cells occupied only one or two layers on the outer rim of large spheroids. Obviously, central necrosis was due to insufficient diffusion of nutrients through the surface of the spheroid.

Cell multiplication on various substrata

When anchorage-dependent cells do not make contact with a substratum, they do not proliferate. One can also expect that when these cells contact a substratum, their proliferation will depend on the adhesiveness and area of

---

* The number of colonies with a diameter larger than 0.1 mm per 10^5 seeded cells.

this substratum. The role of the area of the substratum available for cell contact was shown by Stoker *et al.* (1968). In their experiments, glass fibrils 500 μm in length, but not silica particles about 5 μm in diameter, supported the growth of BHK 21 cells in agar. In similar experiments, Maroudas (1972) tested glass beads of various sizes and found that 50 μm was the critical diameter supporting growth. In other experiments Maroudas (1973*b*) added glass fibres of different lengths to suspension cultures of 3T3 and BHK cells in agar. The minimum fibre length that promoted division of suspended cells was 30 μm. Colony formation increased sigmoidally with fibre length from 30–250 μm; about 50% of the suspended cells formed colonies on the fibres that were 115 μm long. These figures are similar to the observed lengths of substratum-spread cells in normal cultures.

Folkman & Moscona (1978) described a method in which the adhesiveness of the substratum surface could be gradually varied. The showed that the adhesiveness of the surfaces of plastic dishes can be reduced in a graded manner by applying increasing concentrations of poly(2-hydroxyethyl methacrylate) which forms films of increasing thickness. In experiments with bovine endothelial cells, a range of 12 cell conformations from spherical to flat were obtained by this method. Cells plated on the thickest layer of the polymer (35 μm) retained an almost spherical shape whilst on thinner polymer layers the cells became flatter. The rate of DNA synthesis was inversely proportional to the thickness of the cell layer on the substratum; that is, directly proportional to the degree of spreading. Paranjpe, Boone & Eaton (1975) showed that 3T3 cells do not proliferate on teflon, presumably because they cannot spread well on this substratum; in contrast, virus transformed derivatives of these cells grow well on teflon. In future experiments comparing cell growth on substrata with varying adhesiveness it would be important to check the possible effects of cell aggregation on poorly adhesive surfaces (see chapter 7). This aggregation may lead to the formation of areas with a high local cellular density even in cultures with a low average density. Local high density can inhibit cell multiplication and thus obscure the real effects of decreased spreading. Martin & Rubin (1974) showed that cultivation of chick fibroblasts on poorly adhesive bacteriological plastic dishes leads to the formation of cell clumps which are correlated with the inhibition of cell growth.

The 'usual' substrata of cell cultures may also vary in their ability to support cell growth. Those who work with cell culture know that certain cells will grow better on plastic than on glass although the reason for these differences is not clear. Coating of the plastic surface with polylysine or other basic polymers increased the cloning efficiency of human and chick fibroblasts but not human fibroblasts transformed with SV-40 virus; the growth rate of cell colonies on the polylysine-coated substrata was also higher than those of control cultures (McKeehan & Ham, 1976). Presumably, the effect of polymers was associated with an increase in cell adhesiveness to the substratum. When a cell culture was removed from the substratum and fresh cells were seeded at a low density on the same surface, the cells grew faster on this 'microexudate-coated' surface than on the control clean plastic (Weiss *et al.*, 1975). Possibly, in these experiments, the stimulation of cell growth was also a result of increased adhesiveness of the surface precoated with cell products such as fibronectin or collagen.

Possible mechanisms of anchorage dependence

The data presented in the previous section suggest that in order to start proliferation an anchorage-dependent cell should not only contact the substratum but should also become spread on it. It was suggested that the stimulation of proliferation might be correlated with an increase of the total surface area in the course of spreading (Folkman & Greenspan, 1975). However, the areas of the surface of spherical and spread cells are of the same order (see chapter 4). Cell spreading is accompanied by a complex structural reorganization of the cell and some aspects of this reorganization are probably important for the initiation of proliferation. For instance, it is possible that this initiation is correlated with a redistribution of surface receptors leading to the formation of focal contact structures or with the alterations of the architecture of the cortical layer. Formation of pseudopodia and of lamellar cytoplasm by the spread cell may be of special importance. As we know, the surface of pseudopodia and the lamellar cytoplasm have special properties, e.g. the ability to become clear of patched receptors. Possibly, many other properties of the surface of pseudopodia (e.g. permeability, the state of fibrillar elements, etc.) are also different from those of the other cell parts. We do not know which of these alterations, if any, is essential for initiation of proliferation. The loss of anchorage dependence by neoplastic cells is also an unexplained phenomenon. It would be important to find out any differences between the suspended normal and transformed cells that could be correlated with different growth. At present no reliable differences of this type are known.

Cell survival and proliferation in aggregates

Several methods for the cultivation of multicellular aggregates unattached to any solid substratum have recently been described (Yuhas *et al.*, 1977; Sutherland, MacDonald & Howell, 1977; Ng & Inch, 1978; Haji-Karim & Carlsson, 1978; Yuhas & Li, 1978). In a number of experiments, the behaviour of normal and transformed cells in aggregates was compared: suspended cells were first aggregated and then their growth was studied. Different authors used various cells and somewhat different methods of aggregation. Steuer *et al.* (1977*a*) obtained cell aggregates by placing the cell suspension on a non-adhesive agar surface. The cells of certain non-tumorigenic liver cell lines from rat rapidly degenerated in this state while tumourigenic lines survived 3–4 days and even proliferated. The same group of authors observed similar differences in the aggregate survival between minimally transformed human osteosarcoma and its strongly transformed sub-lines (Steuer *et al.*, 1977*b*). Several types of normal or minimally transformed rat, mouse and hamster fibroblast cells also did not survive in the aggregate state in contrast to their strongly transformed analogues (Putman *et al.*, 1977). Carrino & Gershman (1977) obtained aggregates of normal 3T3 cells and those transformed with SV-40 virus either mechanically, by placing flasks with cell suspensions on gyratory shakers, or by sedimenting the cells from the suspension. 3T3 cells in the aggregates incorporated [^{3}H]thymidine and divided but the rate of cell death was higher than the growth rate so that the total number of cells in the aggregates gradually decreased. In contrast, in aggregates of strongly transformed cells the growth rate exceeded the death rate.

Thus, in both types of experiments strongly transformed cells showed higher viability and more active growth in the aggregates than their minimally transformed and normal analogues. This different behaviour might be due to a different sensitivity of the cells to high density, to the absence of substratum, to the shortage of nutrients in the inner parts of the aggregates, etc. Separate assessments of the role of each of these factors have not been made because of obvious technical difficulties. Therefore, aggregate cultures are not very convenient for analytical studies of the distinctive growth properties of transformed cells.

Nevertheless, the study of cell growth in aggregate cultures is possibly interesting because it reproduces certain features of the growth of neoplastic cells in three-dimensional small non-vascularized tumour nodules *in vivo*.

9

Density-dependent inhibition of growth and its alterations in transformed cultures

Assessment of density-dependent inhibition

The rate of proliferation of normal substratum-spread cells decreases as the local density of the cell population per unit area of substratum increases. This phenomenon is designated 'density-dependent inhibition' (DDI) or 'topoinhibition'. The term 'contact inhibition of growth' is not synonymous because it implies a certain unproven mechanism for DDI, namely, its association with the formation of cell–cell contacts.

A culture which is sensitive to DDI stops growing when it reaches a certain density, the so-called saturation density. The stationary state of a density-inhibited culture is not due to increased cell death but to the cessation of proliferation – to the transition of the population into the resting state. A decreased proportion of DNA-synthesizing cells and of those undergoing mitosis are the main manifestations of this inhibited proliferation. A decreased proliferation in dense culture is an essential symptom of DDI but is not sufficient for its diagnosis; besides DDI this inhibition may be caused by various changes in the humoral medium such as the accumulation of growth-inhibiting substances, and exhaustion of nutrients or serum factors. These act on all the cells in the vessel while DDI, by definition, acts locally.

There are several ways of distinguishing between growth inhibition caused by general changes in the medium and that caused by a local increase in density.

(1) *Calculation of labelling indices.* Cells are labelled with [³H]thymidine for autoradiography and various areas of the culture are examined for their local population densities by microscopy. Under the microscope, the number of cells and labelled cells are counted in many different fields of view. This allows labelling indices to be calculated for various areas of the culture (density-adjusted indices). If the culture is sensitive to DDI, these indices decrease as local population density increases. Discrete colonies of cells on the substratum are particularly convenient for these counts: due to DDI, labelling indices in sparse peripheral zones of the colony will be higher than in the dense central zones of the same colony (Fisher & Yeh, 1967).

(2) *Wounding of the culture*. Mechanical removal of a part of the culture (wounding) produces conspicious variation in local densities in different areas of the culture. After wounding, the cells start to migrate into the wound; that is, on to the free surface of the substratum. During the first days after wounding, the local population density in the wound is much lower than that in the surrounding areas of the culture. Accordingly, in DDI-sensitive cultures the labelling index and the mitotic index increase in the wound area considerably at certain times after wounding (Fig. 9.1), while in the surrounding dense areas these indices remain the same as in control, non-wounded cultures. Local activation of proliferation by wounding was revealed long ago in experiments with explants (Fischer, 1946) and later in the experiments with dense cell cultures (Todaro, Lazar & Green, 1965; Vasiliev *et al.*, 1966*a*, 1969; Gurney, 1969; Dulbecco & Stoker, 1970). The degree of stimulation of DNA synthesis is similar in large wounds and in small wounds (about 1 mm in diameter) surrounded by large dense areas (Vasiliev *et al.*, 1969). Thus, factors responsible for density dependent inhibition are effective within a distance of less than 1 mm. The exact ratio of the labelling indices in the wound and the dense area may, of course, vary depending on the density of the culture in which the wound was made: the higher the density of the sheet, the lower the ratio. These variations should be taken into account when the wound:sheet ratios obtained in cultures of different types are compared; these comparisons become more meaningful, when wounds are made in cultures of different densities and two series of ratios are compared.

(3) *Comparison of proliferation*. Cells seeded on the upper surface of pre-established dense cultures are compared with those seeded on the cell-free substratum in the same medium. Often in these experiments one half of a dense culture is removed mechanically and then the suspended cells are added to the medium. These cells become attached both to the dense culture surface and to the substratum. The seeded cells may be pre-labelled with [³H]thymidine or with carmine but this is not necessary if the seeded cells are clearly different morphologically from the cells of the dense culture. Proliferation of seeded cells can be assessed by various methods, for instance, by counting the colonies formed by these cells or by counting the densities of pre-labelled cells. If seeded cells are sensitive to DDI their multiplication on the cell layer will be decreased as compared with the free substratum (Stoker, 1964). Obviously, this method may be used only in a system where seeded cells are attached equally well to the cell layer and to the substratum. For instance, it is senseless to use, as a lower cell layer, non-adhesive epithelial sheets. An advantage of this method is that it permits the assessment of sensitivity to the local inhibiting action not only of the cells of the same type (homotypic DDI) but also to that of the cells of other types (heterotypic DDI).

The methods of assessment listed above can be regarded as various operational definitions of the DDI phenomenon. All the methods probably reveal the same phenomenon but different, additional changes are induced in cultures in various types of experiments: cell injury and migration in the wounding experiments, cell detachment and spreading in the experiments with seeding on the pre-established dense layer, etc. Therefore it is advisable if possible to use several methods for the assessment of DDI in each type of culture and to compare the results obtained. Sometimes the value of the saturation density is regarded as a measure of DDI: the lower the density, the higher the cell sensitivity to DDI. However, density inhibition, as defined above, is a result of local effects of neighbouring cells. Besides these local factors the rate of cell growth is affected by many non-local humoral factors such as serum, nutrients, etc. The value of saturation density depends not on the density inhibition alone but on the complex equibrium of DDI and non-local factors (see chapter 11).

Density-dependent inhibition in normal and minimally transformed cultures

When cells are cultivated in excess medium with 10% serum, the saturation density of mouse embryo fibroblasts may reach about 400000–600000 cells/cm², that of primary hamster embryo cells 600000–800000 cells/cm². Certain continuous lines have a similar saturation density; for instance, that of BHK cells was reported to be 500000–1000000 cells/cm². However, other lines may have a much lower saturation density; for instance, that of 3T3 cells has been reported to be about 50000–200000 cells/cm². Normal mouse kidney epithelium also has a low saturation density, about 150000–200000 cells/cm² (Zetterberg & Auer, 1970; Pitts, 1971; Martz & Steinberg, 1972; Pletyushkina, Vasiliev & Gelfand, 1975). It should be remembered (see chapter 2) that 3T3 lines are specially selected as cells that are the most sensitive to DDI. As mentioned in the previous paragraph, the saturation density of the same line grown in different media may vary considerably.

In experiments with wounding, DDI (increased labelling index in the wound compared with that in surrounding dense areas) was revealed in all types of primary fibroblastic cultures examined: mouse, rat, hamster, chick; in the cultures of certain continuous lines such as 3T3 and BHK (Todaro *et al.*, 1965; Vasiliev, Gelfand & Erofeeva, 1966, 1969; Gurney, 1969; Dulbecco & Stoker, 1970; Pletyushkina *et al.*, 1975). In undisturbed cultures of normal fibroblasts from man, hamster and mouse the labelling index decreases as the local population density increases (Defendi, 1966; Vasiliev *et al.*, 1969; Pletyushkina *et al.*, 1975; Harel & Jullien, 1976; Ellem & Gierthy, 1977). The same is true for mitotic indices in the cultures of chick and guinea pig fibroblasts (Abercrombie, Lamont & Stephenson, 1968). When pre-labelled homologous cells are seeded on the dense sheets of normal fibroblast cultures of primary mouse or hamster cells, multiplication of these cultures is inhibited compared with that of the same cells attached to the free substratum (Domnina *et al.*, 1972; Pletyushkina *et al.*, 1975). Thus by several criteria DDI is revealed in cultures of all the examined types of normal and minimally transformed fibroblastic cells.

The increased labelling index in the area near the wound compared with the more dense areas far from the wound was also revealed in primary cultures of mouse baby kidney (Pitts, 1971) and green monkey kidney epithelial line BSC-I (Pitts, 1971; Holley *et al.*, 1977). These data show that DDI is also characteristic for certain types of epithelial cells. Probably, it will be revealed in all types of normal epithelial cells but at present the number of cell types examined is too small for any final conclusions. We do not know how long a cell must remain in a density-inhibited area before DDI becomes effective. Analysis of time lapse films of 3T3 cells taken in the course of their transition from rapid growth to the stationary phase has shown that a high local cell density present throughout G_1 phase often did not produce an immediate inhibition of cell division (Martz & Steinberg, 1972). Possibly, DDI becomes effective only after some minimal time of cell residence in a high density area.

In summary, homotypic DDI seems to be effective in all types of normal fibroblastic and epithelial cultures examined. Is this dependence species-specific or type-specific? In other words, can heterotypic DDI between cells of various types or species be revealed in cultures? The main method used for studies of heterotypic density effects was the seeding of cells on an established sheet of cells of another type. In the experiments of Westermark (1973*a*), stationary glia-like cells inhibited the proliferation of human skin fibroblasts. Different strains of human fibroblasts usually inhibit each other in mixed cultures (Eagle & Levine, 1967; Bard & Elsdale, 1971). In a large series of experiments human, mouse and monkey fibroblasts inhibited each other, regardless of species (Eagle, Levine & Koprowski, 1968). Thus, heterotypic DDI was effective in these experiments.

An interesting phenomenon of non-reciprocal heterotypic DDI was described by Njeuma (1971*a, b;* see also Weiss & Njeuma, 1971) in Abercrombie's laboratory. In these experiments primary mouse fibroblasts inhibited the multiplication of chick fibroblasts; however, dense cultures of chick fibroblasts were unable to inhibit multiplication of mouse cells. The same cultures effectively inhibited the multiplication of homologous chick cells. This interesting phenomenon certainly deserves further study. Little is known about DDI between the cells of different histological types; for instance, between epithelial cells and fibroblasts. Rabbit lens epithelium was inhibited by human and mouse fibroblasts (Eagle, Levine & Koprowski, 1968) but mutual inhibition was not observed between primary epithelial cultures of human amnion and human diploid fibroblasts (Eagle & Levine, 1967; Eagle *et al.*, 1968).

Alterations of density-dependent inhibition in strongly transformed cultures

Two main types of DDI were assessed in cultures of transformed fibroblasts: DDI by parent normal cells and DDI by homologous transformed cells. Many types of transformed cells lose their sensitivity to DDI by parent normal cells. When seeded on the top of normal dense cultures these cells grow at almost the same rate as on the free substratum. The formation of transformed colonies on the top of a pre-established normal sheet is one of the standard tests for transformed cells (Eagle & Levine, 1967; MacIntyre & Pontén, 1967; Weiss, 1970; Aaronson *et al.*, 1970; Eagle *et al.*, 1970; Weiss, Vesely & Sindelarova, 1973). Transformed cells may also form colonies on the top of dense fibroblastic cultures of other species. For instance, chick cells transformed by Rous sarcoma virus acquire the ability to proliferate on the upper surface of dense cultures of mouse fibroblasts (Weiss, 1970). The growth of other transformed cells, especially those transformed by polyoma virus, was inhibited by normal cells of the same or other species (Stoker, Shearer & O'Neill, 1966; Pontén & MacIntyre, 1968; Weiss, 1970). Probably, different degrees of a loss of sensitivity to DDI by normal cells are characteristic for various lines.

Wounding was used to assess homotypic DDI in a variety of transformed cultures. Certain lines had a high labelling index both in the wound and in dense areas: the ratio of these indices was near to 1.0 even in high density cultures. In particular, this was the case with many virus-transformed lines (see review in Pitts, 1971). The rate of DNA synthesis in these cultures did not decrease with increasing density. However, certain other neoplastic cultures, although reaching high saturation densities, nevertheless retained all the manifestations of sensitivity to DDI: their labelling index in undisturbed cultures decreased with increasing density; homologous prelabelled cells seeded on the top of dense culture did not proliferate; the labelling index in wounds was much higher than in the dense areas. This sensitivity to homologous DDI was found to be characteristic of a number of spontaneously, virally and chemically transformed tumourigenic rodent lines (Domnina *et al.*, 1972; Vasiliev *et al.*, 1975*b*; Pletyushkina *et al.*, 1975; O'Neill, 1978; Moses *et al.*, 1978). In certain transformed lines DNA synthesis was inhibited at a high density but the degree of this inhibition was less complete than in the parent lines: a fraction of the cells continued to multiply even at maximal densities (Lindgren & Westermark, 1977*b*). Thus, the sensitivity to homotypic DDI may be decreased to different degrees in various lines. It is stressed that the saturation density value of the culture, when used alone, cannot be a reliable index of sensitivity to homotypic DDI. Saturation densities of many transformed cultures are not much higher than those of primary fibroblastic cultures. Only when specially selected continuous lines with a very low saturation density such as 3T3 cells undergo additional transformations, may their transformed derivatives have much higher saturation densities than parent cells.

Comparison of the data obtained by different authors suggests that sensitivities to DDI by homotypic transformed cells and by parent normal cells may be lost independently in the course of neoplastic evolution. One can distinguish at least three groups of transformed lines with regard to these characters. (*a*) The first group is not sensitive to homotypic inhibition but retains sensitivity to inhibition by normal cells. BHK cells transformed by polyoma virus (Stoker *et al.*, 1966) and bovine fibroblasts similarly transformed (Pontén & MacIntyre, 1968) apparently belong to this group as they did not show DDI in isolated cultures but stopped growing when seeded on a layer of normal parent cells. (*b*) The second group is formed by cells which retain sensitivity to homotypic cells but lose sensitivity to normal cells. Certain lines of transformed mouse cells and of hamster cells studied in our experiments (Domnina *et al.*, 1972; Pletyushkina *et al.*, 1975) belong to this group. Labelled cells of these lines seeded on the unlabelled dense cultures of similar transformed cells do not multiply; however, these transformed cells seeded on the dense culture of parent normal fibroblasts multiply at the same rate as those seeded on the free substratum. (*c*) The third group is formed by cells which are insensitive both to homotypic and normal cells. Lines of bovine fibroblasts transformed by Rous sarcoma virus (MacIntyre & Ponten, 1967) apparently belong to this group. The criteria of sensitivity to DDI were not identical in the different papers quoted above. In particular, the loss of sensitivity to homologous DDI was sometimes implied from the high saturation densities of transformed cells without additional studies. Therefore, the conclusion that there are three groups of transformed cells with different sensitivities to DDI should be regarded as a preliminary one.

Possible mechanisms of density-dependent inhibition

According to the operational definitions discussed above, DDI is a local effect. Experiments with wounding show that DDI is not effective when a cell is at a distance of more than 1 mm from the dense area. Possibly, the upper limit of effectiveness of DDI is much smaller than this distance.

What morphological type of local cell interactions is characteristic of density-inhibited cultures? Our experiments (Cherny *et al.*, 1975 and unpublished) show that density-inhibited cultures of different cell types may have a diverse morphology: (*a*) normal fibroblasts in dense cultures form multi-layered sheets of well-spread cells (see chapter 4); (*b*) morphologically transformed hamster HEK-40 cells retaining sensitivity to homologous DDI form multilayered structures consisting of the cells poorly spread over each other in dense cultures. (*c*) transformed L cells retaining sensitivity to DDI form monolayered dense cultures consisting of poorly spread, almost spherical cells (see chapter 5).

As seen from this short list, DDI is not correlated with the formation of a monolayer by the cells, and is not correlated with the presence of specialized cell–cell contacts. It may be revealed in cultures of L cells in which electron microscopy does not reveal specialized contacts. Examination of the morphology of density-inhibited cultures of mouse embryo fibroblasts does not support the suggestion (Folkman & Moscona, 1978) that DDI is due to decreased cell spreading in dense culture. As discussed in chapter 4, normal mouse fibroblasts in dense cultures have large areas of lamellar cytoplasm and their mean projection area on the plane of the substratum is not smaller than in sparse cultures. Possibly, the only common morphological feature of various density-inhibited cultures is that the cells in these cultures are closely packed: a large part of the surface of each cell in these cultures is at a small distance from the surface of some other cell.

One may suggest that the rate of cell proliferation depends not on the number of specialized junctional structures but on the fraction of cell surface located at a distance which does not exceed a certain critical limit from the surface of another cell. Close packing of cells leading to an increase of this fraction may be achieved by various means in different cultures. Unfortunately, we cannot say at present what the exact value of the critical distance is. One possibility is that it is close to the membrane thickness, that is, direct contacts between the cell membranes are essential for DDI. Another possibility is that the critical distance is much wider than the membrane thickness. In this case growth inhibition would be due to some substance(s) secreted by the cells into the intercellular milieu or eliminated from it. To explain the local character of growth inhibiting effects it would be necessary to assume that an essential substance has a sharp concentration gradient in the immediate vicinity of the cell surfaces. Hypotheses of both types have been actively discussed in the recent literature but the data available at present to not permit a final choice to be made.

To model the possible growth inhibiting effect of contacts between membranes, Whittenberger & Glaser (1977) added preparations of plasma membranes of 3T3 cells to sparse cultures of homotypic cells. They found that preparations of 3T3 cell membranes have a growth inhibiting effect on 3T3 cell cultures. In contrast, the membranes of virus-transformed SV-3T3 cells did not inhibit the growth of 3T3 cells or of SV-3T3 cells. Further studies of the effects of plasma membranes would be of considerable interest: possibly these would reveal some specific membrane components responsible for growth inhibition. At present, however, it is not clear whether the effects of membranes described by Whittenberger & Glaser are identical to DDI. If DDI is due not to direct cell–cell contact but to alterations of local concentrations of certain substances in the diffusion boundary layer near cell surfaces, then an increase of diffusion between this layer and other parts of the fluid medium should decrease growth inhibition. In fact, two types of experiments have shown that increased circulation of fluid medium can lead to stimulation of growth in density inhibited cultures. (a) Stoker (1973) used a miniature pump to produce a local increase in the velocity of medium across the surface of confluent 3T3 cells. The labelling index of the cells exposed to the fastest steam from the pump increased locally to 76% compared with 10% in other areas of the same culture. Local cell density also increased in the 'jet area' after 26 h of pumping. (b) Shaking cultures of 3T3 cells (Stoker & Piggott, 1974) and human fibroblasts (Froelich & Anastassiades, 1975) led to stimulation of DNA synthesis.

These results, especially those of the ingenious 'pumping' experiments performed by Stoker, certainly strengthen the case for the diffusion boundary hypothesis. However, they do not prove this hypothesis; more exactly, they do not prove that the growth stimulating effect of pumping is due to alterations of diffusion rates. It remains possible that the mechanical movement of the fluid acting on the cell surface may activate DNA synthesis in quiescent cells. In this connection it would be interesting to study the effects of shaking or pumping on cells made quiescent not by a high local density but by other means, e.g. by a low serum medium.

Thus, DDI is associated with unknown diffusible or contact-mediated signals effective only when two cell surfaces are in a close proximity to each other. Until we identify these signals we cannot say what type of change is responsible for altered DDI in the cultures of transformed cells. Theoretically, these alterations can be due to at least three types of changes: (a) to altered production of DDI-inducing signals; (b) to altered cell sensitivity to these signals; (c) to altered distribution of cells in dense cultures leading to decreased proximity between cell surfaces.

10

Growth control by components of the humoral medium and its alterations in transformed cultures

Introduction

At least three groups of components affecting cell growth are present in the culture media: growth factors, nutrients and ions. Growth factors are molecules that can stimulate cell proliferation without being utilized by the cell for its metabolic purposes. It is supposed that in the absence of an essential growth factor the cell stops growing but will not die. If the absence of some molecule causes cell death, this molecule may be termed a 'survival factor'. Most identified growth factors are polypeptides (see review in Holley, 1975; Gospodarowicz & Moran, 1976), while all known nutrients are substances of low molecular weight. Of course, by defining a particular molecule as a 'growth factor' we describe only one of its effects. This designation does not mean that growth stimulation is the only or even the main effect of this molecule in culture or in the organism. Obviously, the same is true for 'survival factors', or 'adhesion factors', etc.

By definition, a growth factor may stimulate the proliferation of certain cells in certain culture conditions. This set of conditions may include the presence of a number of other molecules ('permissive factors'). The distinction between growth factors and permissive factors may be rather artificial: depending on the experimental design molecule A may act as growth factor which stimulates proliferation only in the presence of permissive molecule B or vice versa. In the first part of this chapter we will describe the effects of various growth factors, nutrients and ions on non-transformed cells. In the second part of the chapter we will discuss alterations of the sensitivities to these factors characteristic for strongly transformed cells. All experiments described in this chapter were performed with cells attached to the usual culture substrata.

The effect of humoral factors may depend on cell population density. Three main types of experimental designs may be used to reveal the growth-controlling effects of medium components.

(1) *Examination of the effect of the tested component on the cloning efficiency of cells;* that is, the assessment of its ability to stimulate cell growth in the condition of very low cell population density.

(2) *Examination of the effect of the tested component on cell proliferation in sparse mass cultures;* that is, in cultures where population density is much lower than saturation density. Often multiplication of cells in these cultures is inhibited by lowering the serum concentration in the medium for some time before the addition of the examined factor.

(3) *Examination of the effect of the tested component on cultures in the stationary phase;* that is, examination of the ability of this factor to counteract the density-dependent inhibition of proliferation.

Obviously, the effect of the same factor may be different in experiments of various types. Most experiments discussed in this chapter belong to the groups (2) and (3). It is not always easy to distinguish between designs (2) and (3) from the data given in the published papers. This distinction becomes even more difficult if we remember that DDI acts locally and that even sparse cultures may contain local high density areas with inhibited proliferation. Components of humoral medium, especially macromolecular growth factors, are very specific in their action on different cell types. Each cell type, either in the organism or in culture, probably has its specific set of growth-stimulating regulatory molecules. Lists of cell types stimulated by two different molecules may overlap only partially or not overlap at all. Here, as in other parts of this book we will describe only the data related to commonly used fibroblastic and epithelial cells. We will not describe the growth factors essential for other cell types, e.g. erythropoietin, 'colony-stimulating activity' and other factors controlling the growth of haematopoietic cells.

Growth factors and other medium components controlling proliferation of normal and minimally transformed cells
The effects of whole serum

Serum is the source of macromolecular growth factors in most culture media. High concentrations of serum (5–20%) are usually added to culture media. The sera obtained from animals of various ages have different properties. In particular, growth-stimulating properties are characteristic of fetal sera. The nature of the differences between sera is not clear. Serum concentrations needed for the active growth of various cell types are different. Primary fibroblastic cultures and most minimally transformed permanent lines obtained from these cultures need high concentrations of serum. Epithelial cells may have lower serum requirements than fibroblasts (Dulbecco, 1970; Dulbecco & Elkington, 1975). For instance, calf lens epithelium was grown in a medium with 1% fetal calf serum supplemented with thymidine (Taylor-Papadimitrou, Shearer & Walting, 1978). Cultured vascular endothelial cells were reported to be much less sensitive to growth stimulation by serum than smooth muscle cells of the vessel wall (Haudenschield *et al.*, 1976). Systematic comparisons of the serum requirements of various epithelia have not yet been done. Routine cultures of epithelia are usually grown in media with high (5–10%) concentrations of sera.

Transfer of fibroblastic cultures into a medium with a low serum content (0.5–1.0%) leads to the cessation of proliferation; the cells, however, may remain viable for a long time in these media. The low serum medium may be described as supplying the cells with a sufficient amount of survival factors but not of growth factors. The addition of fresh serum to the medium of these low-serum sparse cultures leads to the stimulation of proliferation. The addition of fresh serum may also stimulate proliferation in density-inhibited cultures. In both cases the proportion of [^{3}H]thymidine-labelled cells increases after a certain latent period, the average length of this period depending on the culture type; usually it varies between six and eighteen hours. When various amounts of serum are added to cultures, the minimal length of the latent period remains the same but the fraction of cells stimulated to enter the S phase and mitosis may vary (Todaro *et al.*, 1965; Temin, 1966; Holley & Kiernan, 1974*a*; Bartholomew, Neff & Ross, 1976*a*; Ellem & Gierthy, 1977; Okuda & Kimura, 1978). When the cells are continuously grown in a medium with a constant serum concentration, the rate of proliferation is proportional to the serum concentration within certain limits. The final stationary density of the cell population also increases in proportion to the concentration of serum (Todaro *et al.*, 1965; Holley & Kiernan, 1971; Bartholomew, Yokota & Ross, 1976*b*; Hassell & Engelhardt, 1977). Fresh serum incubated with the cells gradually loses its stimulating activity. This 'depletion' is due to the presence of living cells: it is not observed in media without cells. The addition of fresh serum to a medium containing depleted serum produces growth stimulation suggesting that depletion is not due to the accumulation of growth inhibitors but probably to inactivation of growth factors (Holley & Kiernan, 1968; Clarke & Stoker, 1971).

The effects of purified growth factors

Platelet factor. Serum usually added to tissue culture media (whole blood serum) is a supernatant fluid obtained after the clotting of whole blood. Platelet-poor plasma may be prepared from the whole unclotted blood by centrifuging out all the cellular elements including platelets and a serum may be produced by the subsequent clotting of this plasma (plasma-derived serum). The clotting of plasma, in contrast to that of whole blood, takes place without the participation of platelets. Consquently, plasma-derived serum, in contrast to whole blood serum, does not contain the products released by platelets which are activated in the course of clotting. It was found that plasma and plasma-derived serum have much lower growth-stimulating activities for cultured cells than whole blood serum (Balk, 1971). This difference suggested that one of the active mitogens present in whole blood serum may be derived from platelets (Balk, 1971; Balk *et al.*, 1973; Ross *et al.*, 1974). This suggestion was confirmed by experiments showing that platelet extracts stimulate the growth of many cell types including 3T3 mouse cells, smooth muscle cells of monkey aortic wall, human glial cells and human fibroblasts (Kohler & Lipton, 1974; Ross *et al.*, 1974; Rutherford & Ross, 1976; Westermark & Wasteson, 1976; Pledger *et al.*, 1977; Antoniades & Scher, 1977; see review in Ross & Vogel, 1978). Endothelial cells that are not significantly stimulated by serum also show low sensitivity to the serum-derived platelet factor (Wall *et al.*, 1978). For maximal and sustained stimulation of growth of sensitive cells, platelet factor should be added together with platelet-poor plasma. Components acting co-ordinately with platelet factor have not yet been identified (Pledger *et al.*, 1977; Vogel *et al.*, 1978; Yen, 1978). Purification experiments indicate that the platelet-derived growth factor is a heat-stable cationic protein with a molecular weight between 10 000 and 35 000. Semi-purified preparations of this factor have growth stimulating activity in doses ranging from 50 to 100 ng/ml when added to 5% cell-free plasma (Heldin, Wasteson & Westermark, 1977; Eastment & Sirbasky, 1978). In-vivo tissue injury is usually accompanied by blood coagulation. Production of platelet-derived growth factors in these conditions may have great physiological importance.

Epidermal growth factor. One variant of epidermal growth factor (EGF) was isolated from male mouse submaxillary salivary glands (mEGF) (Cohen, 1962). Another variant, human EGF or hEGF, with a somewhat different amino acid composition was purified from human urine (Cohen & Carpenter, 1975). Both human and mouse EGF are single-chain polypeptides with a molecular weight of about 6000; they compete for the same surface receptors and have similar biological activities *in vivo* and *in vitro*. Surface receptors of EGF have been partially characterized, they are large protein molecules of about 190 kd (Das *et al.*, 1977). mEGF and hEGF stimulate proliferation in non-dividing low density and confluent cultures of many cell types including human fibroblasts (Hollenberg & Cuatrecasas, 1973; Carpenter & Cohen, 1976*b*; Lembach 1976*b*), mouse 3T3 cells (Armelin, 1973; Rose, Pruss & Herchman, 1975), human glial cells (Westermark, 1976), human mammary epithelium (Stoker, Piggott & Taylor-Papadimitriou, 1976), human epidermis (Rheinwald & Green, 1977) and bovine luteal cells (Vlodavsky, Brown & Gospodarowicz, 1978*a*). A serum protein with a molecular weight of 18 000 enhances the effect of EGF on 3T3 cells (Mierzejewski & Rozengurt, 1978). Bovine endothelial cells do not respond to EGF; human endothelium may be stimulated by combined action of EGF and thrombin but not by thrombin alone (Gospodarowicz, Bialecki & Greenburg, 1978*a*; Gospodarowicz *et al.*

1978*b*). Effective concentrations of EGF causing growth stimulation in sensitive cultures are about 10^{-9}–10^{-10} M. The concentration of mEGF in mouse plasma is 2.5–10^{-10} M (Byyny *et al.*, 1974).

Fibroblast growth factor. Fibroblast growth factor (FGF) isolated from the brain and from the pituitary is a heat-stable basic protein, or rather a family of proteins,* with molecular weights of about 13000 (reviewed Gospodarowicz, Lindstrom & Benirschke, 1975; Gospodarowicz & Moran, 1976). Two peptides possessing FGF activity were isolated from the brain: FGF-1 and FGF-2 with molecular weights of 13000 and 11700 (Gospodarowicz *et al.*, 1978*a*). FGF stimulates proliferation in quiescent sparse and dense cultures of many cell types: mouse and human primary fibroblasts, mouse 3T3 cells, chondrocytes, myoblasts, and adrenal cells. Bovine and human endothelium are stimulated by FGF in combination with thrombin (Gospodarowicz *et al.*, 1975, 1978*b*; Gospodarowicz, Moran & Braun, 1977; Kamely & Rudland, 1976). All cells stimulated by FGF are of mesodermal origin; this factor was ineffective in experiments with cells of ectodermal or endodermal origin, such as epidermis, liver, pancreatic and pituitary cells (Gospodarowicz *et al.*, 1975; Rheinwald & Green, 1977). Concentrations of FGF needed to obtain a maximal stimulatory effect in sensitive cells are usually in the range 50–100 ng/ml.

Insulin-like growth factors. Somatomedin A, multiplication-stimulating activity (MSA) and non-suppressible insulin-like activities (NSILA) form a family of small proteins (molecular weight about 5000–6000) present in serum and having growth-stimulating activities. They are all structurally related to insulin and to each other. The exact number of 'insulin-like growth factors' is not established. They induce growth stimulation when added to cell cultures in low concentrations. For instance, the partially nomologous peptides NSILA-I and NSILA-II isolated from human plasma caused 4–5-fold stimulation of DNA synthesis in chick fibroblasts at concentrations of 75 ng/ml (Rinderknecht & Humbel, 1976*a*, *b*). Cells of a line derived from Buffalo rat liver release MSA into the culture medium. MSA partially purified from this medium stimulated growth of 3T3 cells at a concentration 1 μg/ml. MSA also had a stimulatory effect on chicken and rat fibroblasts (Dulak & Temin, 1973*a*; Smith & Temin, 1974; Dulak & Shing, 1976). Insulin itself also had a stimulatory effect in experiments with chick fibroblasts (Temin, 1967; Vaheri *et al.*, 1973; Teng, Bartholemew & Bissel, 1976) and human liver cells (Leffert, Koch & Rubaclava, 1976). The stimulating effect of insulin in experiments with rodent fibroblasts has also been observed (Jimenez de Asua *et al.*, 1973) but seems to be lower than that obtained with chicken fibroblasts.

---

* Proteins with similar biological properties from different sources, which have slightly different molecular weights and amino acid compositions.

Insulin may potentiate the effect of other growth factors, e.g. of FGF in rodent cultures (Gospodarowicz & Moran, 1974*a*). Even in the experiments with chicken fibroblasts, the growth-stimulating properties of insulin may be lower than those of insulin-related growth factors. For instance, in the experiments of Rinderknecht & Humbel (1976*a*) insulin produced the growth-stimulating effect on chicken fibroblasts only at concentrations 50–100 times larger than NSILA.

Other growth factors present in serum. Glucocorticoids form a group of compounds with potential growth-stimulating activity. Addition of cortisol stimulated the growth of dense cultures of 3T3 cells (Thrash & Cunningham, 1973). Dexamethasone and insulin potentiated the effect of FGF on the same cultures (Gospodarowicz *et al.*, 1975; Holley & Kiernan, 1974*a*). Among other experiments studying the effects of sera, those revealing the growth-stimulating effects of globulins (Janchill & Todaro, 1970) and of a sialoprotein with a molecular weight 12000 (Houck & Cheng, 1973) may be mentioned.

The factors present in these fractions have not yet been studied in detail. Of course, certain specialized cells that have been dependent on some special hormones *in vivo* may retain this sensitivity after their transfer in culture. For instance, rat mammary epithelial cells needed a combination of prolactin, insulin and corticosterone for maximal stimulation of DNA synthesis; however, this combination did not stimulate the growth of fibroblasts from mammary glands (Hallowes *et al.*, 1977).

In summary, it is clear that serum contains a variety of growth factors. Each cell type needs, probably, not one single factor but a group of factors. These groups are not identical for different cell types. Serum-free media supplemented with various combinations of known growth factors have been described as supporting the proliferation of permanent cell lines (Hutchings & Sato, 1978; Rizzino & Sato, 1978) and also of primary cultures of myogenic cells (Puri & Turner, 1978), of bone cells (Burks & Peck, 1977) and of lymphocytes (Iscove & Melchers, 1978). However, the perfect recipe of a combination of growth factors completely replacing serum in cultures of primary fibroblasts and epithelia has not yet been proposed. Apparently, some serum factors essential for these cells are still to be identified. In addition to polypeptide growth factors, undialyzed serum may also be a source of other growth-promoting factors such as serine (Allen & Moscowitz, 1978*a*, *b*), fatty acids (Nilausen, 1978) and selenium (McKeehan, Hamilton & Ham, 1976).

Does serum contain growth inhibiting molecules? This question has not received much attention. Very low density lipoprotein present in plasma was found to suppress proliferation of liver cells in culture (Leffert & Weinstein, 1976). Recently, the isolation of a partly purified serum-protein of high molecular weight (160000–300000) inhibiting proliferation of a number of cell types was briefly reported (Harrington, Godman & Wilner, 1978).

Growth factors secreted by cultured cells

'Feeder' effects of pre-established cells on the clonal growth of other cells are well known. A medium conditioned by one type of cell may also sometimes exert a beneficial effect on other cells. In particular, it was shown recently that epidermal cells grow in a medium conditioned by fibroblasts much better than in a non-conditioned medium (Rheinwald & Green, 1975a, b). The rate of growth of human mammary epithelial cells and of calf lens epithelium may be increased by feeder layers of fibroblasts (Taylor-Papadimitriou, Shearer & Stoker, 1977; Taylor-Papadimitriou *et al.*, 1978). Conversely, epithelial cultures may produce growth-stimulating substances for fibroblasts: dialysed medium from the cultures of normal and neoplastic mammary, epithelium stimulated the growth of primary mouse fibroblasts (Howard, Scott & Bennett, 1976). Medium conditioned by BALB/3T3 cells stimulated proliferation of cultured bovine endothelial cells (Birdwell & Gospodarowicz, 1977). The nature and specificity of the effects of these growth-stimulating factors produced by normal cells remain to be studied. Cultured cells have also been reported to produce growth-inhibitory substances. A glycoprotein with a molecular weight about 160000 is released into the medium by a hamster melanocytic cell line and inhibits the growth of many types of normal and transformed cells (Lipkin & Knecht, 1976; Knecht & Lipkin, 1977; Lipkin, Knecht & Rosenberg, 1978).

Holley, Armour & Baldwin (1978) have found that the monkey epithelial cell line BSC-1 releases into the medium at least three growth-inhibiting substances: lactic acid, ammonia and an unidentified inhibitor that may be an unstable protein. These diffusible inhibitors, in addition to locally acting DDI, may limit cell growth in overcrowded cultures.

Effects of ions

Alterations of concentrations of ions in the culture medium may profoundly affect cell growth. Three ions most actively studied are Ca^{2+}, Mg^{2+} and H^+. The concentration of calcium in the media is usually 1.2–1.7 mM. If this concentration is decreased to 0.1 mM or less, the proliferation of cells is reversibly inhibited. This effect was observed in normal cultures of chicken, rodent and human fibroblasts as well as in mouse 3T3 lines (Balk *et al.*, 1973; Swierenga, MacManus & Whitfield, 1976a; Boynton & Whitfield, 1976; Boynton *et al.*, 1974, 1977a, b; Mitchel & Tupper, 1978). Dulbecco & Elkington (1975) increased the calcium concentration in serum-starved 3T3 cell cultures to 5.4 mM or more and observed stimulation of growth. It is, however, possible that in this case the stimulatory effect could be partly due to the formation of calcium-containing precipitates (see below). The role of magnesium in the growth control of chicken fibroblasts was studied in detail by Rubin and his group (Rubin & Koide, 1976; Kamine & Rubin, 1976; Rubin, 1977*b*; Sanui & Rubin, 1977; Rubin & Chu, 1978). The concentration of magnesium in a standard medium is 0.8 mM. If this concentration is decreased to 0.2–0.1 mM or lower, inhibition of DNA synthesis and growth was observed. This inhibition is more pronounced when the calcium concentration is also decreased in a medium with low magnesium. In contrast, the inhibition of proliferation in a low-magnesium medium can be reversed by doubling or tripling the normal concentration of calcium. Magnesium deprivation prevents the serum stimulation of DNA synthesis. A decrease of pH was also found to be a growth inhibitory factor. Using a non-volatile organic buffer Ceccarini & Eagle (1971) and Eagle (1973) showed that normal human fibroblasts have a well-defined pH optimum in the range 7.5–7.8. The highest saturation density was reached by these cells at pH 7.6 (Ceccarini & Eagle, 1971). In the experiments of Rubin (1971) the effect of pH on the growth of chick fibroblasts was most marked in dense cultures; sparsely seeded cells grew almost as well at pH 6.7 as at higher pH, while crowded cells were inhibited below 7.4. According to Froelich & Anastassiades (1974) the initial growth rate of human fibroblasts in sparse cultures was similar at pH 7.1 and pH 7.7. However, the saturation density achieved at pH 7.7 was 2–4 times that achieved at pH 7.1. These results confirm that sensitivity to pH changes increases with growing population density.

Media deprived of Zn^{2+} were found to inhibit cell growth (Rubin, 1972, 1973). Selenium was found to stimulate proliferation of human fibroblasts and serum appears to be a source of selenium in most media (McKeehan *et al.*, 1976).

Effects of nutrients

The minimal list of nutrients essential for the growth of the usual mammalian cultures was established by the classical work of Eagle (1955). Later, Eagle & Piez (1962) showed that the requirements for certain metabolites (e.g. serine or cystine) are population-dependent; that is, the addition of these metabolites to the medium is essential only for the growth of cultures at a low population density. At a high population density exogenous metabolites of this type become unnecessary, presumably because they are released into the medium by the cells themselves. Since then there has been little progress in this field. At present we still do not know whether the requirements of cells of different species and different tissue types for nutrients are identical. It was reported (Leffert & Paul, 1973) that hepatic parenchymal cells are able to synthesize arginine and, therefore, in contrast to other cell types, are able to grow in an arginine-deficient medium. Normal cells in the medium lacking certain nutrients cease to proliferate and are usually collected in the post-mitotic phase with diploid DNA content (Holley & Kiernan, 1974*b*; Holley, Armour

& Baldwin, 1978*a;* Allen & Moscowitz, 1978*a, b*). If a medium with one essential nutrient missing was replaced by a medium with another missing component, stimulation of growth was not observed (Pardee, 1974). These data suggest that normal cultures have some mechanism causing their entry into the resting state when some essential nutrient is lacking. However, amino acid deficiency may also block part of the cells in the S phase of the cycle (Yen & Pardee, 1978*b*).

Diverse growth stimulators

Besides the factors discussed above, there are a number of growth stimulatory agents of diverse nature that are not standard components of culture media but can stimulate cell growth when added to these media. Proteases are the best studied of these agents. Proliferation of normal chick embryo fibroblasts may be stimulated by various proteases including trypsin (Sefton & Rubin, 1970; Vaheri, Ruoslahti & Hovi, 1974; Cunningham & Ho, 1975; Blumberg & Robbins, 1975) and thrombin (Chen & Buchanan, 1975*a;* Teng & Chen, 1975). Data about the effects of proteases on mammalian cells are more diverse. Certain investigators obtained stimulation of mouse 3T3 cells by trypsin or pronase (Burger, 1970; Burger, Bombik & Noonan, 1972; Noonan & Burger, 1973; Noonan, 1976). However, other investigators did not observe protease-induced stimulation in these cells (Glynn, Thrash & Cunningham, 1973). Pohjanpelto (1977, 1978) observed stimulation of DNA synthesis in serum-starved human fibroblasts by thrombin. In the experiments of Zetter *et al.* (1977), thrombin alone was not a very effective stimulator of a variety of mammalian cell lines (BHK 21, 3T3 etc.); however, thrombin potentiated the mitogenic action of serum and other growth-promoting factors such as EGF or prostaglandin $F_{2\alpha}$. In the experiments of Gospodarowicz *et al.* (1978*b*) thrombin also potentiated the proliferative response of human endothelial cells to FGF or EGF. Some of the discrepancies of the results obtained by different authors may be due to variations of culture conditions affecting the stimulatory action of proteases. In their detailed study of the action of trypsin and thrombin on human, mouse and chick fibroblasts Carney, Glenn & Cunningham (1978) revealed two factors changing the cellular response to these proteases. (*a*) A stimulatory response of all types of fibroblasts was significant only after pre-incubation of the cells in a serum-free medium. The addition of calf serum inhibited the response, especially in mouse cells. (*b*) The responsiveness of chick embryo cells to proteases rapidly decreased in the course of serial passages of these cells.

These results suggest that in order to obtain a maximal stimulatory effect of proteases one has to use only primary cultures and not established lines.

What is the mechanism of the growth-stimulatory action of proteases? Carney & Cunningham (1977) showed that trypsin immobilized on the polystyrene beads retains its ability to stimulate chick fibroblasts; this suggests that growth stimulation is a result of the action of trypsin on the cell surface. The same seems to be true for the growth stimulating effect of thrombin (Glenn & Cunningham, 1978). Thrombin is bound by receptors on the fibroblast surface (Carney & Cunningham, 1978; Perdue *et al.*, 1978). It is not quite clear whether the stimulating effect of proteases is always correlated with their specific proteolytic activity. In the experiments of Brown & Kiehn (1977), inactivation of trypsin by specific inhibitors did not prevent its stimulating action on BHK cells. However, in similar experiments, a number of other authors observed significant inhibition of growth stimulation suggesting that this stimulation may be due to direct proteolytic activity (Sefton & Rubin, 1970; Burger, 1970; Carney & Cunningham, 1977).

Many other substances have been reported to stimulate growth in different cultures. 12-*O*-Tetradecanoyl-13-phorbolacetate (TPA) was reported to stimulate the growth of the BALB/c-3T3 line (Sivak, 1973, 1977) in confluent culture. In the experiments of Boynton, Whitfield & Isaacs (1976) TPA stimulated DNA synthesis in 3T3 cell cultures in which growth had been inhibited by pre-incubation in low-calcium medium. On the other hand, O'Brien & Diamond (1977) did not obtain stimulation of DNA synthesis in dense cultures of hamster fibroblasts treated with phorbol esters. TPA and certain other phorbol esters were able to stimulate considerably the growth of primary cultures of mouse epidermal cells (Slaga *et al.*, 1976).

Digitonin stimulated DNA synthesis in density-inhibited cultures of mouse fibroblasts (Vasiliev *et al.*, 1970), and protaglandin $F_{2\alpha}$ stimulated growth of 3T3 cells (Jimenez de Asua, Clingan & Rudland, 1975).

Agents that selectively destroy microtubules (colchicine, colcemid and vinblastine) were found to cause a marked stimulation of DNA synthesis in dense primary cultures of mouse fibroblasts (Vasiliev, Gelfand & Guelstein, 1971). The minimal concentrations of these agents producing growth stimulation in mouse cultures were similar to the minimal concentrations producing a complete metaphase block of mitotic cells and inhibition of cell migration into wounds; that is, effects specific for microtubule-destroying drugs (Guelstein & Stavrovskaya, 1972). Microtubule-destroying agents, antitubulins, were also reported to potentiate the growth stimulating action of fresh serum and insulin in chick embryo cultures (Teng *et al.*, 1977). However, in dense cultures of primary hamster and rat fibroblasts even the large doses of colchicine and colcemid that produced metaphase block and inhibition of migration were unable to stimulate DNA synthesis (O. Y. Pletyushkina, unpublished). Colchicine inhibited serum-induced initiation of DNA synthesis in human WI-38 fibroblasts, in the mouse C3H10T1/2 line (Walker, Boynton & Whitfield, 1977), in a mouse neuroblastoma line (Baker, 1976) and in 3T3 cells (Baker, 1977). Possibly, cells of different species and lines have very different sensitivities to the growth-stimulating action of colchicine and related drugs and this question needs further investigation.

Commercial preparations of testicular hyaluronidase and ribonuclease stimulated DNA synthesis in dense cultures of mouse fibroblasts (Vasiliev *et al.*, 1970). As shown by Greenberg & Cunningham (1973), the growth-stimulating component in hyaluronidase preparations can be dissociated from the fraction containing the main enzymatic acitivity. The nature of this component remains unknown.

Concanavalin A and other lectins stimulated proliferation and increased saturation density in cultures of BHK-21 hamster fibroblasts (Aubery, Bourrillin & O'Neill, 1975). Plant lectins also stimulated DNA synthesis in cultures of embryonic cells of chick neural retina (Kaplowitz & Moscona, 1976).

Sub-toxic concentrations of certain metal ions such as Zn^{2+}, Ca^{2+} and Hg^{2+} stimulated the rate of DNA synthesis 5–20-fold in cultures of chick embryo fibroblasts deprived of serum (Rubin, 1975*a*).

Insoluble complexes of Ca^{2+} with inorganic pyrophosphate and/or ortho-phosphate stimulated thymidine incorporation and cell multiplication in 3T3 cultures (Barnes & Colowick, 1977; Rubin & Sanui, 1977).

Conclusion

The facts presented above demonstrate the multiplicity of growth-controlling factors in culture media. Alterations of concentrations of many macromolecules, nutrients or ions may induce the transition of a culture from a resting into a growing state and vice versa. Specific protein growth factors are of particular importance for proliferation control. The number of purified factors of this type to be characterized is steadily increasing. Certain macromolecular growth factors were shown to be bound by specific receptors at the cell surface. Such factors as EGF, FGF and MSA are bound by different receptors, so that at least the first stages of the process by which growth stimulation is induced may be different for various factors. It is interesting that most known growth-controlling macromolecules stimulate cell growth; examples of growth-inhibiting macromolecules are very scarce although the search for growth-inhibiting chalones has been in progress for many years (see Houck & Attallah, 1975; Houck *et al.*, 1977). Under appropriate conditions, growth stimulation may be caused not only by growth factors but also by a miscellany of agents. These facts suggest that growth stimulation can be a result not only of the binding of specific macromolecules at the cell surface but also of less specific alterations of this surface or of other cell components. We will discuss this question in chapter 12.

Altered reactions of transformed cells to growth factors, ions, nutrients and growth-affecting drugs

Introduction

The rate of multiplication of normal or minimally transformed cells in optimal media at low density may be as high, or even higher, than that of strongly transformed cells. To mention but two examples, in the experiments of Bartholomew *et al.* (1976*b*) the doubling times of a population of BALB/c 3T3 cells and of its derivative line transformed with SV-40 virus were similar (16 h) in media containing serum at concentrations of 8% or more. Malignant cells from mouse mammary carcinomas had a larger doubling time than normal mammary epithelial cells (Buehring & Williams, 1976).

To reveal characteristic growth differences between normal and transformed cells one has to cultivate them in non-optimal media; that is, in media with a decreased content of one or several components essential for the growth of normal cells. In certain suboptimal media normal cells stop growing while transformed cells continue growth. In this section we will discuss the reactions of transformed cells to media deficient in serum, calcium or nutrients. We will also compare the reactions of normal and transformed cells to various purified growth factors and to growth-stimulating and growth-inhibiting agents which are not standard components of the culture media.

Alterations of serum requirement

In the course of neoplastic evolution transformation of serum requirement may take place: cells may acquire the ability to grow in a medium with a low serum concentration which cannot support the growth of the parent, non-transformed cells. To test the serum requirement one has to compare the growth of sparse mass cultures and cloning efficiencies in media with decreased serum concentrations. A line which can grow actively in the medium with 1–2% serum can be regarded as a line with a strongly transformed serum requirement (see numerous examples of these lines in Holley & Kiernan, 1968; Temin, 1968; Clarke *et al.*, 1970; Dulbecco, 1970; Pitts, 1971). In media with a low serum content, the rate of proliferation of transformed cells may decrease but these cells, in contrast to non-transformed cells, are not blocked in any particular stage of the cycle. For instance, in the experiments of Bartholomew *et al.* (1976*b*), in a medium containing 1% serum about 95% of BALB/3T3 cells were blocked in the post-mitotic phase of the cell cycle and only about 5% cells were in S, G_2 or M phases. In contrast, when 3T3 cells transformed with SV-40 virus were grown in a medium containing 1% serum about 50% of the population had a DNA content corresponding to the S and G_2 phases, although the population doubling time in this low serum medium increased to 51 h by comparison with 16 h in the medium with 10% serum. When standard culture methods are used, even strongly transformed lines will not grow for a long time in a medium that contains neither serum nor a set of purified growth factors. However, several lines which grow actively in serum-free media have been obtained by special selection (McQuilkin, Evans & Earle, 1957; Shodell, 1972; Katsuta & Takaoka, 1973; Gelfand 1974; Banks, Bhavanada & Davidson, 1977). Bush & Shodell (1976) added the soybean inhibitor of trypsin to the culture medium in order to decrease the damaging effect of trypsin during sub-cultivation of cells. They showed that two 'usual' strongly transformed lines (polyoma-transformed BHK cells and SV 40-transformed 3T3 cells) could grow in a serum-free medium containing this inhibitor without any special previous selection. In these cultures, the rate of cell passage through the cycle was low in a serum-free medium and increased after the addition of serum. Thus, in these conditions even the complete absence of serum did not produce a post-mitotic block although it decreased the rate of proliferation. Serum-restricted transformed cells may die much earlier than normal cells in similar conditions. For instance, human WI-38 fibroblasts plated in 1–3% serum without subsequent medium changes ceased to proliferate; their number remained unchanged for 11–13 days. In contrast, WI-38 cells transformed with SV-40 virus in similar conditions reached saturation density at 9 days but later their number gradually decreased (Schiaffonati & Baserga, 1977).

The features described above are characteristic of cell lines with strongly transformed serum requirements. Certain other lines have less changed growth requirements. A detailed analysis of the growth behaviour of one particular line of this type, 3T3 cells transformed by benzo(a)pyrene (BP-3T3), was made by Holley *et al.* (1976). These cells had a lower serum requirement than parent cells: they grew in medium with 2–4% serum but they became blocked in the post-mitotic phase when transferred into medium with 0.2% serum. DNA synthesis in BP-3T3 cell cultures incubated with 0.2% serum could be stimulated by the addition of more serum or FGF; the minimal effective concentrations of these agents were about two to three times less than those needed for the stimulation of 3T3 cells The ability of BP-3T3 cells to deplete the serum-containing medium of growth-stimulating activity was considerably lower than that of parent 3T3 cells. These results confirm that the threshold concentrations of serum factors needed for the initiation of proliferation may be decreased at 'intermediate' degrees of transformation of serum requirement. In more strongly transformed cells these threshold concentrations may decrease to zero level.

By special selection, serum revertants can be isolated from strongly transformed cultures (see chapter 2). These revertants lose the ability to grow in media containing 1% serum (Vogel & Pollack, 1973). Different serum revertants have been found to have different mechanisms for the cessation of proliferation in low serum media: certain lines stop DNA synthesis (like non-transformed cells); other lines continue to synthesize DNA but are blocked before entry into mitosis (Vogel & Pollack, 1975). Thus, in the course of selection these latter revertants did not restore the normal characteristic of blocking in the post-mitotic phase but instead attained the ability to become blocked in other phases of the cell cycle.

Reactions to individual growth factors

Almost by definition, cells with decreased serum requirements should have altered sensitivities to growth factors contained in the serum but the available data bearing on the effects of individual factors are few. 3T3 cells infected with SV-40 virus in contrast to non-infected cells could enter the S phase in media without platelet factor (Scher *et al.*, 1978; Stiles, Pledger & Scher, 1978). In the experiments of Hallowes *et al.* (1977), cells of transplantable rat mammary tumours that were hormone-independent *in vivo* were grown in culture. As might be expected, proliferation of tumour cells, in contrast to that of normal mammary epithelium (see above), was not affected by mammotropic hormones. Rudland *et al.* (1974*a*) studied the effect of FGF on clones selected from the BALB/3T3 line with a temperature-sensitive expression of the transformed phenotype. It was found that DNA synthesis was stimulated by FGF only at a non-permissive temperature; at a permissive temperature DNA synthesis in the low serum medium proceeded at the same rate with or without FGF. In the experiments of Moses *et al.* (1978) density inhibited, chemically transformed C3H10/T1/2 mouse cells, in contrast to the similarly inhibited parent line, were not stimulated by EGF or FGF. Possibly, in some cases insensitivity to exogenous growth factors may be associated with the formation of endogenous growth factors by neoplastic cells: these factors may then block corresponding receptors (see below).

Thus, transformed cells may lose both the ability to stop proliferation without growth factors and the ability to initiate proliferation in the presence of these factors.

Alterations of calcium requirement

Media with a decreased calcium content (0.01–0.02 mM) inhibit the growth of normal cells (see above) but have no effect on the growth of homologous transformed cells such as cells of chick embryos transformed with Rous sarcoma virus, 3T3 cells transformed with viruses or chemical carcinogens, human fibroblasts transformed with SV-40 virus, or adenovirus-transformed hamster cells (Balk *et al.*, 1973; Boynton & Whitfield, 1976; Swierenga, Whitfield & Gillan, 1976*b*; Boynton *et al.*, 1977*a*, *b*; Levinson & Levine, 1977). Normal rat kidney (NRK) cells transformed with temperature-sensitive Rous sarcoma virus had a decreased calcium requirement only when grown at a permissive temperature (Boynton & Whitfield, 1978). Probably, a lowered calcium requirement should be regarded as a special transformed character. Its relation to other transformed characters remains to be studied. A lowered dependence of growth of cells transformed with Rous sarcoma virus on pH changes was observed by Rubin (1971, 1973).

Sensitivity to alterations in the concentrations of nutrients

Little is known about the alterations of nutritional requirements accompanying transformation. It was shown that a number of strongly transformed lines, unlike their parent cells, are methionine auxotrophs: these lines did not grow when methionine in the medium was replaced by its immediate precursor, homocysteine (Chello & Bertino, 1973; Hoffman & Erbe, 1976; Wilson & Poirier, 1978). This example shows that a systematic comparison of nutritional requirements of normal, minimally transformed, and strongly transformed cells may possibly provide some interesting results.

Cultivation of strongly transformed cells in media deficient in essential nutrients, e.g. in isoleucine, in contrast to normal cells, does not result in transition into the resting state. Cells in these conditions may remain distributed in various phases of the cell cycle and their rate of death increases considerably (Pardee, 1974; Martin & Stein, 1976). These experiments suggest that transformed cultures, in contrast to normal ones, may be unable to enter the resting state in response to the absence of essential nutrients.

Altered sensitivity to miscellaneous growth-stimulating agents

The growth of density-inhibited normal cells can be stimulated by various agents. Transformed lines that retain the ability to stop proliferating at high population densities may lose sensitivity to these growth-stimulating agents. For instance, DNA synthesis in dense cultures of L and S-40 lines of transformed mouse fibroblasts could be stimulated by wounding but not by colcemid or by commercial preparations of testicular hyaluronidase (Vasiliev *et al.*, 1975*b*). As mentioned in the previous section, both colcemid and hyaluronidase caused considerable stimulation of DNA synthesis in dense cultures of normal mouse fibroblasts. It would be interesting to study the reactions of strongly transformed cells to the action of other growth-stimulating agents such as proteases and phorbol esters

Altered sensitivity to the action of agents which block entry into the growing state

A number of chemicals with different mechanisms of action block non-transformed or minimally transformed cells in the post-mitotic (G_1) phase but do not produce a similar block in strongly transformed cells. For instance, the blocking of BHK cells in G_1 was induced by caffein, streptovitacin, 5-fluorouracil and puromycin aminonucleoside but these compounds did not block the proliferation of polyoma-transformed BHK cells (Pardee & James, 1975). Dibutyryl-cAMP in combination with an inhibitor of cyclic nucleotide phosphodiesterase prevented proliferation in cultures of 3T3 cells. The labelling index after 24 h incubation with [³H]thymidine decreased from 75 to 13%. However, the same drugs had no effect on the labelling index of 3T3 cells transformed with SV-40 virus (Rozengurt & Po, 1976). Similarly, rat liver epithelial cells were blocked by dibutyryl-cAMP outside the S phase, while malignant liver cells were slowed in the S phase (Williams, 1977).

Picolinic acid (a metal-chelating agent structurally related to nicotinic acid) was reported to block NRK and BALB/3T3 cells in G_1; these cells remained viable in the drug-containing medium for extended periods. In contrast, various strongly transformed sub-lines of 3T3 cells were blocked by this agent in G_2 or randomly in different stages of the cycle; these cells died after 90–120 h of incubation in the drug-containing medium (Fernandez-Pol, Bono & Johnson, 1977; Fernandez-Pol & Johnson, 1977). The drugs used in these experiments primarily affect different metabolic processes. Cell shift into the quiescent state (post-mitotic block) in these conditions may be an adaptive reaction induced by various metabolic abnormalities. It seems that transformed cells lose the ability to perform this reaction.

Interesting effects are produced in cultures of normal and transformed cells by small concentrations (about 1–3 μg/ml) of cytochalasin B which block cytokinesis but not the other stages of mitosis. As a result mitosis leads to the formation of binucleated cells (Carter, 1967b; Defendi & Stoker, 1973). When normal cell cultures are incubated for several days in a cytochalasin-containing medium, the proportion of binucleated cells may reach 40–80%, but then the average number of nuclei per cell is not further increased: the proportion of cells having more than two nuclei remains very small even after many days of incubation. In contrast, in strongly transformed cultures the number of nuclei per cell continues to increase as the time of incubation with cytochalasin increases. This loss of 'limitation of multinucleation' was observed in experiments with many types of cells transformed by different agents (Wright & Hayflick, 1972; Kelly & Sambrook, 1973; O'Neill, 1974, 1976; Hirano & Kurimura, 1974; O'Neill et al., 1975; Steiner et al., 1978). For instance, in our experiments (J. Vasiliev, unpublished) with cultures of NRK cells transformed by a temperature-sensitive mutant of mouse sarcoma virus, after seven days incubation with cytochalasin B at a non-permissive temperature (39 °C), 75% of the nuclei were located in binucleated cells and only 9% in cells with more than two. A similar incubation of the same line at a permissive temperature (32 °C) led to the formation of a culture containing 45% of nuclei in multinucleated cells.

The limitation of multinucleation in normal cultures incubated with cytochalasin B may be the result of a reversible G_1 block that this agent, like many other agents, can produce in these cells (Westermark, 1973b). The cells that are in G_1 at the moment when the drug is added are immediately blocked while other cells pass into the S phase and mitosis and become binucleated. The binucleated cells are then also blocked in G_1 and to not enter the following mitosis, so that the number of nuclei per cell is not further increased. Transformed cells are not blocked in G_1 and therefore have no 'limitation of multinucleation'. It would be important to find out how insensitivity to drug-induced blocks is correlated with other transformed characters.

The different sensitivities of non-transformed and transformed cells to drug-induced G_1 blocking has been used as a basis for the selective killing of these cells by cytotoxic agents. Non-transformed and transformed cells were first treated with the G_1-blocking agents and then with drugs that are selectively toxic for cells in the S phase or in mitosis (cytosine arabinoside, hydroxyurea, colcemid). In these conditions G_1-blocked non-transformed cells survived while cycling transformed cells were killed (Pardee & James, 1975; O'Neill, 1975; Rozengurt & Po, 1976; Bradley *et al.*, 1977). For instance, in the experiments of Pardee & James (1975) hydroxyurea killed 94% of the unprotected BHK cells and none of the same cells protected with caffeine or streptovitacin. In contrast, in the experiments with BHK cells transformed with polyoma virus, hydroxyurea killed 99% of the cells in unprotected cultures and 88–94% in protected ones. The potential importance of these data for tumour chemotherapy is obvious.

Mutual stimulation of transformed cells

Theoretically it is possible that selective growth of neoplastic cells may be due to mutual stimulation. This stimulation could be mediated either by the factors secreted by these cells into their micro-environment or by the stimulating action of cell–cell contacts.

The cells of certain plant tumours (crown galls) produce growth factors of low molecular weight (cytokinins and cytokinesins) which are not synthesized in similar conditions by corresponding normal cells. Therefore these tumour cells may grow in media without added growth factors (see review in Braun, 1978). Certain animal cells also secrete growth factors into their media (see review in Shields, 1978). In particular, L cells growing in a serum-free medium secrete unidentified proteins that are able to stimulate growth of BHK cells (Shodell, 1972).

In three systems, growth factors produced by transformed cells have been purified. (*a*) MSA is produced by a buffalo rat liver cell line growing in serum-free medium (Dulak & Temin, 1973*a*, *b*). The properties of this insulin-like factor were described earlier. Certain human fibrosarcomas in culture produce MSA-like activity (Todaro & De Larco, 1978). (*b*) 'Migration factor' is produced by a line of BHK cells transformed by SV-40 virus. It is a protein with a molecular weight of 18000. This protein stimulates the migration of BHK cells into the wound made in confluent culture. It also stimulates DNA synthesis and growth in cultures of 3T3 line as well as in primary cultures of mouse and human fibroblasts (Bürk, 1976; Bourne & Rozengurt, 1976). (c) De Larco & Todaro (1978a) isolated several polypeptides from a serum-free medium of 3T3 cells transformed by sarcoma virus. These sarcoma growth factors (SGF) stimulated DNA synthesis in serum-deprived cultures of an NRK rat fibroblastic line being effective in nanogram quantities. The addition of SGF also stimulated colony formation by NRK cells in methylcellulose. Partial purification of SGF revealed three major peaks of activity with apparent molecular weights of 25000, 12000 and 7000. All three molecular species competed with EGF for membrane receptors; they were not, however, identical to EGF in their properties. 3T3 cells transformed with SV-40 virus did not produce SGF. In all these cases, corresponding non-transformed cells in contrast to strongly transformed ones, did not produce growth factors. Systematic studies of the production of various growth factors by many types of non-transformed and transformed cells will be probably made in the near future.

Besides the secretion of growth factors, mutual stimulation of neoplastic cells may be a result of growth-enhancing influences transmitted to neighbouring cells via cell–cell contacts. The possible existence of this 'contact stimulation' is suggested by experiments with ascitic mouse hepatoma (Vasiliev *et al.*, 1966b). A suspension of hepatoma cells in ascitic fluid comprises single cells and groups of two and three cells. The cells of each floating group are linked with each other by firm specialized cell–cell contacts. It was shown that the mean duration of the mitotic cycle of single cells is longer than that of cells forming two-cell groups. This prolongation of the cycle in single cells was due to the prolongation of the G_1 phase: the mean period spent by the cells in this phase was 17–18 h for single cells and 10–11 h for cell doublets. The mechanism of this 'contact stimulation' is not clear. Fournier & Pardee (1975) showed that the G_1 phase may be shorter in binucleate cells compared with mononucleate cells of the same culture. Probably, two nuclei located in the same cytoplasm or in two contacting cells may mutually stimulate their entry into the S phase. It would be very interesting to study further the mechanism of these mutual stimulatory effects as well as their occurrence in cultures of normal and transformed cells.

Conclusion

Transformations may be manifested by alterations of cell reactions to several different medium components such as growth factors, calcium and nutrients. The general characteristic of a transformed culture is, probably, its decreased ability to shift into the resting state when the medium becomes deficient in certain components. Even if the cells are unable to grow in very deficient media they may still continue to cycle and eventually may die due to their inability to lower their metabolism in adverse conditions. A similar situation may arise when transformed cells are incubated with certain drugs that disturb their metabolism: transformed cells, in contrast to normal ones, may be unable to react to these disturbancies by a transition into the quiescent state. Alterations of the requirements for medium components, like other transformed characters, may have different degrees of expression. For instance, certain lines have a decreased serum requirement but still retain the ability to become blocked when transferred to a medium with a very low serum concentration; other cells decrease their growth in these media but are not blocked in G_1; certain other lines may continue to proliferate in serum-free media.

Possibly, transformation may be accompanied by the acquisition of a decreased sensitivity not only to growth-inhibiting conditions but also to certain growth-stimulating factors. Certain macromolecular growth factors and diverse stimulators of growth were described as being ineffective in cultures of transformed cells. It is probable that transformed cells promote mutual growth by secreted growth factors or by 'contact stimulation'.

11

Inter-relationships between the effects of different growth-controlling factors

In the previous chapter we described the effects of various components of the humoral medium on the growth or normal and transformed cells. Do these effects depend on other components of the cell environment; namely, on the availability of the substratum and on cell density? In this chapter we will discuss the scanty experimental data related to this question.

Serum and anchorage dependence

Clarke *et al.* (1970) showed that the anchorage dependence of BHK21 cells can be partially overcome by a high serum concentration: in the usual medium containing 10% serum, the proportion of [³H]thymidine-labelled cells was lower than 5% in methylcellulose suspension but when the serum concentration in the medium was increased to 50%, proportion of labelled cells increased to 12%. The proportion of substratum-attached cells labelled by [³H]thymidine in the same conditions was about 70–80% in the medium with 10% serum. Suspended cells in these experiments were estimated to be about 60 times less sensitive to the serum than spread cells. These data suggest that attachment to the substratum and serum factors act co-ordinately. Sensitivity to serum stimulation is increased when the cells are attached to the substratum. The same relationship may be expressed in another form: sensitivity to stimulation by the substratum grow with increasing serum concentrations. Another good example of an alteration of anchorage dependence by a humoral factor was published by De Larco & Todaro (1978): in the presence of sarcoma growth factors isolated by these authors (see p. 213) the proportion of colonies formed by NRK cells in agar increased to 50% or more compared with none in the control. It would be very important to continue investigation of the relationships between anchorage dependence and humoral growth factors.

Density dependence and humoral growth factors

As described in chapter 10, density inhibited non-transformed cells can be stimulated by fresh serum and by purified growth factors such as EGF and

FGF. The fraction of cells stimulated to enter into the S phase by a given concentration of fresh serum decreased with increasing population density (Ellem & Gierthy, 1977). In the experiments of Mierzejewski & Rozengurt (1977) the proportion of 3T6 mouse cells stimulated by low concentrations of EGF (1 ng/ml) and insulin (0.1–0.2 μg/ml) decreased with increasing density. When the insulin concentration was increased up to 5–10 μg/ml and that of EGF up to 3–10 ng/ml maximal stimulation was obtained both in sparse and dense cultures. The percentage of cells stimulated by colcemid in mouse embryo cultures was also inversely proportional to the local population density (Vasiliev *et al.*, 1971). As first shown by Holley & Kiernan (1971), cultures grown in media with various serum concentrations reach different saturation densities. For instance, in the experiments of Bartholomew *et al.* (1976*a*), the saturation density of 3T3 cells varied from about 10^4 cells/cm^2 with 6% calf serum to about 4×10^4 cells/cm^2 with 20% serum. The concentration of low molecular weight nutrients may also affect the saturation density. For instance, in the experiments of Holley *et al.*, (1978*a*) with the BSC-1 line of African green monkey kidney epithelium, if the glucose concentration in the medium was increased four-fold, the saturation density increased approximately 50%. These facts suggest that sensitivity to stimulation by serum, purified growth factors and probably increased concentrations of nutrients, decreases with growing cell density. The same phenomenon may be described in a somewhat different way: humoral factors and reduced density stimulate cell growth in a co-ordinated fashion. These results do not exclude the possible existence of special growth stimulators that are active only in dense cultures but not in sparse ones. Presumably, these hypothetical stimulators counteract the action of unknown factors inducing density-dependent inhibition; therefore it would be interesting to find agents with this type of effect.

Co-ordinated reaction norm to growth-controlling factors

The results described above suggest that the action of humoral growth factors depends on the availability of the substratum and on cell density and vice versa. For instance, less serum is needed to produce the same stimulating effect in a spread culture at a low population density than in either a suspended culture or one at high population density. Various growth-stimulating factors added to the humoral medium also interact in controlling cell reactions. For instance, in the experiments of Holley & Kiernan (1974*a*), the combination of insulin and FGF caused considerably greater stimulation of DNA synthesis in serum-restricted or dense 3T3 cell cultures than each of these factors alone. A number of other examples of mutual potentiation of the effects of various stimulators have been mentioned earlier (see for example the description of the effects of platelet factors, glucocorticoids and proteases in chapter 10).

Thus, the degree of reaction of cells to a given concentration of a particular growth factor (norm of reaction to this factor) depends on the presence of all the other growth controlling factors. In other words, stimulation of growth is the result of the co-ordinated assessment by the cell of various parameters. The cell has a co-ordinate reaction norm to many controlling factors. Depending on the conditions one or other controlling factor (serum, ions, nutrients, density, substratum) may become growth limiting.

Alterations of reaction norm in transformed cultures

The reaction of certain genetically transformed cells to the substratum depends on the amount of serum in the medium. In experiments with the LSF subline of L cells (Bershadsky *et al.*, 1976) it was shown that the cells are anchorage independent only in a serum-containing medium: when LSF cells were suspended in methylcellulose 77–80% formed colonies in a medium with 10% serum and only 2–11% in a serum-free medium. The same cells attached to the substratum grew well both in medium with 10% serum and without serum. In the experiments of Paul (1978), cells of the human carcinoma-derived line HeLa-S3 grown in suspension culture demonstrated serum dependence: proliferation of these cells stopped when the stimulating serum factors became depleted. Suspended cells in depleted medium were shown to be arrested in the post-mitotic phase of the cycle. However, when these resting cells obtained from suspension cultures were transferred into the standard tissue culture dishes in their depleted medium, they attached to the substratum and initiated DNA synthesis about 20 h later. Thus, independent experiments with two different transformed lines, LSF and HeLa, have shown that either serum or substratum is essential for initiation of DNA synthesis. In other words, sensitivity to serum depends on the presence of the substratum and vice versa. This situation is qualitatively similar to that observed in the experiments of Clarke *et al.* (1970) with BHK cells (see p. 215). However, the reactions in LSF cells in our experiments and of BHK cells in the experiments of Clarke *et al.* seem to be quantitatively different. 10% Serum was sufficient to induce the proliferation of most LSF cells not only on the substratum but also in suspension but BHK cells were much less reactive.

One may suggest that transformed cells are characterized by a shift of the co-ordinated reactivity norm to various growth-controlling factors: the sum of the stimulating environmental factors sufficient for the initiation of proliferation is smaller for transformed cells than for normal ones. Is such a shift of co-ordinated reactivity a characteristic of all transformed cells? On the face of it, the answer is no because we know that certain transformed lines seem to lose their sensitivity to one growth controlling factor but retain their sensitivity to other factors: certain cells grow in a lower serum medium but retain anchorage dependence and density dependence; other lines grow in methylcellulose but not in a low serum medium, etc. (see chapter 2). Thus, it is possible that in the course of transformation one or several independent losses of sensitivity to various controlling factors may take place. An alternative suggestion is that transformed cells may always have altered sensitivities to all parameters, but normally not all the alterations are revealed by presently used tests. For instance, when we label a particular transformed cell line as 'anchorage dependent' but 'serum independent', this means that it will not form colonies in suspension but will form colonies on the substratum in a medium containing 1–2% serum. But perhaps its anchorage sensitivity is also somewhat altered; for instance, if we had tested the growth of these cells on a series of substrata of varying adhesiveness (see the work of Folkman & Moscona described in chapter 8), this cell line might have started growing on a less adhesive substratum than normal cells. Or, if we had cultivated suspended cells with increased concentrations of serum or growth factors, these cells might have started to form colonies with a lower threshold of stimulation than normal cells. Possibly, each transformation changes the sensitivity of cell growth to all environmental parameters, although the degree of altered sensitivity to each parameter may be different in various lines. To find out whether transformations induce several independent changes of sensitivity to various parameters or induce shifts of co-ordinated reaction norms to all parameters, we will have to design a more extended and more exact system of tests assessing cell reactions to different environmental factors and to apply this system of tests to many transformed lines. The experiments described in this chapter show that 'growth-related' transformed phenotypic characters may undergo apparent reversion when environmental conditions are changed. For instance, anchorage dependence of transformed cells may be restored in serum-free or serum-depleted media. On the other hand, genetically normal cells may acquire certain traits characteristic of the transformed phenotype. For instance, anchorage-dependent cells may become anchorage-independent when stimulated by large concentrations of serum or certain growth factors.

12

Mechanisms of action of proliferation-controlling factors

Kinetics of growth stimulation
Lag phase of growth stimulation

We have described the growth-controlling effects of various environmental factors. Now we will discuss the general characteristics of the effects produced by these factors. The usual method of studying these effects is to take a normal resting culture (e.g. a sparse culture in a low serum medium, or a dense culture), to add the stimulating agent (e.g. fresh serum or purified growth factors) to its medium and to examine the alterations preceeding DNA synthesis and mitosis. These transitions of cultures from the resting into the growing state are, despite many variations, characterized by several common kinetic features.

(1) The cells of normal or minimally transformed resting cultures are usually arrested in the phase of the mitotic cycle that is located between mitosis and DNA synthesis. Accordingly, stimulated cells first enter the S phase and only later, mitosis. Certain exceptional populations, with a large proportion of cells blocked in the G_2 phase, have been described *in vivo* and *in vitro* (see review of Gelfant, 1977). When stimulated, these cell first enter mitosis and only later enter the S phase. However, these situations are rare and later in this chapter we will discuss only the growth stimulation processes observed in the large majority of typical resting populations residing in the interval between mitosis and S phase. All the stimulating agents in the culture change the mean duration of cell residence in this interval: none of the stimulating agents has been observed to change the duration of the S phase.

(2) When a stimulating agent begins to act on resting cells, the rate of cell entry into the S phase, measured by [³H]thymidine incorporation, does not start to increase immediately but only after a certain lag-period. Several exceptions to this rule have been summarized in the review of Gelfant (1977). In these exceptional cases, stimulated cells *in vivo* were observed to enter the S phase without an appreciable lag (1–2 h after stimulation). Such exceptions have not been observed in cultures of tissue cells and will not be discussed here. In most cultures the minimal duration of the lag phase is within the range of 6–20 h. This duration may depend both on the cell type and probably on

219

the type of stimulation (see review in Gelfant, 1977). A relatively short latent period (4 h) was observed in stimulated chick fibroblasts (Rubin & Steiner, 1975). The minimal part of the lag period is followed by a varying part. The duration of this second part may vary for individual cells from 0–24 h or more. The minimal length of the latent period is not dependent on the concentration of the stimulating agent (see e.g. Jimenez de Asua *et al.*, 1977). An alteration of this concentration may, however, change the rate of cell entry into the S phase during the second, variable part of the lag-period. As a result, the total fraction of cells entering the S phase or mitosis during the fixed time interval may be changed. For instance, within certain concentration limits, the more serum is given to the resting cultures, the more cells will enter the S phase (see for example, Ellem & Gierthy, 1977). The length of previous residence in the resting state may affect the mean duration of the latent period (Augenlicht & Baserga, 1974; Rossini & Baserga, 1976).

(3) To activate DNA synthesis the stimulating agent should act upon the cells during the whole latent period or, at least, during a large part of it. For instance, in the experiments of Brooks (1976), the lag period preceeding the activation of DNA synthesis by quiescent cultures of 3T3 cells exposed to fresh serum was 14 h. However, the reverse transfer of cells from a medium with fresh serum into a low serum medium led to a decrease in the rate of DNA synthesis after only 5 h. In a similar way, exponentially growing 3T3 cells with a median G_1 duration of 5.4 h were prevented from entering the S phase by reducing the amount of serum to such a level that their median residence in G_1 was less than 3.2 h (Yen & Pardee, 1978*a*). In cultures of human glial cells DNA synthesis stimulated by fresh serum or EGF after a lag period of about 17 h; the minimal time of exposure to these growth factors needed to produce activation was about 8–12 h (Lindgren & Westermark, 1976, 1977*a*). The entry of chick fibroblasts into the S phase was reduced almost immediately after the removal of serum and the lowering of pH (Rubin & Steiner, 1975). In cultures of chick fibroblasts stimulated by MSA, the transfer of cells into stimulator-free medium during any part of the latent period prevented activation of DNA synthesis (Bolen & Smith, 1977). Thus, in this system, the activation of DNA synthesis takes place only when the stimulating agent has been present in the medium during the whole lag period. After removal of the simulator in the middle of the lag period, the cells gradually slip back into their original stage and if this stimulator is administered again after an interval of several hours, the length of the second lag period is not decreased compared with that of the usual lag (Lindgren & Westermark, 1977*a*).

(4) Certain environmental factors added during the lag period can modulate the final effect of an activating agent. For instance, in the experiments of Jimenez de Asua *et al.* (1977), prostaglandin $F_{2\alpha}$ (PGF$_{2\alpha}$) activated DNA

synthesis in quiescent 3T3 cells after a minimal lag period of 15 h. Addition of insulin to the medium at the same time as $PGF_{2\alpha}$ or 8 h later did not alter the minimal duration of the lag period but considerably increased the rate of cell entry into the S phase after a 15 h-lag. Insulin alone did not stimulate DNA synthesis. In other experiments, $PGF_{2\alpha}$ was added to the medium several hours after insulin in which cases the 15-h lag period began only from the moment of the addition of $PGF_{2\alpha}$. Thus, in this system, insulin added during the lag period acted as a modifying agent which increased the stimulating effect of $PGF_{2\alpha}$.

In summary, growth-stimulating agents do not immediately and irreversibly commit the exposed cells to DNA synthesis and division. The transition through the lag period is a lengthy and reversible process requiring external stimulators for its progress and completion.

Models of cell kinetics in resting and growing cultures

Differences between the kinetic characteristics of resting and growing cultures have been described above. To interpret these differences several theoretical models of cell kinetics have been proposed.

(1) *A model with varying* G_1 (Prescott, 1968). The cells in growing culture move through G_1, S, G_2 and M phases of the mitotic cycle. The cells in resting cultures continue to pass through these stages but the rate of their passage through the first part of the G_1 phase is decreased while the duration of the second part of G_1 and of other phases remains unaltered. Stimulating agents shorten the duration of the first part of G_1.

(2) *A model with* G_0 *phase* (Lajtha, 1963; review in Epifanova & Terskikh, 1969). The cells in growing culture move through all the phases of the mitotic cycle. In contrast, most cells of resting cultures do not move through the cycle at all; they enter a special phase designated G_0. Stimulating agents activate cell reversion from G_0 to the G_1 phase of the cycle.

(3) *A transition probability model* (Burns & Tannock, 1970; Smith & Martin, 1973). After mitosis each cell enters state A: the cells residing in this state do not progress to mitosis. Each cell in the A phase has a certain probability to make a transition into the B phase which includes part of G_1, S, G_2 and M phases. Both in resting and growing cultures, the probability of a cell's transition from the A into the B phase (transition probability) does not depend on the time of previous residence in the A phase. The duration of passage through the B phase is similar in resting and growing cultures, but the transition probability is much lower in resting cultures. Stimulating agents increase this probability.

All these models have one common feature in that the inter-mitotic interval is divided into two parts: a variable regulated part and a constant, non-regulated part. The cell residence time in the first part of the interval may vary widely and is regulated by external stimulators; cell residence time in the second part of the interval is relatively constant and is not regulated by external stimulators. The difference between the models is in the interpretation of the action of regulating factors on the variable phase of the inter-mitotic interval. The regulated parameter is the rate of cell progress through the variable phase (model 1) or the probability of cell exit from the variable phase (model 3) or both the probabilities of the entry into that phase and of the exit from it (model 2). Of course, it is possible that regulating factors change not one parameter but all three parameters (entry, rate of progress and exit), so that these models are not mutually exclusive. None of the three models takes into account the special features of the latent period preceeding entry into the S phase after activation of resting cultures: these are its relatively constant minimal duration and its reversibility. As we have seen in the last section, stimulating factors are needed not only for cell entry into that period but for cell transition through it. If the cell is deprived of stimulating factors in the middle of the latent period its progress through that period not only stops but eventually the cell returns to the point preceeding the beginning of the latent period. Thus the latent period is a regulated part of the cycle. We do not know at present whether environmental stimuli are needed continuously throughout the whole latent period. More probably detailed analysis of the latent period would reveal that it includes several 'regulated' and 'non-regulated' sub-periods. It is also possible that complexes of essential regulating factors are not identical in different parts of the latent period. On the basis of these considerations we think that regulation of cell kinetics may be more adequately described by a model dividing inter-mitotic intervals into three parts: two regulated parts and one non-regulated part. This model is shown in fig. 12.1. It contains the following three stages of the life cycle:

(1) *The Resting stage* (synonyms: G_0, A-state). This is the first part of the M–S phase interval. The time of cell residence in this stage may vary from near zero to an indefinite period. Regulating factors control the duration of this phase. What particular parameters (rate of exit, rate or progress, etc.) are controlled, remains unknown.

(2) *The Activation stage* (lag period of activation, pre-replicative phase). This is the second part of the M–S phase interval. Its duration can vary considerably but is never lower than a certain minimum. Regulating factors control the cells' passage through the whole of this stage or at least through many parts of it: absence of these factors may cause the return of the cells to the resting stage.

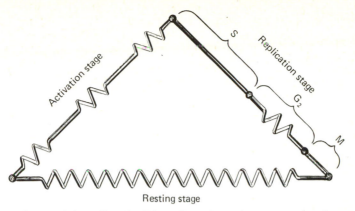

Fig. 12.1. Scheme of the cell cycle. Direct lines show the non-regulated parts of the cycle; zig-zag lines, the regulated parts. As discussed in the text, the exact distribution of regulated and non-regulated parts of the activation stage remains unknown.

(3) *The Replication stage* including S, G_2 and M phases. Usually the duration of this stage is not regulated by external factors but one cannot exclude that short regulated periods are contained in G_2 (Gelfant, 1977).

Most cells in growing cultures are at the activation or replication stages. Most cells in resting cultures are at the resting stage, although one cannot exclude that some of these cells may occasionally enter the early parts of the activation stage and then slip back into the resting stage. Is the resting stage an essential part of the life cycle or is it entered by the cell only in special conditions? In other words, can duration of this phase reach zero in growing cultures? This question remains open at present. Possibly, as discussed by Baserga (1978), it can be solved with the aid of cell mutants that have temperature-sensitive defects of the exit from the resting stage; at present perfect mutants of this type have not yet been found.

The frequency distribution of inter-mitotic times of cells within populations of each cell type is also unknown. Considerable data on these distributions in cultures of several permanent mouse lines (3T3, 3T6 and SV 3T3) have been collected by Shields & Smith (1977). In their experiments, time-lapse films of cultures were used to measure the inter-mitotic times of individual cells; that is, the time interval between mitosis giving birth to a cell and another mitosis finishing the life of that cell. As could be expected, the inter-mitotic times of various cells within the same culture had a certain minimal duration. The experimentally observed frequency distributions of the fractions of inter-mitotic times exceeding this minimum could be closely fitted by exponental functions. The differences between the inter-mitotic times of sister cells born in the same mitosis were also distributed nearly exponentially. As suggested by Shields & Smith, exponential distributions can be easily explained by postulating that in the course of cell progress towards mitosis some transition from one stage into another occurs at random and its probability per unit time is constant. This is obviously the simplest explanation of the distributions obtained. However, other theoretical distributions may have nearly exponential asymptotic parts so that, possibly, these experimental data will not contradict some other statistical hypothesis. This question needs further analysis. In contrast to the cultures studied by Shields & Smith, an analysis of inter-mitotic times in cultures of WI-38 human fibroblasts has revealed a pattern obviously deviating from an exponential distribution (Mets & Verdonk, 1978). It would be important to compare the experimental distributions of inter-mitotic times in many types of cultures before making any generalizations. In any case, as discussed above, mechanisms of the regulation of proliferation, probably cannot be reduced to alterations of the probability of some single-step transition.

Early cellular changes accompanying growth activation

Transition of cultures from the resting to the growing state induced by various stimulators is accompanied by a co-ordinated set of cellular reactions designated surface-induced responses (Vasiliev & Gelfand, 1968); co-ordinate growth responses (Rubin, 1975b, 1976) or pleiotropic reactions (Herschko *et al.*, 1971). Early changes developing during the first hours of the activation stage are of special interest with regard to the mechanisms of activation.

Table 12.1 gives a number of examples of experimental results revealing these early changes of metabolic parameters accompanying growth stimulation; reverse alterations of the same parameters accompanying density inhibition of growth are also included. As seen from table 12.1, the complex of early alterations includes: (*a*) simulated uptake of certain ions (K^+, PO_4^{3-}, possibly Ca^{2+}); (*b*) increased uptake of certain nutrients (glucose, some amino acids, uridine); (*c*) altered intracellular concentrations of cyclic nucleotides; (*d*) activation of glycolysis probably associated with an activation of its key enzymes, such as phosphofructokinase; (*e*) increased turnover of certain membrane phospholipids is possibly also a characteristic early change.

All these alterations occur within the first hour of activation. The dynamics of the later alterations of these parameters have not yet been studied in detail. Some of the changes, especially alterations of cyclic nucleotides, have been observed to return to normal levels in spite of continued stimulation. The increased level of the uptake of nutrients seems to persist for many hours.

Similar early changes have been observed in various cell systems. Certain exceptional results are also included in table 12.1; for instance, the absence of stimulation of glucose uptake by glucocorticoids and the small decrease of cyclic AMP in cells stimulated by pure FGF. These exceptions give reason to suggest that in some cases growth stimulation can occur in spite of the absence of certain symptoms of the 'syndrome of early changes'.

The increased uptake of nutrients characteristic of this syndrome may be due either to their increased transport through the membrane or to an increased utilization of nutrients in metabolic reactions. Analysis of the kinetics of the serum-stimulated uptake of uridine has led to the conclusion that it may be caused by increased phosphorylation rather than by increased transport (Rozengurt, Stein & Wigglesworth, 1977; Koren *et al.*, 1978; Goldenberg & Stein, 1978). Increased glucose uptake also may be associated with increased phosphorylation (Colby & Romano, 1975; Romano, 1976). However, serum has been observed to increase the uptake of the non-metabolizable glucose analogue, 3-*O*-methylglucose (Weber, 1973; Lang & Weber, 1978). Thus, stimulators probably induce not only the activation of phosphorylation but also change in sugar transport. Many other manifestations of early changes are possibly also due to changes of plasma membrane components such as the altered uptake or ions, altered concentrations of cyclic AMP and cyclic GMP, and altered turnover of phospholipids. Certain early changes are found in the nuclei and may be manifestations of the first steps of a re-programming of cellular synthesis.

The list of the early changes is probably still far from its final form. In particular, it would be important to find out which alterations of cortical structures and of cell morphology may be characteristic of this syndrome. In chapter 4 we mentioned certain effects of growth factors on cell morphology. Among those described was the formation of numerous microvilli on the surface of insulin-stimulated 3T3 cells; microvilli were formed during the first hour of stimulation (Evans *et al.*, 1974). As described in chapter 10, the migration factor of Bürk stimulates not only cell proliferation but also migration into wounds. Increased ruffling induced by EGF and morphological alterations caused by FGF were also observed. These fragmentary data give an impression that stimulation may be accompanied by increased pseudopodial activity and possibly by decreased spreading. It would be very important to study systematically the early effects of growth stimulators on morphogenetic reactions.

Early changes induced by stimulators are followed by a sequence of intermediate change. These changes become evident only after several hours of stimulation. They include: activation of protein synthesis, increased synthesis of ribosomal RNA and of other types of RNA (Johnson *et al.*, 1974, 1975, 1976; Abelson *et al.*, 1974; Mostafapour & Green, 1975); increased formation of polysomes (Levine *et al.*, 1965; Levine, Jeng & Chang, 1974); decreased rate of protein degradation (Castor, 1975; Hendil, 1977); alterations of the state of chromatin and of the properties of isolated DNA (Baserga, 1976; Collins, 1977). One particularly characteristic change is an increase in the activity of ornithine decarboxylase, the rate limiting enzyme in the synthesis of polyamines (Haselbacher & Humbel, 1976; D'Amore & Shepro, 1978). Most of these changes are probably prerequisities for cell growth and for DNA synthesis (see reviews in Baserga, 1976 and in Epifanova, 1977). It

Table 12.1. *Alterations accompanying transitions of cultures into the growing state*

Parameter	Cells	Treatment	Effect	Sources
K$^+$ uptake (measured by the uptake of ^{86}Rb$^+$)	3T3	Serum stimulation	Increase within minutes	Rozengurt & Heppel, 1975; Tupper, Zorgniotti & Mills, 1977
	3T3	Stimulation by prostaglandins E$_1$ and E$_2$ or by phorbol ester (TPA)	Increase within 15 min	Moroney et al., 1978
K$^+$ content within the cell	Chick fibroblasts	Stimulation by insulin	14% Increase after 16 h	Sanui & Rubin, 1978
	3T3	Serum stimulation	75% Increase	Tupper et al., 1977
Electric trans-membrane potential	3T3 and CHO	Density inhibited compared with growing sparse cultures	Increase	Cone & Tongier, 1973
Li$^+$ uptake	3T3	Serum stimulation	Two- to three-fold increase within minutes	Smith & Rozengurt, 1978
Intracellular Na^{2+} content	CHO	Medium change	Increase after 15 min	Terskikh & Malenkov, 1973
Ca^{2+} uptake into intracellular compartment	3T3	Medium change	Increase within 1 h	Tupper et al., 1978
Ca^{2+} content	Chick fibroblasts	Stimulation by insulin	No change after 16 h	Sanui & Rubin, 1978
Mg^{2+} content	Chick fibroblasts	Stimulation by insulin	22% increase after 16 h	Sanui & Rubin, 1978

	Cell type	Stimulation	Increase within 15 min	References
P_i content	3T3	Density-inhibited versus growing cultures	Five-fold decrease	Cunningham & Pardee, 1969; Jimenez de Asua, Rozengurt & Dulbecco 1974; Jullien & Harel, 1976 Weber & Edlin, 1971
	3T3	Stimulation by prostaglandins or phorbol ester	Increase	Moroney et al., 1978
	3T3	Stimulation by glucocorticoids	No change	Greenberg et al., 1976
	Chick fibroblasts	Stimulation by trypsin or insulin	No change	Greenberg et al., 1976
P_i content	3T3	Density inhibited cultures compared with growing ones	Five-fold decrease	Gray et al., 1976
	Chick fibroblasts	Density inhibited cultures compared with growing ones	Decrease	Sefton & Rubin, 1971; Weber, 1973; Kletzien & Perdue, 1974a, b
Glucose uptake (measured by the uptake of analogues, 2-deoxy-D-glucose or 3-O-methyl glucose)	Human glioma	Density inhibited cultures compared with growing ones	Decrease	Edström, Kanje & Walum, 1976
	3T3	Density inhibited cultures compared with growing ones	Decrease	Bose & Zlotnick, 1973;
	Revertant transformed 3T3	Density inhibited cultures compared with growing ones	Decrease	Schultz & Culp, 1973; Dubrow, Pardee & Pollack, 1978
	Rat liver cells	Density inhibited cultures compared with growing ones	Decrease	Siddiqi & Jype, 1975

Table 12.1. (*cont.*)

Parameter	Cells	Treatment	Effect	Sources
	Chick fibroblasts	Inhibition of growth in Mg^{2+}-deprived medium	Decrease	Rubin, 1976; Bowen-Pope & Rubin, 1977
	Chick fibroblasts	Stimulation by serum, insulin or proteases	Increase within minutes	Sefton & Rubin, 1971; Vaheri, Ruoslahti & Nordling, 1972; Kletzien & Perdue, 1974*b*; Hale & Weber, 1975; Rubin, 1977*a*
	Rat NRK line	Stimulation by serum, insulin neuraminidase or proteases	Increase within minutes	Gregory & Bose, 1977
	Rat NRK line	Stimulation by elevated pH or excess Zn	Increase	Rubin & Koide, 1975
	3T3	Stimulation by serum	Increase	Bradley & Culp, 1974
	3T3 cells	Stimulation by glucocorticoids	Decrease	Thrash & Cunningham, 1974
	Serum revertants from 3T3 transformed by SV-40 virus	Stimulation by serum	Increase	Dubrow *et al.*, 1978
Uptake of amino-acid analogue, α-aminoisobutyric acid	3T3	Stimulation by FGF of serum-deprived cultures	Increase after 1 h of stimulation	Quinlan & Hochstadt, 1977
	Serum revertant lines ob-	Stimulation by serum	Increase	Dubrow *et al.*, 1978

	formed by SV-40 virus			
Uptake of amino-acid analogue, cycloleucine	Chick heart cells	Serum stimulation	Increase	Frelin & Padieu, 1976
	WI-38 human diploid fibroblasts	Stimulation by serum	Increase after 3 h	Costlow & Baserga, 1973
Uptake of a mixture of ^{3}H-labelled amino acids	3T3	Stimulation of serum-deprived cultures by FGF and hydrocortisone or by serum	Increase without an appreciable lag period	Rudland *et al.*, 1974*b*; Jimenez de Asua *et al.*, 1974
Intracellular concentration of cyclic AMP	Mouse mammary epithelial lines	Serum stimulation	50% decrease within 5–10 min	Howard, Scott & Manter, 1977
	Human fibroblast	Serum stimulation	Decrease to 40% of control within 2–5 min	Rechler *et al.*, 1977*a*
	Human fibroblasts	Serum stimulation in the presence of dibutyryl-cAMP	No change of cAMP (activation of DNA synthesis not inhibited)	Rechler *et al.*, 1977*a*
	Human fibroblasts	Serum stimulation	Decrease	Froelich & Rachmeler, 1972; Ahn *et al.*, 1978
	3T3	Serum stimulation	Decrease	Heidrick & Ryan, 1971; Seifert & Rudland 1974*b*

Table 12.1. (*cont.*)

Parameter	Cells	Treatment	Effect	Sources
	3T3	Serum and insulin stimulation after incubation in Ca-deprived medium	No change	Boynton *et al.*, 1978
	3T3	FGF	No change or small change	Gospodarowicz *et al.*, 1975
	BHK	Serum stimulation	Three-fold decrease during the first hour	Pledger *et al.*, 1979
Activity of cyclic nucleotide phos-phodiesterase	BHK	Serum stimulation	Increase	Pledger *et al.*, 1979
Uptake of uridine	3T3	Stimulation by FGF or serum of serum-deprived cultures	Increase after 15 min lag	Rudland, Seifert & Gospodarowicz, 1974; Jimenez de Asua, *et al.*, 1974; Quinlan & Hochstadt, 1977
	3T3	Serum stimulation	Nine- to 11-fold rise within minutes	Seifert & Rudland, 1974*a, b*; Yasuda *et al.*, 1978
Intracellular concentration of cyclic GMP	3T3	FGF	Ten- to 15-fold rise within 20 min	Gospodarowicz *et al.*, 1975
	Mouse mammary epithelium	Serum stimulation	Four- to five-fold rise within minutes	Howard *et al.*, 1977

Property	Cell type	Stimulation	Effect	Reference
The rate of turn-over of phosphatidyl-inositol	Human fibroblasts	Serum stimulation	Five-fold increase within 2 min	Ahn et al., 1978
	3T3	Serum stimulation	Fall within 4–6 min	Boynton et al., 1978
	3T3	Serum stimulation	24-fold increase within 1 h; turnover of other phospholipids was not increased	Ciechanover & Herschko, 1976
Rate of glycolysis	Chick fibroblasts	Stimulation by serum or increased pH	Increase	Fodge & Rubin, 1975
	3T3	Stimulation by serum or EGF	Increase (EGF stimulates glycolysis by 12-fold after 2 h)	Diamond et al., 1978
Activity of phospho-fructokinase	Rat NRK	High density	Decrease	Gregory & Bose, 1977
	3T3	Serum stimulation	Increase	Schneider, Diamond & Rozengurt 1978
Labelling of certain non-histone chromosomal proteins	Human fibroblasts	Serum stimulation	Selective increase after 10 min of stimulation	Rovera & Baserga, 1971; Augenlicht & Lipkin, 1977
Binding of acridine orange to nuclear chromatin	3T3	Serum	Increase after 30 min	Smets, 1973

is worth mentioning that growth stimulation may be accompanied by activation of synthesis not only of internal cell proteins but also of specific cell products secreted into the medium, such as collagen (Kamine & Rubin, 1977) and hyaluronic acid (Moscatelli & Rubin, 1975, 1977; Tomida, Koyama & Ono, 1975; Lembach, 1976*a*; Vannuchi & Chiarugi, 1976).

Mechanisms of growth activation
Occupancy of receptors as a factor inducing growth

A complex of stimulating environmental factors is needed to induce the transition of a population into the growing state. This complex includes the substratum, growth factors, nutrients, ions, and low density. The reactivity of the cell to each factor seems to depend on the presence of other factors (see chapter 11). What is the nature of the effect of growth stimulating factors on the cell? At least two groups of growth-inducing factors, the substratum and growth factors, probably act as ligands attaching to receptors of the cell membrane. It is obvious that attachment to the substratum is somehow involved in anchorage control of growth, although we do not know the exact nature of the membrane receptors involved in this attachment. Each group of macromolecular growth factors (EGF, FGF, insulin-like factor, etc.) is bound to specific surface receptors present at the cell membrane. Competition experiments show that different groups of growth factors (EGF, FGF, insulin-like factors) interact with different receptors (Hollenberg & Cuatrecasas, 1973; Carpenter *et al.*, 1975; Raizada & Perdue, 1976; Todaro, De Larco & Cohen, 1976; Todaro *et al.*, 1977; Das *et al.*, 1977; Rechler *et al.*, 1977*b*; De Larco & Todaro, 1978*b*). Probably, reaction with these external receptors is essential for growth stimulation although some growth factors may also penetrate into the cell and interact with internal structures; for instance, insulin was described to be bound by nuclear receptors (Goldfine & Smith, 1976).

We do not know how many different types of receptors present in the membrane can induce growth stimulation by interaction with corresponding ligands. It is possible that there are no specialized 'growth stimulation receptors' but occupancy of almost every type of receptor present in the membrane by its corresponding ligands may activate growth. Growth stimulation of lymphocytes may be produced by many different types of ligand–receptor interactions. In particular, many lectins and antibodies against different surface components may act as growth-inducing factors in this system (see later). Certain lectins seem to be growth-stimulatory also for tissue cells (see above), but systematic studies of the effects of antibodies and lectins on the growth of the cells remain to be done.

What fraction of receptors of one type has to be occupied by a ligand for the induction of growth stimulation? Raizada & Perdue (1976) have found that half-maximal stimulation of DNA synthesis is produced in serum-starved chick fibroblasts when insulin occupies 42% of its surface receptor sites. The total number of insulin receptors per cell in these conditions is estimated to be 6×10^4. Each 3T3 cell is able to bind approximately 94000 molecules of ^{125}I-labelled EGF at 37 °C and the apparent dissociation constant of the binding reaction is 1.1 nM (Das *et al.*, 1977). A similar concentration of EGF is needed to stimulate DNA synthesis maximally in these cells (Rose, Pruss & Herschman, 1975). Half-maximal stimulation is observed at 0.1 nM EGF, a concentration at which 10% of the receptors are occupied (Das & Fox, 1978). These results indicate that growth stimulation requires an occupancy of a large fraction of receptors. Many more studies are obviously needed to elucidate the quantitative characteristics of various receptor–ligand interactions involved in growth stimulation. It would be also important to find out the valencies of receptors and ligands involved in growth stimulation and whether cross-linking and redistribution of receptors are essential for this stimulation.

The binding of growth factors to corresponding receptors may be accompanied by a redistribution of these receptors. In particular, the incubation of 3T3 cells with EGF leads to a substantial decrease in the number of EGF receptors present at the cell surface (Carpenter & Cohen, 1976a; Das *et al.*, 1977). Analysis of this phenomenon suggests that it is due to ligand-induced aggregation and endocytosis of EGF receptors (Das & Fox, 1978; Haigler *et al.*, 1978; Schlessinger *et al.*, 1978). It remains to be established whether this redistribution is essential for growth stimulation. This question is closely related to another unresolved question: that of the inter-relationships between pseudopodial reactions and growth activation reactions. At least some external factors, such as the substratum surface, can induce both types of reactions. As we have seen, growth factors, probably also induce pseudopodial reactions leading to internalization of the corresponding receptors. Possibly, both groups of reactions are initiated by similar surface changes and have some common steps.

Besides factors acting as ligands, there may be another group of growth stimulating factors: agents that alter the membrane not by binding to specific receptors but by destroying or modifying some surface structures. This group might include proteases and phorbol esters.

The primary targets of growth regulatory effects of ions and nutrients remain a mystery. Possibly, these components interact with some surface receptors or alter receptor binding of other growth factors. Alternatively, ions or nutrients may not act at the first step of a set of growth activation reactions; certain concentrations of these components inside the cells may be essential for the development of intracellular changes induced by other receptor-binding growth factors.

It would obviously be useless to discuss the possible targets of action of the factors responsible for density dependent inhibition until we learn more about the nature of these factors. A decrease in the numbers of surface receptors for growth factors in dense cultures (Holley *et al.*, 1977) may possibly contribute to density-dependent regulation of proliferation.

The problem of the 'chief executives' of pre-replicative changes

As we have seen above, the interactions of external molecules with surface receptors can initiate a complex set of intramembraneous and intracellular changes eventually leading to the initiation of DNA synthesis. We do not know how an alteration of one membrane component is transformed into multiple early changes of intramembraneous and intracellular cell components. It is natural to suggest that an initial growth-stimulating event leads to the alteration of one particular cellular parameter which, in its turn, is responsible for induction of many other alterations. At different times, various alterations have been nominated as candidates for this role of 'chief executive' of pre-replicative changes'.

(1) *Calcium.* The leading role of calcium in the activation of proliferation and of other intracellular processes was suggested along ago (Mazia, 1937; Heilbrunn, 1956). Calcium deprivation may prevent cell activation (see chapter 10). Calcium uptake seems to increase at the early stage of the pre-replicative phase (table 12.1). There are, however, some data indicating that pre-replicative alterations can be induced by fresh serum in calcium-deprived medium although DNA synthesis is not initiated in this medium (Boynton *et al.*, 1978).

(2) *Magnesium.* Rubin (1975*b*) (see also Sanui & Rubin, 1977, 1978) proposed that free intracellular magnesium is the central agent responsible for the co-ordinated control of metabolism and growth in animal cells. The heuristic value of this hypothesis has been confirmed by the accumulation of experimental facts on the role of magnesium content in the external medium for the activation of proliferation (see chapter 10) and on the changes of intracellular magnesium in the pre-replicative phase (table 12.1).

(3) *Changes of transmembrane electric potential associated with alteration of potassium and/or sodium concentrations inside the cell.* The role of these changes has been advocated by Cone & Tongier (1973). Increased potassium uptake is one of the most conspicuous early changes in the pre-replicative phase (table 12.1). Little is known about sodium uptake and about the possible role of sodium and potassium in the alterations of other cellular parameters. This role may be significant, as sodium–potassium gradients may be involved in the regulation of many intracellular parameters; for instance, in the

regulation of sugar and amine acids transport (Parnes, Carvey & Isselbacher, 1976; Lever, 1976; Bader, 1976; Bader, Sege & Brown, 1978).

(4) *Cyclic nucleotides.* These, especially cyclic AMP, are now among the veteran candidates (see discussion in Ryan & Heidrick, 1974; Whitfield *et al.*, 1976; Pastan, Johnson & Anderson, 1974; Aboe-Sabe, 1976; Puck, 1977). A decrease in cyclic AMP is observed at the early stage of stimulation although certain exceptional cases have also been described in which stimulation of DNA synthesis was not preceeded by an early fall in cyclic AMP (see references to Rechler and to Boynton in table 12.1). The addition of dibutyryl cyclic AMP (Bt$_2$cAMP) to the medium has been observed to inhibit the growth of many types of culture (see review in Pastan *et al.*, 1974; Aboe-Sabe, 1976), but the mechanism of this inhibition is not clear. Possibly, this agent does not block the early stages of growth initiation but acts at some later step of pre-replicative events (Rechler *et al.*, 1977a). Possible non-specific toxic effects of Bt$_2$cAMP and of butyrate present in these preparations should also be taken into account. In experiments with hepatocytes cyclic AMP and Bt$_2$cAMP inhibited rather than stimulated cell proliferation (Armato, Draghi & Andreis, 1977). Both the rise and fall of cyclic GMP were observed by different authors at the early stage of growth stimulation (see table 12.1), so that the nature of changes of this nucleotide and their role remain obscure.

(5) *Nutrients.* It was suggested that cell proliferation may be regulated through an alteration in the transport rates of nutrients (Holley, 1972). A deficiency of nutrients may lead to a proliferation block (see chapter 10) and the uptake of nutrients is increased at an early stage after stimulation (see table 12.1). However, intracellular concentrations of glucose were not observed to decrease in quiescent cells (Naiditch & Cunningham, 1977) as would have been expected if changes of these concentrations were primary regulators of proliferation.

(6) *Changes in microfilaments and/or microtubules.* It is possible that the alterations of fibrillar structures induced by ligand–receptor interactions somehow transmit the proliferation-initiating signal into the cell. Suggestions of this type were made by Edelman (1976; see also McClain *et al.*, 1977) and by Puck (1977). At present we know little about the concrete alterations of fibrillary systems at the early stages of growth stimulation. This is obviously one of the most promising directions for further studies.

Thus, at present each of the alterations listed above deserves to remain on the list of supposedly important early changes but there is no evidence that would permit to single out one crucial event responsible for all other alterations. Possibly, this single crucial event does not exist at all. Due to the mobility of membrane components, alteration of one type of receptor may induce alterations of many other membrane structures (see Rozengurt, 1976; Jacobs & Cuatrecasas, 1976). As a result, several intercellular parameters may be changed simultaneously. In other words, the set of activation changes is probably not produced by a single 'chief' alteration but by a complex of alterations arising simultaneously or almost simultaneously. Alterations occurring during a large part of the latent period are reversible: they disappear when the cell is deprived of external stimulator. This suggests that early membrane changes do not only trigger the first step of the pre-replicative process but are essential for the maintenance and progress of these processes. The whole set of pre-replicative processes up to the initiation of DNA synthesis does not progress spontaneously but requires continuous 'confirming' signals from the cell environment. Possibly, the nature of essential signals may be different at various stages of the pre-replicative phase. Recently, a number of mutant cell lines with temperature-sensitive blocks in the G_1 phase have been obtained (see reviews in Basilico, 1978; Pringle, 1978; Levine, 1978; Stanners, 1978). The use of these mutants is obviously of paramount importance for mapping the intracellular events responsible for cell progression from the resting state to the S phase.

Other processes related to growth activation of cultured tissue cells

It may be useful to compare the growth activation in fibroblastic cultures with blast transformation of lymphocytes. Both processes lead to the transition of a population from the resting to the growing state. Blood lymphocytes remain in a quiescent state unless they are treated with certain agents called mitogens. Mitogens activate a set of pre-replication changes including many macromolecular syntheses. These changes are followed by DNA synthesis and mitosis. In contrast to the growth activation of fibroblasts, transformation of lymphocytes does not require their attachment to a solid surface: mitogen-stimulated lymphocytes may proliferate in the suspended state. Mitogens include many different proteins which are able to attach themselves to different types of receptors present on the surface of lymphocytes; in particular, lectins and antibodies against various component of the membrane. When an organism is immunized by a certain antigen, this antigen becomes mitogenic for a sub-population of immune lymphocytes which carry the corresponding receptors on their surfaces. A mitogen binding to a cell surface often induces patching and capping of corresponding receptors; multivalency of ligands essential for the induction of patching and capping seem also to be essential for the mitogenic action of these molecules (Wands, Podolsky & Isselbacher, 1976; Ravid, Novogrodsky & Wilchek, 1978; Prujansky, Ravid & Sharon, 1978). Besides ligands reacting with certain receptors a number of agents that possibly cause chemical modifications of some surface molecules are also mitogens. These are proteases, including thrombin and trypsin (Chen, Callimore & McDougall, 1976), lactoperoxidase and periodate. The last two agents act presumably by oxidizing carbohydrate groups of surface glycoproteins or glycolipids (Ravid & Novogrodsky, 1976). Ionophores increasing permeability of cell membrane for calcium are also mitogenic for lymphocytes (Luckasen, White & Kersey, 1974; Maino, Green & Crumpton, 1974). The stimulating action of some mitogens on a population of lymphocytes is not always direct: mitogen-treated cells were reported to acquire the ability to stimulate growth of other lymphocytes that had no direct contact with mitogen (Beyer & Bowers, 1975). Mitogens acting on lymphocytes induce a number of early membrane changes including the activation of uptake Na^+, K^+ (Averdunk & Lauf, 1975; Negendank & Collier, 1976) and Ca^{2+} (Whitfield, MacManus & Gillan, 1973; Freedman, Raff & Gomperts, 1975; Ozato, Huang & Ebert, 1977; see Hesketh *et al.*, 1977, for conflicting evidence), increased uptake of glucose (Hume & Weidemann, 1978) and of amino acids (van den Berg & Betel, 1973) as well as alterations of cyclic nucleotides (Hadden *et al.*, 1976; Haddox *et al.*, 1976; Wang, Sheppard & Foker, 1978). Although these early changes are developed within minutes after the beginning of stimulation, the process of blast transformation may remain reversible for a long time. Initiation of DNA synthesis does not take place unless the cell was incubated with mitogen for 6 h or more (Ravid & Novogrodsky, 1976).

Thus, although the many concrete factors inducing growth stimulation in fibroblasts or blast transformation in lymphocytes are not identical, many common features of these processes are obvious. One may regard blast transformation of lymphocytes and growth stimulation of fibroblasts as two different variants of one class of 'surface-induced growth activation reactions' initiated by the action of external agents on the surface, accompanied by a complex of similar early membrane changes that eventually lead to DNA synthesis and mitosis. The diversity of agents inducing activation reactions indicates that they may be initiated by alterations of different components of the membrane. Possibly, the nature of alterations initiating these reactions is not identical in different cases; induction may be associated with cross-linking of certain membrane components, with conformational changes of certain receptors, or with chemical modification of membrane components. In other words, similar activation reactions may be initiated by various types of disturbancies in the normal structure of the membrane. In this connection it would be interesting to study in more detail the effects of 'non-specific' stimulating agents and to find out what other agents may have a similar action.

As we suggested some time ago (Vasiliev & Gelfand, 1968) growth stimulation of fibroblasts and blast transformation of lymphocytes can be regarded as special variations of a more general group of cellular reactions, of surface-induced restorative responses. We will present here an up-dated version of this hypothesis. In the course of their life, eukaryotic cells come into contact with many external agents which may interact with different receptors of their membranes. These interactions are potentially dangerous for a cell as they may alter its membrane structure and diminish the isolation of the cell interior from the medium. At the same time certain disturbances of the membrane structure are essential for normal cell life; these disturbances inevitably occur when the cell phagocytoses food, moves on the substratum, or otherwise reacts to external stimuli. Thus, the cell is confronted with two problems: it has to keep its membrane structure and internal functions intact and at the same time it has to retain the sensitivity of the membrane to many external stimuli. The problem posed by this dichotomy was possibly solved in the course of evolution by the development of cellular reactions designed to restore the normal state of the membrane and the whole cell after surface-induced disturbancies. Pseudopodial reactions, discussed in chapter 3, form one group of these restorative processes: they are aimed at the removal of locally altered groups of receptors from the membrane. Another group of restorative responses may be designated 'surface-induced metabolic activation'. This set of responses is directed towards the replacement of altered membrane components and of intracellular structures and consists of many co-ordinated metabolic changes including exteriorization of internal membrane structures, activation of many syntheses, etc. The set of surface-induced metabolic activation reactions possibly proceeds in several sequential steps: if surface-induced disturbancies of the cell structure have been erased after the first step of the reaction, then the course of the reaction stops. In contrast, if disturbances persist after the first step, then the second step begins and so on. The intensity of restorative changes may overshoot the intensity of cellular alterations induced by external agents. Protracted metabolic activation can eventually lead to cell growth and division.

The ability of cells to give surface-induced metabolic activation could be a basis for the evolutionary development of many types of surface-associated control reactions in multicellular organisms. Each cell type in multicellular organisms has its specific set of surface receptors reacting with certain physiologically important ligands such as hormones, mediators, the surface of the substratum, etc. These interactions may lead to membrane alterations that have a signalling function. They set in motion metabolic activation reactions. If an external ligand acts for a short time, only the first stages of restoration reactions (secretion, activation of specific metabolic processes) will be observed. If, however, receptor–ligand interaction is continued for a longer time, development of the activation reaction will continue and eventually it will lead to cell growth and division. Depending on the duration of membrane stimulation, the same external agent may induce short-term physiological activation, cell hypertrophy and proliferation. Certain special cell types may have become unable to perform the last steps of surface-induced metabolic activation. For instance, polymorphonuclear leucocytes react to phagocytosis by the activation of a member of metabolic processes but not by initiation of DNA synthesis (Romeo *et al.*, 1975; Korchak & Weissmann, 1978). On the other hand, phagocytosis has been reported to induce DNA synthesis and division in the cells of reticular cells of liver and spleen (Kelly, Brown & Dobson, 1962; Jandl *et al.*, 1965). Surface-induced metabolic activation reactions in the cells of different types may vary depending on: (*a*) the spectrum of inducing ligands; (*b*) the presence or absence of blocks in the late stages of reactions; (*c*) the spectrum of metabolic activities stimulated in the course of the reaction.

Many types of physiological responses apparently unrelated to each other may turn out on closer examination to be different variants of the same archetype – the surface-induced restorative metabolic activation. To establish possible relationships between various membrane-related responses it would be important to compare in detail the sequences of early and late metabolic changes associated with the processes of growth stimulation by different growth factors and mitogens in various cell types. Such processes would include phagocytosis, the excitation–secretion response in secretory cells, the excitation–contraction response and functional hypertrophy of muscle cells, and effects of single and sustained excitation in nervous cells. Comparative analysis of many membrane-related phenomena might help to choose those responses which have a homologous sequences of metabolic changes and therefore presumably have a common evolutionary origin.

The ability to perform metabolic activation is essential for cell survival after membrane alterations. However, it is equally important to prevent excessive and unwarranted metabolic activation that would put too great a stress on the cell's economics and destroy the homeostasis of the multicellular system. Each normal cell type has its own spectrum of stimuli which result in metabolic activation. Several stimuli may act simultaneously or consecutively on the surface of the same cell inducing different initial disturbancies. The cell needs, and probably has mechanisms to assess the combined effects of all these stimuli and to establish the final metabolic restorative response; in other words the cell probably has a co-ordinated reaction norm to all these stimuli.

As mentioned before, the activation of DNA synthesis and division occurs at the late stages of restorative responses. According to this interpretation, the resting state of a cell is not necessarily a state in which the cell does not receive any stimuli but a state in which the disturbances induced by these stimuli are not large enough, and do not exist for long enough, to induce the full set of restorative metabolic responses. Only if the level of these disturbancies exceeds a certain threshold will DNA synthesis and mitosis be induced. These considerations suggest that some cells in resting cultures probably enter the first parts of the activation stage from time to time but then return to the resting state. This possibility deserves investigation.

Possible mechanisms for alterations of growth control in transformed cells

The proliferation of transformed cells depends less on external stimulation than the proliferation of normal cells. The range of environmental conditions in which a culture undergoes transition from the resting into the growing state becomes broader in the course of neoplastic evolution. Strongly transformed cultures do not undergo transition into the resting state even in conditions where they cannot grow; for instance, in a medium without essential nutrients or in a medium containing various potentially harmful drugs. As we have seen in chapter 10, these cells may continue futile (and even suicidal) efforts to cycle in these conditions. The increased sensitivity of transformed cells to growth stimulation may be due either to a series of independent changes of sensitivities to different environmental parameters or to a shift of the co-ordinated reaction norm to all the controlling factors. At present it is difficult to choose between these two possibilities, although we favour the second suggestion (see chapter 11). Many metabolic features of transformed cultures are common with those of growth-stimulated normal cultures. For instance, transformed cultures, with few exceptions, usually have a high rate of glucose uptake (Hatanaka, Huebner & Gilden, 1969; Hatanaka & Hanafusa, 1970; Sefton & Rubin, 1971; Eckhart & Weber, 1974; Oshiro & DiPaolo, 1974; Hatanaka, 1976; Perdue, 1976; Gionti & Lawrence, 1977; Lang & Weber, 1978). The rate of glycolysis is also increased in these culture (Morgan & Ganapathy, 1963; Bissel, Hatié & Rubin, 1972; Gregory & Bose, 1977). The intracellular level of cyclic AMP is often decreased although there are exceptions (Otten *et al.*, 1972; Burstin, Renger & Basilico, 1974; see review in Pastan & Willingham, 1978). Although the quantitative value of each of these parameters may vary in different transformed lines, the directions of metabolic shifts accompanying transformation are usually qualitatively similar to those accompanying growth activation in normal cultures. Naturally, growth-related metabolic differences between normal and transformed cultures are maximal in the conditions in which normal cells become quiescent; that is, in high density or low serum cultures. These differences become much less pronounced in conditions in which both normal and transformed cells grow actively. Thus, transformed cultures are those in which growth stimulation reactions are more easily switched on than in normal cultures; in strongly transformed cultures, those reactions may remain permanently switched on regardless of cultural conditions.

On the basis of the discussion of normal growth regulation presented above it is easy to construct a number of hypotheses that might explain the sustained stimulated state of transformed cells: (*a*) these cells may form endogenous stimulating ligands which interact with the receptors of the same or neighbouring cells (this question of auto-stimulation was discussed in chapter 10); (*b*) certain receptors of transformed cells may be structurally altered in such a way that they remain permanently 'switched-on'; (c) new proteins may be incorporated into the cell membrane and permanently keep the membrane in the 'switched-on' state; (d) absence of certain normal membrane components may alter the state of the membrane and 'switch-on' restorative metabolic activation; (e) the mechanism transmitting activation signals from the membrane into the cell interior may be altered: the cells may have an altered ability to form cyclic nucleotides, altered permeability to some ions, etc. (f) internal cell proteins that are regulated by activation signals (e.g. phospho-kinases) may be altered quantitatively or qualitatively. At present there is no decisive evidence in favour of any of these or other hypotheses.

Part IV

General discussion

13

Comparison of characteristics of transformed cells *in vitro* and neoplastic cells *in vivo*

Introduction

Growth *in vivo* and *in vitro*, morphogenesis and differentiation in different types of normal cells are regulated by complexes of environmental factors. Both *in vivo* and *in vitro* these complexes comprise three main groups of factors: components of the humoral medium, the substratum (stroma, tissue territory) and homotypic neighbour cells. However, many of the concrete factors forming each of these groups are different *in vitro* and *in vivo*. Neoplastic cells *in vivo* and transformed cells *in vitro* react to these environmental factors abnormally so that they form altered multicellular structures and may grow selectively in comparison with normal cells. In this chapter, we will compare in more detail the behaviour of transformed cells in culture conditions with that of neoplastic cells in the organism. We will first compare alterations in the formation of multicellular structures; then we will discuss two aspects of alterations in growth control mechanisms: alterations in the reactions to components of the humoral media and alterations of cell interactions with the substrata or with tissue territories. Finally, we will briefly discuss the factors involved in the neoplastic evolution of cells *in vivo* and *in vitro*.

Alterations in the formation of multicellular structures
Morphogenesis

In the course of embryogenesis *in vivo*, normal organ-specific mesenchymal and epithelial populations, interacting with each other, build various complicated organo-typic structures. Neoplasms growing in the organism are morphologically atypical to various degrees. The formation of organ structures is deficient in all neoplasms; the formation of simple tissue structures (e.g. epithelial alveoli, stretches of mutually parallel fibroblasts) also becomes deficient in neoplasms with advanced degrees of atypicalness. In cell cultures, even normal fibroblasts and epitheliocytes form only simple tissue-like structures and are unable to make organoid structures. Accordingly, alterations observed in morphologically transformed cultures probably correspond to deficiencies of tissue structures seen in morphologically atypical neoplasms. Alterations in the simple tissue-like organization in culture are usually diagnosed as morphological transformations only when these alterations have reached an advanced level, e.g. when fibroblasts have lost the ability for parallel orientation or when epithelial cells have lost the ability to form coherent sheets. Smaller alterations may pass unnoticed during light-microscopic examination. The development of more exact and more objective methods for assessing cellular morphogenetic potentialities may help to reveal minimal degrees of morphological transformation. Some of these methods (e.g. measurement of lamellar cytoplasm area, assessment of cell reactions to substrata with grooves or cylinders, measurement of cell aggregation) have already proved useful for diagnosing transformation in fibroblastic cultures, but thus far they they have been applied only to a few lines. It is not clear whether any single method is sufficient for this diagnosis. Reliable objective methods revealing minimal transformations in epithelial cultures are yet to be developed. The analysis of morphological alterations in strongly trans-formed fibroblastic and epithelial cultures suggests that these alterations are due to a deficiency in one group of basic morphogenetic reactions: pseudopodial attachment reactions. It is highly probable that advanced atypicalness of morphogenesis in neoplasms growing *in vivo* is also due to the same deficiency of pseudopodial attachment. The reduced ability of invasive epidermoid carcinomas to form cell–cell adhesions was shown long ago (Coman, 1944). The same is probably true for many other neoplasms. For instance, a decreased number of desmosomes is a characteristic feature of cervical carcinomas *in situ* (Shingleton *et al.*, 1968; Shingleton & Wilbanks, 1974). Possibly, small degrees of atypicalness in tissue structures are also associated with minimal alterations of pseudopodial attachment. Cell–cell and cell– substratum attachments in various types of benign and malignant neoplasms deserve further systematic studies.

Due to the simplified conditions we are able to reproduce in culture only a few simple tissue structures and their abnormal neoplastic variants. To give but one example, normal cervical epithelial cells are attached to the basal membrane via vertical outgrowths containing numerous hemidesmosomes at their surfaces (see fig. 13.1). At the basal surfaces of neoplastic cells of carcinoma *in situ* these outgrowths disappear and the number of hemidesmo-somes is diminished. To reproduce these normal and abnormal cell–substratum interactions in culture we have to learn how to cultivate cervical epithelial cells on a special substratum that has properties similar to those of the basal membrane and, in particular, can be evaginated by attached cells.

There seem to be three general approaches to the problem of development of more natural and more specific substrata for different cell types *in vitro*.

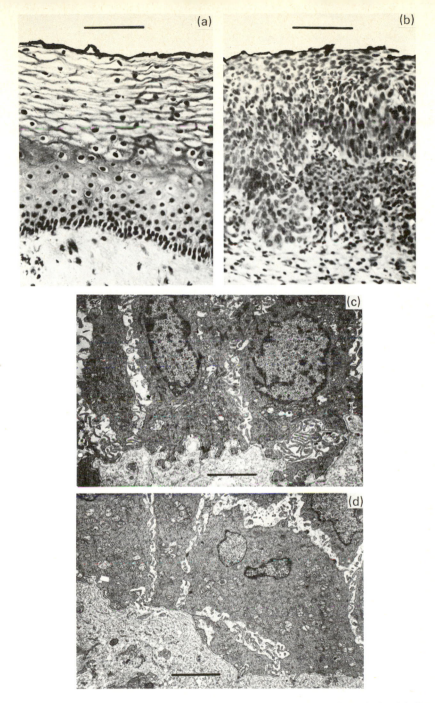

Fig. 13.1. A carcinoma of human uterine cervix *in situ*. (a) Normal cervical epithelium. (b) Carcinoma *in situ;* vertical stratification of the epithelium is almost absent. (c) The lower surface of basal cells of normal cervical epithelium. (d) The lower surface of basal cells of carcinoma *in situ*. In contrast to normal epithelium, basal cells have no cytoplasmic processes 'rooted' in the derma. (a, b) Light micrographs of haematoxylin and eosin stained preparations. (c, d) Electron micrographs. Bars: (a, b), 100 μm; (c, d), 2 μm.

(1) *Reconstruction of substrate from normal components of connective tissue matrices and of basal membranes*. Collagenous surfaces (Ehrman & Gey, 1956) may be simplified prototypes of substrata of this type.

(2) *Preparation of substrata with complex geometrical shapes which imitate the surfaces of structures existing in the organism*. Epithelial cells growing on sponge-like three-dimensional substrata may form complex tissue-like structures such as alveoles and tubules (Leighton, Siar & Mahoney, 1962). Comparative studies of the morphogenetic potentialities of normal and transformed epithelial cells on these substrata are yet to be made.

(3) *Preparation of 'living' substrata: cultivation of cells on layers of cells of another tissue type*. These 'living' substrata imitating specific stromal territories may provide specific micro-environments essential for the attachment and growth of special cell types. An example of the successful use of this principle is provided by the experiments of Dexter, Allen & Lajtha (1977) who cultivated haematopoietic stem cells on a layer of adherent bone marrow cells containing phagocytic mononuclear cells, 'epithelial' cells and 'giant fat' cells. Obviously, in experiments of this type both contact interactions with 'living' substrata and/or humoral factors produced by these substrata can be essential for the attachment and growth of specific cells. The effects of factors produced by fibroblastic feeder layers have already been mentioned (chapter 10). One hopes that the development of more sophisticated methods of cultivation will soon permit the study of various degrees of alteration in tissue morphogenesis accompanying transformations of many cell types.

Differentiation

Alterations in the formation of multicellular structures are manifested not only by deficient morphogenesis, but also by abnormalities of cell differentiation (maturation). We do not know anything about the maturation processes in normal fibroblastic cultures or those of most epithelia. Therefore this aspect of neoplastic transformation has not been studied in these systems. The only exception is the abnormal differentiation of certain transformed epithelia (chapter 10). In cultures of certain other cell types, neoplastic transformation was accompanied by considerable alterations of differentiation. For instance, when chick embryo muscle cultures were transformed with a temperature-sensitive mutant of Rous sarcoma virus, specific differentiation of cells (formation of multinucleated myotubes) was inhibited at the permissive temperature. However, when transformed cultures were transferred to a non-permissive temperature, within 72 h many cells fused and formed contracting myotubes (Holtzer *et al.*, 1975). Thus, the blocking of myogenesis, like other transformed characters, depended on the sustained activity of the viral oncogen.

Certain neoplastic cells derived from specialized cell types *in vitro*, like *in vivo*, retain the ability to undergo maturation. In particular, alterations of environmental conditions (attachment to the substratum, increased cell density, decreased serum content) have been shown to induce differentiation in cultures of teratocarcinomas (Topp *et al.*, 1976; Gearhart & Mintz, 1974, 1975; Jack, Rajewsky & Kapp, 1975; Herman & van den Berg, 1978) and of neuroblastomas (Schubert *et al.*, 1971; Prasad & Sinha, 1978).

Comparative investigations of the conditions of maturation in normal and neoplastic cell cultures have been performed only with certain haematopoietic precursor cells. These experiments have shown that conditions essential for maturation may be altered in transformed cultures. For instance, a specific protein, erythropoietin, is essential for the growth and maturation of normal erythroid precursor cells in culture (Iscove, 1978). The growth of murine erythroleukemia cells becomes independent of erythropoietin which does not induce differentiation in these cells. Certain low molecular weight polar compounds (dimethylsulphoxide, dimethylformamid, butyric acid, etc.) may, however, induce terminal maturation of these leukaemic cells (Friend *et al.*, 1971; Singer *et al.*, 1974; Filbach *et al.*, 1977).

In experiments with cultures of normal myeloid precursors it was found that a specific protein factor (macrophage and granulocytic inducer, MGI) is essential for viability, growth and maturation of these cells. The cells of certain clones of mouse myeloid leukaemia, in contrast to their normal counterparts, remain viable and continue to grow in the absence of MGI, but still need this factor for differentiation into mature macrophages and granulocytes. In other clones of leukaemic cells maturation is not induced by MGI but can be induced by factors of a different nature, e.g. steroids. There are also clones in which neither MGI nor steroids induce maturation (see review in Sachs, 1978).

These facts suggest that neoplastic cells, at least at the first stages of their evolution, may retain the ability for full-scale maturation but their reactivity to the factors inducing this differentiation is changed compared with normal precursors. It would obviously be important to compare environmental factors controlling maturation in cultures of other types of normal and neoplastic cells.

Positional information

In complex normal organ systems *in vivo* the cells of one tissue type located in different areas may have unequal abilities for differentiation and morphogenesis. For instance, cytologically similar cartilage cells located in different parts of limbs form structures of different shapes in the course of limb regeneration. These differences have been described by Wolpert and collaborators (Tickle, Summerbell & Wolpert, 1975; Wolpert, 1976; Lewis & Wolpert, 1976; Lewis, Slack & Wolpert, 1977) as differences in 'positional information'. 'Positional information' of cells in morphologically abnormal neoplastic tissue should be considerably altered compared with normal structures, but at present we know nothing about these alterations.

Alterations in growth dependence on components of the humoral medium

Transformations *in vitro* may lead to a decreased dependence of cell growth on the composition of the humoral medium and, in particular, on the concentrations of protein growth factors, ions and nutrients. A number of facts suggest that similar alterations of sensitivities to humoral controlling factors may be partly responsible for the selective growth of neoplastic cells *in vivo*.

Sensitivity to growth factors

Altered proliferation control by hormones (growth factors) has been revealed in neoplastic populations both *in vivo* and *in vitro*. However, the cell types and growth factors used for these investigations have differed in cultures compared with animals. In culture, most studies have been performed with fibroblasts and with unfractionated serum or, during the last years, with purified platelet factor and other growth factors (see chapter 10). But little is known about the sensitivities of normal and neoplastic cells *in vivo* to such factors as FGF, MSA or platelet factor. The in-vivo evolution of hormone dependence of neoplastic tissues has been described in numerous studies reviewed by Furth (1953), Gardner (1953), Huggins & Yang (1962), De Ome (1965), De Ome & Medina (1969), Foulds (1969, 1975), Clifton & Sridharan (1975), Nandi (1978). Most of these studies were concerned with cell sensitivity to 'usual' hormones including pituitary hormones, ovarian steroids and thyroid hormone. Accordingly, neoplasms arising from the tissues normally sensitive to these hormones (e.g. mammary gland, pituitary, thyroid, ovary, testis) were examined. Systematic studies of these normal and neoplastic tissues in culture are only just beginning.

Despite all the differences between experimental systems, the results obtained are remarkably similar. The step-wise, progressive decrease in the dependence of proliferation on external growth factors is a general feature of neoplastic evolution both *in vivo* and *in vitro*. In the first stages of evolution, *in vivo* neoplastic cells may retain their sensitivities to hormones although quantitatively this dependence may be altered in comparison with normal precursors. Correspondingly, transformed cells in culture may retain some degree of sensitivity to serum factors. In the advanced stages of this evolution these dependences may be lost completely. Animal experiments show that in certain cases neoplastic cells may lose their sensitivity to one hormone but remain sensitive to another growth factor. For instance, the mouse pituitary tumours secreting thyroid stimulating hormone (TSH), may lose sensitivity to thyrotropin-releasing factor formed in hypothalamus which acts locally on pituitary cells. Therefore, these neoplastic cells acquire the ability to grow outside the pituitary e.g. in subcutaneous tissue. The same tumours retain sensitivity to the growth-inhibiting effect of thyroid hormone; therefore, they grow only in thyroidectomized mice (Clifton & Sridharan, 1975). It would be important to study in detail the sequences of alterations in the sensitivities to various individual growth factors accompanying different consecutive transformations of fibroblasts *in vitro*.

Another interesting problem raised by studies *in vivo* is that of the nature of regressions and recurrencies of hormone-dependent tumours. By definition, hormone deprivation leads to regression of these tumours; however, a fraction of the neoplastic cells usually survives in these conditions. For instance, Noble (1977) examined a large series of oestrogen-dependent transplantable rat tumours of 14 different types e.g. adrenal carcinomas, uterine leiomyoma, mammary carcinomas. The tumours were grown in rats with implanted ostrogen pellets. Removal of the hormone-containing pellet from the rat usually led to regression of the tumour. Reintroduction of the pellet was followed by a re-growth of the regressed tumour even when regression had been as long as 39 weeks; the recurrent tumour remained oestrogen-dependent. These observations raise a number of questions: what are the differences between the cells killed by hormone deprivation and those surviving deprivation? Both cell groups are apparently hormone-dependent: what is the state of the 'dormant' neoplastic cells present for many a month in hormone-deprived animal? Reproduction of this intriguing phenomenon in culture may be essential for the solution of these problems.

We do not know the nature of many of the growth controlling factors which stimulate proliferation of normal cells and 'conditional' neoplasms. Among them are unidentified factors stimulating the growth of 'conditional' skin papillomas near the wounded area (see chapter 1) as well as those stimulating the proliferation of transplanted tumour cells in an area of inflammation and in the vicinity of embryonic cells (see chapter 2). As discussed in chapter 10, 'autostimulating' growth factors may be produced by neoplastic cells and the nature of these factors has been established only for few cell lines *in vitro*. We do not know how common autostimulation phenomena are *in vivo* or *in vitro*.

Deficiency of nutrients.

Many transformed cultures are characterized by an increased uptake of glucose. This increased uptake is probably a result of the permanently activated state of transformed cells (see chapter 12). A rapid uptake and utilization of glucose and probably of other nutrients is also characteristic for many experimental tumours *in vivo* (Henderson & Le Page, 1959; Shapot,

1972, 1979). As a result, the concentration of glucose in the intercellular fluid of solid tumour nodules is much lower than in the intercellular fluid of normal connective tissue (Gullino, Clark & Grantham, 1964; Gullino, 1966).

Ascitic fluid surrounding the suspended cells of ascites tumours also contains very low concentration of glucose; all the labelled glucose injected into this fluid can be taken up by the cells within a few seconds (Gorozhanskaya & Shapot, 1964). If neoplastic and normal cells are located in the same area, the former can utilize metabolites more actively than the latter. Food deprivation may cause the death of normal cells in an area invaded by neoplastic cells. Eventually, deficiency of nutrients may also become deleterious to neoplastic cells themselves. The cells in overcrowded tumour nodules live in a condition of severe nutrient deprivation. Experiments in culture have shown that strongly transformed cells, in contrast to normal ones, do not react to a deficiency of nutrients by a transition into the resting state which is characterized by decreased food requirements. Transformed cells in a deficient medium may continue to move through the mitotic cycle; eventually these cells may die due to an inability to reduce their requirements (chapter 10). Similar processes probably take place in tumour nodules. The analysis of proliferation kinetics of a number of experimental and human tumours has shown that the actual rates of growth of tumour nodules are much lower than those calculated on the basis of the assumption that all the new cells formed by division of the cycling tumour cells remain alive. This means that a considerable proportion of the tumour cells die in each generation. From the differences between calculated and real rates of tumour growth one can calculate the rate of cell loss. Although these calculations are subject to many limitations and the obtained values vary considerably from one tumour to another (see critical discussion in Hill, 1978), their results indicate that the rates of cell loss are usually high in solid tumours. In certain tumours, 80–90% of new cells were estimated to die in each generation (Steel & Lamerton, 1966; Tubiana, 1971; Guelstein & Klempner, 1973; Frankfurt, 1975; Klempner & Guelstein, 1976; Schertz & Marsh, 1977). Various cells within the same tumour nodule have a different supply of nutrients and oxygen depending on the distance of these cells from blood vessels. In certain transplanted tumours the fraction of [³H]thymidine-labelled tumour cells was found to decrease with increasing distance from the capillaries (Tannock, 1968). Central parts of the nodules have the worst supply of nutrients and their necrosis is a common occurence. Thus, it seems probable that neoplastic cells *in vivo*, like strongly transformed cells *in vitro*, may die due to the inability to decrease their requirements in response to a nutritional deficiency. This mechanism is probably the main cause of cell death in many malignant tumours that are at the advanced stages of progression. In addition, cell maturation may be a significant factor in differentiated tumours such as epidermoid carcinomas with keratinization (Frankfurt, 1975).

Another aspect of the same problem is that of the state of non-proliferating cells in tumours. Many animal tumours contain a considerable proportion of cells which remain unlabelled after continous or repeated labelling with [^{3}H]thymidine (Mendelsohn, 1960; see reviews in Frankfurt, 1975; Hermens & Barendsen, 1978). It is often assumed that at least part of these cells are in the non-proliferating resting state (G_0) similar to that of normal cells in serum-deprived or high density cultures. However, existing methods do not distinguish between slowly proliferating and non-proliferating resting cells (see review in Hill, 1978). Also we cannot distinguish between cells that have reversibly or irreversibly stopped proliferation. In other words, at present we cannot determine the true 'resting' fraction of cells in tumours. By analogy with culture systems, one may suggest that the ability to enter the resting state is decreased in the course of neoplastic evolution. Strongly transformed cells in unfavourable conditions *in vitro* and *in vivo* are likely to enter not the 'true' resting state but the state of slow suicidal cycling.

Interactions of cells with their territories
Cell territories in vitro *and* in vivo

Selective growth of neoplastic cells is associated with spreading; that is, with the progressive expansion of tissue territories occupied. Here we will compare interactions of cells with their territories *in vitro* and *in vivo*. A normal cell in usual conditions is able to proliferate only when it has a particular local territory; that is, a certain surface provided by a non-cellular substratum and/or by cells of another type. The territory of a normal anchorage-dependent cell in culture is the substratum on which it spreads. In an organism, the cells of each type that are able to proliferate have their specific territories. These territories are formed by intercellular structures made by cells of the same type or other types. For instance, fibroblasts and other mechanocytes form their own territories by secreting collagen and other components of intercellular scaffoldings upon which they spread. Mechanocytes also participate in the formation of territories for other cells, e.g. for epithelial cells. Networks of blood capillaries are also important parts of cell territories. Specific stromal cells are probably essential components of the territories of many cell types, e.g. haematopoietic stem cells. Of course, as discussed above, the availability of the territory affects not only cell proliferation but also cell morphology and differentiation.

The spread of normal cells on new territories

If cell-free territory is available, normal cells may expand into it and eventually occupy it. Two processes participate in this expansion: the migration of cells from previously occupied territories and the proliferation of cells on new territories. Spreading and multiplication of suspended cells on a free substratum as well as wound healing in dense cultures are obvious examples of these expansions *in vitro*. Similar expansion *in vivo* can be observed when a specific cell territory has been freed from occupying cells. Epithelial cells and fibroblasts migrate on the free territories in the wounds and proliferate there. Transplanted or surviving haematopoietic stem cells can re-populate specific stromal territories of bone marrow and other blood-forming tissues after irradiation. Normal cells *in vivo* can also invade stromal territory formed *de novo*. Haematopoietic cells of the host may re-populate fragments of bone tissue formed *de novo* in the subcutaneous tissue by grafted bone marrow fibroblasts (Friedenstein, 1973). Proliferation of connective-tissue cells may lead to the formation of new stromal territory further colonized by epithelium. For instance, skin epithelium and other epithelia may invade underlying tissue in which inflammation and proliferation of young connective tissue cells have been induced by some stimulus. These so-called inflammatory epithelial proliferates regress completely when pro-liferation in connective tissue stops (see reviews in Garschin, 1928, 1939; Vasiliev, 1958, 1961; Vasiliev & Guelstein, 1966). Hormonal stimulation in the course of pregnancy induces the growth of mammary stroma; growing epithelium grows further into this stroma (Toustanovsky & Vasiliev, 1957). Parenchymal–stromal relations in this and probably in many other cases are specific: normal mammary epithelium grows only in a specific stromal territory; the mammary fat pad. An alteration in the state of specific stromal cells may cause the death of parenchymal cells. An elegant analysis of one example of specific stromal effects on parenchymal cells has been performed by Dürnberger *et al.* (1978).

An androgenic hormone, testosterone, induces necrosis of epithelium from the mammary gland rudiment of mouse embryos. Dürnberger *et al.* (1978) isolated mesenchyme-free epithelium and mesenchyme from the mammary glands of the embryos of normal androgen-sensitive mice and androgen-insensitive mutant mice. Various epithelia and mesenchymes were then recombined and incubated with testosterone in organ cultures. Testosterone-induced necrosis of epithelium was observed only in cultures containing sensitive mesenchyme; epithelium could be taken either from sensitive or from insensitive mice. The direct contact of sensitive mesenchymal cells with the epithelial surface was essential for the induction of necrosis. Thus, the effect of hormone on the epithelium was mediated by a specific mesenchyme. The specificity of parenchymal–stromal interactions may be determined by chemical parameters (the chemistry of the surfaces to which the cell is attached, the production of local 'conditioning factors', etc.) and by the geometrical parameters of the territory. The dependence of cell proliferation on the availability of the territory *in vitro* has two forms: (*a*) the cell starts proliferation only when it is attached to the territory (anchorage dependence); (*b*) the cell stops proliferation when all the surrounding territory is occupied by other homotypic cells (density dependence).

Probably, similar territorial control mechanisms are effective *in vivo*, as suggested by the local activation of proliferation in healing wounds, in the course of re-population of bone marrow, etc. Concrete forms of these mechanisms *in vivo* remain to be studied.

The spread of neoplastic cells

Strongly transformed fibroblasts in culture are less selective than normal cells in their interactions with available territories (substrata). These cells can colonize the surfaces avoided by normal cells; for instance, the bottom of grooves or the surface of millipore filters. They may also become attached to the upper surfaces of normal cells (chapter 5). These cells are less dependent on the availability of territories for their multiplication: either or both forms of territorial growth control mentioned above may become deficient in transformed cultures (chapters 8 and 9). In-vivo interactions of neoplastic cells with their territories have two main traits distinguishing them from similar interactions of normal cells: (*a*) neoplastic cells may become less selective in their choice of territories than normal cells. The degree of this decrease in requirement for specific territories may be different in various neoplasms; (*b*) neoplastic cells acquire an ability to form new territories for themselves; that is, they may acquire the ability to induce local proliferation of normal fibroblasts and sometimes of other cell types. Blood vessels and connective tissue structure formed in the course of this proliferation provide new stromal territory for growing neoplastic cells. The growth of the blood vessels (Algire *et al.*, 1945; Folkman, 1975; Knighton *et al.*, 1977) and of fibroblasts (Belyaeva & Vasiliev, 1971*a*) may be activated in a wide zone around a transplanted tumour nodule. The factor which activates the growth of capillaries passes through a millipore filter (Greenblatt & Shubik, 1968). These results suggest that the stromatogenic activity of a neoplastic cell may be due to the secretion of a diffusible factor. Folkman and collaborators (Folkman *et al.*, 1971; Folkman, 1974, 1975) isolated a protein from tumour tissues which induces proliferation of blood vessels; the tumour angiogenesis factor. We do not know whether the growth of fibroblasts is also stimulated by this protein. Relations between the angiogenesis factor and other growth factors produced by neoplastic cells *in vitro* are also unknown. Certain normal cells, e.g. the cells in a focus of inflammation and activated lymphocytes, may also stimulate blood vessel growth (Auerbach, Kubai & Sidky, 1976). Probably, the stromatogenic activity of neoplastic cells is an anomalous variant of similar activities in normal cells. The loss of stromal specificity and stromatogenic activities may be expressed to various degrees in different neoplasms. Among the main variants of neoplasms differing in the expression of these properties are the following.

(1) *Neoplasms that retain stromal specificity and do not induce the formation of new stroma*. These neoplasms grow only on territory previously occupied by homotypic normal cells and do not invade underlying tissues. Carcinomas *in situ* and other variants of 'intra-epithelial neoplasms' (Richart, 1973; Shingleton & Wilbanks, 1974; Friedell, 1976; Wilbanks, 1976; Farrow *et al.*, 1977) belong to this category. They may grow for many years spreading only within the territory previously occupied by normal homotypic epithelium (fig. 13.1).

Another group of neoplasms retaining stromal specificity are the chronic myeloid leukaemias and related haematological neoplasms: their neoplastic cells spread mainly on territories specific for normal haematopoietic cells such as bone marrow and spleen.

Although the stromal requirements of carcinomas *in situ* and of leukaemias remain tissue-specific, these requirements are possibly altered in comparison with normal parent tissues. As mentioned above, the attachment of carcinoma cells to the basal membrane *in situ* is deficient compared with normal cervical epithelium. By analogy with the behaviour of transformed cells *in vitro* one may suggest that poorly attached cells of carcinoma *in situ* may be less stringent in their requirements for substrata than normal cells. As a result, these neoplastic cells may be able to spread selectively in the areas of stroma altered by some pathological process. In fact, human cervical carcinoma *in situ* is accompanied by vascular changes in the connective tissue underlying the epithelial lesion (Wilbanks, 1976). In a few patients, characteristic vascular changes have been observed to precede the development of the epithelial lesion (Stafl & Mattingly, 1975). Pathological alterations of stromal structures may promote the selective proliferation of minimally neoplastic clones in different tissues.

(2) *Neoplasms inducing stromal proliferation and retaining stromal specificity.* Benign epithelial papillomas and polyps apparently belong to this category. The growth of these neoplasms is accompanied by a proliferation of stromal connective tissue; this proliferation increases the territory available for neoplastic epithelial cells. Newly formed stroma is covered but not invaded by neoplastic epithelium. Mouse mammary ductal papillomas have been shown to have angiogenic properties: they elict the formation of new vessels when transplanted onto the rabbit iris (Brem, Gullino & Medina, 1977). 'Hyperplastic precancerous' nodules in mouse mammary glands (see review in Dunn, 1945; Foulds, 1975) can be regarded as another example of the same category of neoplasms. The cells of these nodules form alveolar structures very

similar to those of normal mammary glands but having an altered sensitivity to hormones. At least some cell lines forming nodules are able to induce the proliferation of fibroblasts (Belyaeva & Vasiliev, 1971*b*) and growth of blood vessels (Gimbrone & Gullino, 1976*a*, *b*). The cells of the nodules, like normal mammary epithelium, can be successfully transplanted into the specific mammary stroma (mammary fat pads) but not into other areas of connective tissue (De Ome *et al.*, 1958). Thus, the cells of these neoplasms retain tissue-specific stromal requirements.

(3) *Neoplasms that induce stromal proliferation and do not retain tissue-specific stromal requirements*. Most 'usual' malignant tumours which invade surrounding tissues may be placed in this category. Animal tumours of this type can be successfully transplanted into different tissues of syngeneic hosts. For instance, mouse mammary carcinomas arising from hyperplastic nodules, in contrast to parent neoplasms, can be transplanted not only in mammary-specific fat pads but also in other areas of subcutaneous tissues.

Experiments showing the ability of neoplastic cells to induce the proliferation of fibroblasts and angiogenesis (see above) have been performed mostly with cells in this category. Each particular tumour has its characteristic amount of stromal structures (measured by the amount of collagen) per neoplastic cell (Gullino, 1966, 1975). This amount probably depends on the relative rates of proliferation and death of the neoplastic and stromal cells and also on the intensity of synthesis of intercellular structures by these cells.

Invasion of normal tissues by neoplastic cells may have at least two different variants: (*a*) neoplastic cells induce the proliferation of new stroma in the surrounding tissue and then invade the new stroma; (*b*) neoplastic cells do not induce proliferation of stroma but migrate along the pre-existing structures of normal tissues. In this case nutrition for neoplastic cells will be provided by pre-existing blood vessels.

Each of these variants is probably rarely encountered in a pure form; usually combinations of the two are observed. In each case a poor attachment of neoplastic cells to the substrata and to other cells may possibly facilitate their detachment and movement into new areas. Detachment may be also facilitated by factors released by necrotic cells present in the tumour (Weiss, 1977*b*).

Experiments with mixed cultures showing the mutual invasion of normal and neoplastic fibroblasts and their 'sorting' in different culture layers (see chapter 5) can be regarded as a simplified prototype of the interactions of invading neoplastic cells with pre-existing or newly formed normal stromal cells.

Mechanisms of the destruction of normal tissue structures in the area invaded by neoplastic cells are not yet clear. Normal parenchymal cells present in these areas may die of starvation (see above). As discussed in chapter 6, proteases may be secreted into the medium by some transformed cells *in vitro*. The role of protease-induced lysis of normal intercellular structures in the invasion of cell *in vivo* has been suggested many times (see e.g. Liotta *et al.*, 1977; Poole *et al.*, 1978; Kuettner, Pauli & Soble, 1978) but still remains to be proved. An interesting possibility is that intercellular structures are destroyed not by tumour cells themselves but by surrounding normal cells. For instance, metastatic nodules of the transplantable rabbit Vx2 carcinoma in bones have been shown to produce a diffusible factor activating osteoclasts in the surrounding bone tissue; these activated osteoclasts were partially responsible for the bone destruction around tumour (Galasko, 1976). The lysis of pre-existing structures may be a part of tumour-induced reorganization of surrounding tissues: it makes a 'clear place' for the growth of new stroma.

(44) *Metastasizing tumours.* In order to undergo metastasis a cell of a solid tumour should perform a sequence of different processes: it should penetrate the wall of a blood or lymph vessel, travel in the fluid medium, stop in an organ and grow into a metastatic nodule in this organ (see review in Zeidman, 1957; Wood, 1971; Wood & Straüli, 1973; Fidler, 1975*b*, 1976; Weiss, 1967, 1976, 1977*a;* Pollack, 1977).

Penetration may be due to the same process as invasion. Most normal and neoplastic cells detached from substrata are probably able to survive for some time in fluid media. However, to remain in the capillaries of certain organs, metastasizing neoplastic cells should have some special properties. In certain cases arrest may be associated with the formation of a thrombus around the circulating tumour cells or with the formation of clumps including neoplastic cells and various normal blood cells such as platelets and lymphocytes (Wood, 1971; Wood & Straüli, 1973; Gasic *et al.*, 1973; Fidler, 1974). The mechanisms of these phenomena deserve investigation in cell cultures.

The experiments of Fidler and collaborators (Fidler, 1975*a*, *b*, 1976, 1977, 1978; Fidler & Nicolson, 1976; Fidler & Kripke, 1977) confirmed the suggestion that the ability to form metastases is a genetically controlled property (or a complex of different properties) of neoplastic cells. They obtained by selection sub-lines of transplantable mouse melanoma B-16 which after subcutaneous or intravenous injection formed metastatic nodules in the lungs with much higher frequencies than the parent tumour line. Non-metastasizing variants have been also selected from a metastasizing melanoma (Tao & Burger, 1977). These melanoma sub-lines provide attractive experimental systems for the investigation of cell properties *in vitro* associated with the ability to form metastases *in vivo*. It has already been shown that cells of high metastatic lines of melanoma cells have an increased rate of attachment to confluent monolayers of melanoma cells and 3T3 cells (Winkelhake & Nicolson, 1976) and an increased ability to form heterotypic aggregates with lung cells (Nicolson & Winkelhake, 1975) compared with low metastatic sub-lines of the same tumour. The cellular characteristics responsible for these differences remain to be clarified. Recently, Nicolson, Brunson & Fidler (1978) developed several new variant mouse melanoma B-16 lines: one of them forms metastases preferentially in the brain, another in the ovary. This unique collection of lines opens new possibilities for an analysis of factors responsible for organ specificity of metastases.

It is worth remembering that the ability to travel in blood and to stop in certain organs is not unique for metastasizing tumour cells; certain normal cells have a similar property. Normal blood, besides usual blood cells, may also contain small numbers of haematopoietic stem cells and cells that form fibroblastic colonies *in vitro* (Friedenstein, 1976). Cells of other types may be found in the blood when adequate cultivation methods have been developed. The formation of metastases by neoplastic cells is probably an abnormal expression of physiological properties.

(5) *Ascites tumours*. When the cells of these neoplasms are implanted intraperitoneally, they induce the accumulation of large volumes of ascitic fluid and actively proliferate in this fluid. The ability to grow in ascitic fluid is a genetically determined property of special variants of neoplastic cells. Ascitic tumour strains can be obtained by selection from neoplasms unable to grow in this form. Ascites variants of different tumours (e.g. mammary carcinomas, hepatomas, sarcomas and lymphomas) have been described (Klein, 1953; Klein, G. & Klein, E., 1956). The cells of ascitic tumours can be regarded as extreme examples of cells that in the course of neoplastic progression decreased both their ability to spread upon solid substrata and their anchorage dependence of growth. However, at present we do not know any characteristic differences between the cultural properties of several sub-lines of the same tumours selected for their different abilities to grow in the ascitic form. It is interesting that certain ascitic carcinomas retain to some degree the ability of their parent epithelial cells to form stable cell–cell contacts; these tumours grow in ascitic fluid as complexes of cells firmly linked to each other (fig. 13.2). Systematic comparison of the abilities of these cells to form cell–cell and cell–substratum attachment may be of some interest. It would also be important to find out how ascitic tumour cells induce the accumulation of fluid in the peritoneal cavity. One of the possibilities is that these cells produce some substances increasing the permeability of blood vessels (Fastaia & Dumont, 1976).

In this section we have classified various neoplasms according to the types of interactions of their cells with stromal structures. In most cases the character of these interactions has not been determined directly in various experimental situations but deduced from the pathological features of the neoplasms. At present we do not have the methods to test these interactions, so that any classification of this type is inevitably a preliminary one. It should

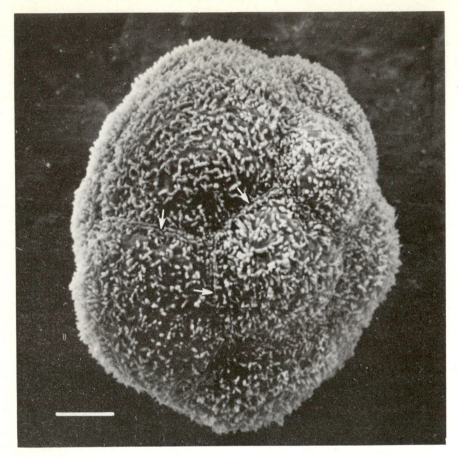

Fig. 13.2. Scanning electron micrograph of a complex of coherent epithelial cells from ascites rat hepatoma (Zaydela line) showing cell–cell contacts (arrows). Bar: 5 μm.

also be remembered that in this section we discussed only those in-vivo properties of neoplastic cells that may be more or less directly related to their cultural characteristics. Certain other factors, especially immune tumour–host interactions, may be very important for tumour spread but their discussion is obviously beyond the scope of this book.

Carcinogenesis without carcinogens: spontaneous transformations *in vitro* and foreign body carcinogenesis *in vivo*

We have already compared several aspects of the abnormal interactions of established neoplasms and transformed cultures with their environments. Another facet of the same problem is the possible role of an abnormal cell environment in the development of neoplastic transformation *in vitro* and *in vivo*. There is reason to believe that exposure of cell populations to abnormal environments may favour the selection of neoplastic clones and possibly even induce their formation. In the course of chemical carcinogenesis the new cell variants are probably induced by the direct action of the inducing chemical on the genetic material of cells (see chapters 1 and 2). However, selective growth of the altered clones may be promoted by a special micro-environment in the foci of carcinogenesis *in vivo*. A direct toxic effect of carcinogens may favour the selection of resistant neoplastic clones (Vasiliev & Guelstein, 1963; Farber, Parker & Gruenstein, 1976). Normal tissue structure may be destroyed by carcinogens. For instance, in the area of subcutaneous carcinogenesis induced by polycyclic hydrocarbons, large amounts of atypical collagen may be accumulated (Vasiliev *et al.*, 1962). These alterations may favour the selective proliferation of neoplastic clones with simplified requirements. Certain promoting agents such as TPA (see chapters 2, 3, and 10) change cell morphology and stimulate cell growth in cultures. Possibly, promoting agents cause similar changes in the foci of skin carcinogensis and these changes somehow stimulate the proliferation of neoplastic clones. 'Pre-cancer' lesions observed in various tissues at the early stages of carcinogenesis (e.g. diffuse and nodular hyperphasias, see review in Shabad, 1967), are possibly the proliferations of minimally changed neoplastic clones. In the course of progression they are replaced by malignant clones.

Of special interest are situations in which alteration of cell environment alone leads to development of neoplasms, although the cells have not been treated with any known agents that are able to induce genetic changes. One example of such 'carcinogenesis without carcinogens' is spontaneous transformation *in vitro*. We have already described these transformations in chapter 2 and we will only recall here that the serial, long-term cultivation of the embryo cells of certain species, e.g. of mouse fibroblasts, may be sufficient to make most cultured populations oncogenic. Foreign-body carcinogenesis is another interesting example of tumour development without chemically active carcinogen. We will briefly describe here this type of carcinogenesis (see review in Oppenheimer, Oppenheimer & Stout, 1948; Oppenheimer *et al.*, 1958; Bischoff & Bryson, 1964; Brand *et al.*, 1975; Brand, 1975, 1976). Foreign bodies implanted subcutaneously or in other tissues of rodents induce local sarcomas (fig. 13.3) in most animals after a long latent period (7–15 months). The tumours can be induced by foreign bodies of diverse chemical nature (different polymers and metals) including those made of chemically inert substances, so that direct chemical interaction of the implanted chemical with the cell is probably not essential for tumour development. The frequency of tumours depends on the size and shape of the implant: unperforated plates or films of different materials induce tumours with the highest frequencies but the same materials in minced or powdered

Fig. 13.3. A mouse with a subcutaneous sarcoma growing around a plastic plate 19 months after implantation; a second implanted plate is seen in the subcutaneous tissue on the other side.

form are inactive. Implanted millipore filters with a pore size of 0.1 μm and less, induced tumours in mice while similar filters with pores larger than 0.1 μm were inactive (Karp *et al.*, 1973).

A collagen capsule is formed around foreign bodies during the first month after implantation. The capsules around the tumour-inducing implants are wider and contain less cells than those around inactive implants.

In the experiments of Brand's group (see reviews in Brand, 1975, 1976; Brand *et al.*, 1975) and Moizhess & Prigogina (1973) polymer films were implanted to CBA-T6T6 mice with chromosome markers. At different times after implantation, the films with attached cells and/or the capsules were removed and transplanted to new syngenic hosts without chromosome markers. Cytogenetic analysis of the tumours arising at the second sites of implantation revealed that their origin was from donor or host cells. These experiments showed that progenitor cells of sarcomas were already present in the capsules at one month after the first implantation; later (after three to five months) these cells appeared also on the surface of the polymer films. In other experiments Brand, Buoen & Brand (1967) removed two-thirds of the film six months after implantation or later, cut this removed part into two equal segments and implanted each segment separately into two new mice. The tumours arising in both mice receiving different parts of the same film had a similar karyotype (similar modal numbers of chromosomes and similar chromosome aberrations), similar morphology and similar growth rates indicating their origin from the same clone. In summary, these results suggest that one 'preneoplastic' (or minimally neoplastic) clone spreads selectively first in the capsule and then on the film surface. This clone later gives rise to malignant cells that invade the surrounding tissues.

Both spontaneous transformation *in vitro* and foreign-body carcinogenesis have one common feature: the appearance of obviously neoplastic cells is preceeded by long-term exposure of the cell population to a drastically changed environment. In the case of spontaneous transformation, the cells are transferred into conditions of culture in which all components of the environment are different from those of the normal tissue *in vivo*. The cells spread on artificial substrata, the relation of the cell volume to that of the intercellular fluid is much lower than *in vivo*, the composition of the intercellular fluid is also different. In foreign body carcinogenesis the focus of the cell environment is also altered in comparison with that of normal subcutaneous tissue: the cells in these foci are attached to the artificial substratum of the film or to atypical collagen of the capsule; the supply of nutrients and of growth factors in the thick collagenous capsule is probably also different from that in the normal tissue. One can suppose that in these conditions neoplastic clones will have selective advantage because they are less sensitive to the environmental changes than normal cells. These variant clones may be present as a minority group in a normal population before the alteration of the environment; alternatively, these variants may be formed *de novo* as a result of some genetic changes produced by the altered environment. It seems that foreign body carcinogenesis and spontaneous transformation may have several stages: the formation of minimally neoplastic clones and the progression of these clones to strongly transformed (sarcomatous) cells.

The similarity between the spontaneous transformation and plastic carcinogenesis is stressed by the experimental results of Boone and collaborators described in chapter 2. Permanent minimally neoplastic cell lines obtained by spontaneous transformation readily gave rise to oncogenic strongly trans-formed variants when these cells were attached to the surface of polymer platelets and implanted subcutaneously to mice; that is, placed in the same conditions as pre-neoplastic clones growing on the surface of a foreign body. This in-vitro–in-vivo system offers interesting possibilities for an analysis of the processes involved in the selection of strongly neoplastic clones. All the present considerations about the role of micro-environment in carcinogenesis inevitably remain hypothetical and too general. Concrete cellular mechanisms of foreign body carcinogenesis and of spontaneous transformation remain almost as enigmatic today as forty years ago when these phenomena were first described.

Conclusion

Both transformed cells *in vitro* and neoplastic cells *in vivo* have altered reactivity to their environments and, in particular, decreased requirements for specific territories (substrata) as well as a decreased requirement for growth factors in the media. Of course, *in vitro* we cannot yet reproduce exactly many components and variants of the environmental factors encountered by normal and neoplastic cells *in vivo*. In particular, we are still unable to reproduce in culture tissue-specific stromal territories. As a result, we cannot yet compare the in-vitro stromal specificities of normal and neoplastic cells.

One can hardly hope to reproduce fully *in vitro* certain processes which play important role in neoplastic growth *in vivo*. Among them are stromal proliferation induced by neoplastic cells and interactions of these cells with the immune system. Cultured cells implanted to a compatible animal will grow into a tumour only when they have strongly neoplastic properties and, in particular, if they are able to spread, to form stroma and to proliferate in the foreign environment of the implantation site. The cell with more specific environmental requirements may not grow selectively in these conditions. We are only beginning to learn how to create selective environments *in vivo* for various types of minimally neoplastic cells.

Despite all the differences, neoplastic evolution *in vivo* and *in vitro* proceed in the same direction and lead to the appearance of basically similar cells. The best proof of this similarity is the oncogenicity of cells transformed by cultivation *in vitro:* these cells form tumours although they have been previously selected for growth in a special culture environment and not for growth in animals.

14

Are different transformed characters inter-related?

Introduction

In previous chapters we have described the transformations of several cellular characters observed in cultures: morphologic transformation, transformations of growth requirements (loss of anchorage, serum and density dependences), unlimited life span and others. These characters can sometimes be dissociated from each other, at least partially, in the course of the evolution of certain cell lines or after special selection procedures (see chapter 2). These dissociations permit the study of each transformed character as a separate unit. However, one cannot exclude the possibility that different characters are inter-related and that their dissociation is a result of secondary alterations. In this chapter we will compare the general nature of the main transformed characters; then we will discuss the inter-relationships between these characters and factors leading to the diversification of the properties of neoplastic clones.

Neoplastic cells have an altered ability to undergo transitions from one integral state to another

Integral states of normal cells

In the course of its life the cell enters various states that can be grouped into several categories: morphology, growth, and differentiation. Each category includes several cellular states. With regard to morphology, in fibroblasts and epithelial cells one can distinguish the isolated (suspended) state and the substratum-spread state with or without cell–cell contacts. With regard to growth one can distinguish the resting state and the growing state; the latter includes activation and replication stages. The cells of certain tissue systems (e.g. haematopoietic cells) may be in several states with regard to differentiation. Within the same category, one state is different from another as judged by many biochemical and morphological characters; therefore we may designate these states as integral cellular states.

A cell may undergo a transition from one state to another within the same category: a suspended cell may become spread on the substratum, a resting cell may undergo transition into growing state, or an undifferentiated cell may undergo maturation. Each of these transitions is a complex multistage process.

Transition between the states in one category is not always accompanied by transitions in other categories. Many transitions are reversible but some are irreversible, e.g. certain maturation processes, and possibly certain transitions from the growing state into the resting state. Many, probably all, transitions are not autonomous but are induced or suppressed by complexes of environmental factors, transition-controlling factors. Therefore, transitions can be regarded as complex cellular reactions to controlling factors. In chapter 3 we discussed morphogenetic reactions in detail, especially pseudo-podial attachments reactions which lead to cell transition from suspension into the spread state. In chapter 12 we discussed surface-induced activation reactions leading to cell transition from the resting into the growing state. Transitions from a less-differentiated into a more-differentiated state have not been discussed in detail in this book because these transitions have been little studied in cultures of fibroblasts or epithelial cells. Several examples of transitions of this type induced in haematopoietic cells by growth factors were given in the chapters 1 and 13. Possibly, cells in the usual fibroblastic cultures may also undergo transition related to the differentiation state but at present the criteria and conditions of these transitions have not been identified.

Altered reactions of transformed cells to transition-controlling factors

The main transformed characters can be regarded as manifestations of altered abilities for transitions from one integral state into another. As discussed in chapters 5 and 6, morphological transformations may be regarded as an expression of the decreased ability of cells to undergo transition from the suspended state to the spread state and to form adequate cell–cell contacts. As discussed in chapter 12, transformed characters related to growth (growth in suspension, low serum, etc.) can be regarded as expressions of a decreased ability of the cells to undergo transition into the resting state or, conversely, as expressions of an increased ability of cells to undergo transition into the growing state. Neoplastic transformations of certain cell types may also involve an altered ability to undergo transition from the undifferentiated state into a series of more mature states. Several examples of these alterations were given in the chapter 13. Transitions from one state into another are induced by environmental changes. An altered ability for transitions may be also be described as an altered reaction norm to transition-controlling factors. In the course of neoplastic progression, the spectrum of environmental conditions in which the cells remain unspread, or do not form cell–cell contacts, or are growing and not differentiated, becomes increasingly broad. Correspondingly, the spectrum of conditions causing transitions into alternative state becomes narrower. The 'ultimate' neoplastic cell would be unable to become well-spread, enter the resting state or start maturation in any environment. Of course, cells with these properties are only the theoretical end-points of progression and are rarely, if at all, reached in human pathology or in experiments. Usually, neoplastic cells retain the ability to undergo transitions into other states, but it is not always easy to find the sets of conditions causing these transitions. For instance, certain transformed cells may gain the ability for good spreading only in serum-free media or in media with artifically increased concentrations of exogenous fibronectins (see chapter 6); that is, in conditions different from those adequate for good spreading of normal cells.

One of the cellular characters that may undergo transformation is growth potential: transformed cells do not become senescent and acquire an unlimited life span. The nature of this transformed character requires special comment. Cell senescence may be a form of cell differentiation or, at least, it may be accompanied by increased cell maturation. Senescent cultures contain a progressively increasing fraction of non-dividing cells which are probably in a resting phase (see chapter 2). Thus, in the course of aging the cells may acquire, by some unknown mechanism, an increased ability to undergo transition into the resting state and possibly to undergo a maturation-like process. In other words, perhaps an aging cell is one whose reactivity to transition-controlling factors has been shifted in the opposite direction to that of a neoplastic cell. If these suggestions are correct, infinite life span can be regarded as one of the expressions of decreased ability of neoplastic cells to undergo maturation and transition into the resting state; by decreasing the ability for these transitions neoplastic transformation prevent or reverses senescence.

Certain other characters of transformed cells may be also related to their abnormal transitions from one state into another. For instance, as discussed in chapter 12, the increased glucose uptake of transformed cells may be a consequence of their permanently 'activated' growing state. It would be important to find out whether other parameters altered in the course of transformation (sensitivity to carcinogenic hydrocarbons, production of plasminogen activators, production of stromatogenic factors, etc.) are also changed during integral transitions of normal cells; for instance, during spreading or activation of cell growth. We do not know what particular alterations of transition or, more probably, what complex of these alterations is essential for oncogenicity *in vivo*.

To sum up, at present we do not know of any 'absolute' or 'static' transformed characters, such as the complete loss of ability to synthesize some normal components or, conversely, the appearance of some completely new components. All the known differences between normal and neoplastic cells are of a dynamic nature: neoplastic cells have altered reaction norms to the factors inducing various categories of integral transitions from one state into another.

Transformations of one or many cellular functions?
Inter-relationships between different transformed characters

Various transformed characters reflect alterations in transitions belonging to different categories (e.g. growth, morphology) or alterations in transitions of a particular category induced by different controlling factors (anchorage dependence, serum dependence, density depence of growth). It is possible that alterations of different characters are independent of one another because they involve different cellular components and functions. Then neoplastic evolution *in vitro* is a series of independent changes of different characters, so that the rule of independent progression of neoplastic characters (see chapter 1) has no exceptions. An alterative possibility is that all neoplastic transformations are different expressions of the changes of one cellular function or at least of one group of functions. Then alterations of different characters should be interdependent, although this interdependence is not easy to reveal. At present it is impossible to make final choice between these two possibilities. As mentioned above, different transformed characters can be dissociated from one another, at least partially which favours the suggestion of independent changes. However, in various parts of this book we have also described a number of facts indicating that different transformed characters are not always independent but may appear or disappear in a co-ordinated fashion. We will briefly recapitulate these facts here.

(1) Experiments with transformation of cells by viruses, especially by oncorna-viruses, show that the co-ordinated expression of different transformed properties may depend on the activity of a single locus in the viral genome. This follows from experiments with temperature-sensitive viral mutants and from other data (chapter 2).

(2) Cell lines with temperature-sensitive expressions of several transformed characters have been also obtained from cultures transformed with chemical carcinogens (Yamaguchi & Weinstein, 1975). Apparently in these lines the expression of several characters is changed by the product of one muted gene.

(3) In the course of spontaneous or induced transformation, several characters may appear simultaneously within the same cell line. To recall just one example, in the experiments of Sanford and her group the development of oncogencity in mouse fibroblastic cultures was found to be correlated with two different transformed characters; with reduced spreading on the substratum and with anchorage independence (chapter 2).

(4) Genetic revertants obtained from transformed cells by selection for one particular character may also have other, non-selected revertant characters (chapter 2).

(5) In certain environmental conditions, transformed cultures have been observed to undergo 'phenotypic reversion'; that is, reversible normalization of several traits of the transformed phenotype. For instance, certain transformed cells become anchorage dependent and better spread on the substratum in serum-free medium (chapters 6 and 11). The restoration of a more normal morphology and of anchorage, serum and growth dependences have been induced by hydrocortisone in a transformed mouse ST1 line (Armelin, M. C. S. & Armelin, H. A., 1978). Other agents of a diverse nature have been also observed to partially normalize the transformed phenotype (chapter 6).
(6) In certain environmental conditions non-transformed or minimally transformed cultures may probably undergo 'phenotypic transformations'; that is they may reversibly develop several traits of strongly transformed cells.

In particular, certain growth factors seem to be able to induce phenotypic transformations of morphology (chapter 6) and anchorage requirements (chapter 11). At present, studies of phenotypic transformations and reversions are still in their infancy. We have yet to learn of the spectrum of agents which are able to induce these effects and to describe in detail the changes produced by the various agents.

Alterations in the co-ordinate reaction norm

The data on the co-ordinated appearance of different transformed characters suggest that the abilities to perform transitions of different categories are inter-related. In chapter 11 we discussed the suggestion that a cell has a coordinated reaction norm to various factors (substratum, serum, density) controlling growth. We may now extend this suggestion and suppose that cell has a co-ordinated reaction norm to the various factors controlling transitions of different categories. Any environmental factor that induces a certain transition or alters the ability for this transition will also affect the abilities for other transitions. For instance, any factor that affects growth will also affect morphology (e.g. spreading) and vice versa. The same may be said about the factors controlling morphology and differentiation or growth and differentiation.

The degrees of shift of reaction norms induced by different factors are under genetic control. The existence of temperature-sensitive transformants indicates that products of certain viral and cellular genes may change the ability of a cell to perform several different transitions; that is, may change its co-ordinated reaction norm. From this point of view, genetic changes leading to neoplastic transformation always involve alterations in the co-ordinate reaction norm. Both genetically normal and genetically neoplastic cells may have either a normal or transformed phenotype depending on the sum of stimuli acting on these cells. Genetically normal cells may acquire a transformed phenotype when the sum of external stimuli exceeds a certain limit. This phenotypic change is associated with a decreased ability of 'hyperstimulated' cells to interact with additional external factors. For instance, a cell treated with serum or growth factors in high concentrations may become less able to undergo spreading after contact with the substratum. For the same reason, it may continue to grow despite the absence of additional stimulators such as the substratum or low density. A genetically neoplastic cell has a decreased limit of external stimuli which can cause its transition into the transformed phenotype; therefore, it has a transformed phenotype even in the usual conditions of culture. However, in certain special environments lacking the 'usual' stimuli e.g. in a serum-free medium, it may temporarily revert to a more normal phenotype. Salyamon (1974) suggested that a common feature of neoplastic cells is their altered 'general reactivity'; that is, altered ability to respond to stimuli of various natures. This suggestion is somewhat similar to the considerations discussed above.

What is the nature of cellular functions regulating the combined reaction norm regulating transitions between the normal and transformed phenotypes? For the authors it is natural to think that these functions are associated with the structures responsible for cell surface reactions with external ligands: the cell membrane, cortical structures or internal proteins interacting with membrane-generated signal molecules and ions. We have discussed several forms of these surface-induced reactions, especially pseudopodial attachment reactions and activations of growth. In evolution, both pseudopodial attachment reactions and growth activation reactions have possibly developed from two sets of integrated responses to various surface disturbances (see chapters 3 and 12). Although all the inter-relationships between these two groups of reactions are not yet clear, they obviously have much in common: they may be induced by the same factors and probably have common stages. For instance, contact with the substratum may induce both spreading and activation of growth. Therefore the same alteration of the surface structure may affect the cell's ability to perform both groups of reactions.

Surface reactions are organized in such a way that the specific interaction of a ligand with one type of receptor molecule may cause alterations in many other membrane components and the cell interior. These pleiotropic changes may be due to propagation of some 'signals' within the plane of the membrane or through the cortical layer. 'Signal' ion and molecules (e.g. calcium, magnesium and cyclic nucleotides) are sent from the membrane into the cell interior where they activate special 'regulatory proteins' (e.g. various kinases). Therefore organization of these structures is well adapted for the assessment of the combined effects of many different ligands and for the co-ordinated control of different categories of transition.

The genetic loci altered in transformed cells may be responsible for the synthesis of some components essential for function of this surface-associated co-ordinating system. For instance, these may be the loci controlling the production of endogenous ligands which may react with surface receptors;

an alteration of these loci may lead to permanent 'hyperstimulation' of cells (see chapter 10). These may also be the loci controlling the synthesis of surface receptors, of the systems responsible for production or transport of signal molecules, of the regulatory intracellular proteins. Recent data on the phosphokinase activity of the product of the *src* gene of avian sarcoma virus (Collett & Erickson, 1978) are of great interest in connection with the last of these suggestions. General considerations about the alterations in the co-ordinated reaction norm may be valid for all genetically transformed cells although the concrete mechanisms of these alterations may be different in various lines.

The transformed phenotype and genetic neoplastic transformation

Considerations discussed in the previous paragraph suggest that the same cell may have either a normal or transformed phenotype depending on the environmental conditions. In certain situations genetically neoplastic cells may behave normally and, conversely, genetically normal cells may in certain conditions demonstrate neoplastic behaviour. Probably, an example of the first situation is that provided by the experiments with fibroblasts from patients having an hereditary adenomatosis of the colon (see chapter 2). The behaviour of these genetically changed fibroblasts remains apparently normal while they remain in their natural surroundings: in the subcutaneous connective tissue. However, when these cells are placed in the conditions of culture they reveal certain transformed characters. We do not yet know of any unequivocal case of neoplastic behaviour of genetically normal cells *in vivo*. Perhaps, as discussed in the chapter 1, stem cells of certain teratoblastomas may be normal cells of an early embryo type. When these cells are removed from their normal surroundings they show neoplastic behaviour and when they are returned to their original environment (the early embryo) their behaviour is normalized. However, whether these cells are perfectly normal genetically remains to be proved. It is possible that normal cell populations contain latent genetically changed cells that will grow into minimally neoplastic cell clones only if they are placed in pathological conditions. Another intriguing and equally unproven possibility is that populations of normal adult tissues may contain genetically unchanged cells belonging to some minority cell type e.g. stem cells of the early embryo, and that these cells may acquire a transformed phenotype and form a neoplastic line when their environment is changed.

These considerations show that the selective growth of a phenotypically neoplastic cell line may theoretically require either genetic changes or environmental alterations or both. The concrete relative roles of each of these two types of changes may be different in various neoplasms and should be studied individually for each particular case.

The origins of diversity in neoplastic cells

In previous sections we have discussed considerations favouring the suggestion that all genetically transformed cells have a common type of alterations; namely, alterations in the reactivities controlling several categories of transition. However, different neoplastic cell lines arising *in vivo* and *in vitro* have very diverse properties; in fact, it is impossible to find two identical neoplastic lines. At least three factors may be responsible for the diversity of various neoplastic lines arising within the same normal populations: various properties of normal progenitor cells, various degrees of common cellular lesions and a selection of variants with additional alterations.

Various normal cells within the same tissue may be different from each other depending on the degree of their differentiation and on their 'positional information' (see chapter 13). Naturally, similar genetic changes of two different normal cells may lead to the formation of two non-identical neoplastic lines.

Another factor contributing to the diversity of neoplastic lines is probably the different degree of the changes characteristic for transformation. Depending on the degree of these changes, transformed characters may be expressed more strongly or more weakly. Certain transformed characters (e.g. anchorage independence) are usually expressed only in strongly transformed cells. The stage of transformation at which each particular altered character will be detected obviously depends on the discriminating power of our methods. Minimally expressed transformed characters are not revealed by the usual methods but may be detected in certain neoplastic lines when more exact quantitative tests are applied. This may be true for morphological transformation (see chapter 5), anchorage independence (see chapter 8) and other characters.

The third factor responsible for the diversity of neoplastic populations is selection. Normal cells are adapted to their environments by means of regulating mechanisms controlling the transitions of different categories. In particular, morphogenetic reactions lead to the formation of well-organized tissue structures with a relatively stable microenvironment. Transitions from the growing to the resting state limit the intensity of proliferation and probably also reduce the consumption of nutrients. These mechanisms decrease the competition of cells within the normal tissue. In the tissues containing rapidly proliferating cells 'suicide maturation' prevents the accumulation of genetic variants in this cell group. In contrast, neoplastic cell populations have a decreased ability for cellular adaptation but probably an increased ability for production of genetic variants as well as an increased intensity of the selection of these variants. The accumulation of genetic variants may be facilitated by permanent proliferation and decreased maturation. In addition, as transformed cells are less diversified in their reactions, many genetic changes could be less harmful for them than for normal cells. For instance, the changes that can decrease cell viability in the resting state will not be eliminated from the population if the cells are unable to enter this state. These variant cells will remain in the neoplastic population and increase its genetic heterogeneity. Competition between the cells of a neoplastic population is probably more intense than that between normal cells due to the altered micro-environments and the inability of transformed cells to limit their requirements.

The progression of neoplastic populations may involve the selection of at least three groups of variant cells.

(1) *Cells more strongly transformed than parent cells.* Many examples of additional transformations have been described in this book. According to the suggestions discussed above, these are the cells with additional genetic alterations of the integrating mechanism responsible for the co-ordinated reaction norm.

(2) *Cells with additional isolated changes of transformed characters, that is, with changes affecting only one character.* According to the considerations discussed above, true transformations alter some integrating mechanism regulating the cellular ability to perform several categories of transition. Each of these transitions is a complex multicomponent process. Certain genetic changes may not affect the central integrating mechanism but alter one of the transitions regulated by this mechanism. For instance, such a change may alter one of many molecules participating in cell spreading but not essential for growth regulation and vice versa. An example of mutants of this type is probably that obtained by Pouysségur & Pastan (1976). They selected mutants of 3T3 cells that were defective in their attachment, but had no abnormal growth properties, e.g. anchorage or serum independence. It would be important to study in detail many different variants with 'isolated' transformed characters. A comparative analysis of the properties of these 'partial quasi-transformants' and of true transformants might give much information about the common and unique components of different transitions.

(3) *Cells with different mutations of properties not related directly to transformed characters.* Numerous variants with different metabolic alterations isolated from transformed cultures probably belong to this group. In-vivo selection of certain variants of this type may be the basis for the development of resistance to anti-tumour drugs in neoplastic populations.

To sum up, the same group of cellular functions may be altered in all genetic neoplastic transformations. The nature of these alterations is such that they make cell populations more heterogeneous and promote selection of variant cells with additional changes. As a result, neoplastic populations become so diversified that their common features are recognized only with difficulty. Alterations, common for different neoplastic lines, probably, involve cell functions controlling the co-ordinated reaction norm to factors inducing cell transitions from one state to another. At present we know little about the mechanisms of these functions. In fact, their very existence in normal cells remains a suggestion based mainly on the results of the analysis of neoplastic transformations.

References

Aaronson, S. A. & Todaro, G. J. (1968). Development of 3T3-like lines from Balb/c mouse embryo cultures: transformation susceptibility to SV-40. *J. Cell. Physiol.* **72**, 141–8.

Aaronson, S. A., Todaro, G. J. & Freeman, A. E. (1970). Human sarcoma cells in culture. Identification by colony-forming ability on monolayers of normal cells. *Exp. Cell Res.* **61**, 1–5.

Abelev, G. I. (1971). Alpha-fetoprotein in ontogenesis and its association with malignant tumors. *Adv. Cancer Res.* **14**, 295–358.

Abelev, G. I. (1976). α-fetoprotein as a marker of embryospecific differentiations in normal and tumor tissues. *Transplant. Rev.* **20**, 3–37.

Abelev, G. I. (1978). Experimental study of alpha-fetoprotein re-expression in liver regeneration and hepatocellular carcinomas. In *Cell Differentiation and Neoplasia*, ed. G. F. Saunders, pp. 257–69. New York: Raven Press.

Abelson, H. T., Johnson, L. F., Penman, S. & Green, H. (1974). Changes in RNA in relation to growth of the fibroblast. II. The lifetime of mRNA, rRNA and tRNA in resting and growing cells. *Cell* **1**, 161–5.

Abercrombie, M. (1961). The bases of the locomotory behaviour of fibroblasts. *Exp. Cell Res.* Suppl. **8**, 188–98.

Abercrombie, M. (1970). Contact inhibition in tissue culture. *In vitro* **6**, 128–42.

Abercrombie, M. & Dunn, G. A. (1975). Adhesions of fibroblasts to substratum during contact inhibition observed by interference reflection microscopy. *Exp. Cell Res.* **92**, 57–62.

Abercrombie, M., Dunn, G. A. & Heath, J. P. (1976). Locomotion and contraction in non-muscle cells. In *Contractile Systems in Non-muscle Tissue*, ed. S. V. Perry, A. Margreth & R. S. Adelstein, pp. 3–11. Amsterdam: Elsevier/North-Holland Biomedical Press.

Abercrombie, M., Dunn, G. A. & Heath, J. P. (1977). The shape and movement of fibroblasts in culture. In *Cell and Tissue Interaction*, ed. J. W. Lash and M. M. Burger, pp. 57–70. New York: Raven Press.

Abercrombie, M. & Heaysman, J. E. M. (1953). Observations on the social behaviour of cells in tissue culture. I. Speed of movement of chick heart fibroblasts in relation to their mutual contacts. *Exp. Cell Res.* **5**, 111–31.

Abercrombie, M. & Heaysman, J. E. M. (1954). Observations on the social behaviour of cells in tissue culture. II. 'Monolayering' of fibroblasts. *Exp. Cell Res.* **6**, 293–306.

Abercrombie, M. & Heaysman, J. E. M. (1976). Invasive behaviour between sarcoma and fibroblast populations in cell culture. *J. Natl. Cancer Inst.* **56**, 561–70.

Abercrombie, M., Heaysman, J. E. & Karthauser, H. M. (1957). Social behavior of cells in tissue culture. III. Mutual influence of sarcoma cells and fibroblasts. *Exp. Cell Res.* **13**, 276–92.

Abercrombie, M., Heaysman, J. E. M. & Pegrum, S. M. (1970*a*). The locomotion of fibroblasts in culture. I. Movements of the leading edge. *Exp. Cell Res.* **59**, 393–8.

Abercrombie, M., Heaysman, J. E. M. & Pegrum, S. M. (1970*b*). The locomotion of fibroblasts in culture. II. 'Ruffling'. *Exp. Cell Res.* **60**, 437–44.

Abercrombie, M., Heaysman, J. E. M. & Pegrum, S. M. (1970*c*). The locomotion of fibroblasts in culture. III. Movements of particles on the dorsal surface of the leading lamella. *Exp. Cell Res.* **62**, 389–98.

Abercrombie, M., Heaysman, J. E. M. & Pegrum, S. M. (1971). The locomotion of fibroblasts in culture. IV. Electron microscopy of the leading lamella. *Exp. Cell Res.* **67**, 359–67.

Abercrombie, M., Heaysman, J. E. M. & Pegrum, S. M. (1972). The locomotion of fibroblasts in culture. V. Surface marking with concanavalin A. *Exp. Cell Res.* **73**, 536–9.

Abercrombie, M., Lamont, D. M. & Stephenson, E. M. (1968). The monolayering in tissue culture of fibroblasts from different sources. *Proc. Roy. Soc. Lond.* B **170**, 349–60.

Aboe-Sabe, M. (ed.) (1976). *Cyclic Nucleotides and the Regulation of Cell Growth.* New York: Halsted Press..

Absher, P. M. & Absher, R. G. (1976). Clonal variation and aging of diploid fibroblasts. Cinematographic studies of cell pedigrees. *Exp. Cell Res.* **103**, 247–55.

Adamo, S., De Luca, L. M., Akalovsky, I. & Bhat, P. V. (1979). Retinoid-induced adhesion in cultured transformed mouse fibroblasts. *J. Natl. Cancer Inst.* **62**, 1473–8.

Adams, S. L., Sobel, M. E., Howard, B. H., Olden, K., Yamada, K. M., de Crombrugghe, B. & Pastan, I. (1977). Levels of translatable mRNAs for cell surface protein, collagen precursors, and two membrane proteins are altered in Rous sarcoma virus-transformed chick embryo fibroblasts. *Proc. Natl. Acad. Sci. USA* **74**, 3399–403.

Adelstein, R. S. (1978). Myosin phosphorylation, cell motility and smooth muscle contraction. *Trends Biochem. Sci.* **3**, 27–30.

Ahn, H. S., Horowitz, S. G., Eagle, H. & Makman, M. H. (1978). Effects of cell density and cell growth alterations on cyclic nucleotide levels in cultured human diploid fibroblasts. *Exp. Cell Res.* **114**, 101–10.

Albertini, D. F. & Anderson, E. (1977). Microtubule and microfilament rearrangements during capping of concanavalin A. Receptors on cultured ovarian granulosal cells. *J. Cell Biol.* **73**, 111–27.

Albrecht-Buehler, G. (1977). The phagokinetic tracks of 3T3 cells. *Cell* **11**, 395–404.

Albrecht-Buehler, G. & Lancaster, R. M. (1976). A quantitative description of the extension and retraction of surface protrusions in spreading 3T3 mouse fibroblasts. *J. Cell Biol.* **71**, 370–82.

Alexander, E. & Henkart, P. (1976). The adherence of human Fc receptor-bearing lymphocytes to antigen–antibody complexes. II. Morphologic alterations induced by the substrate. *J. Exp. Med.* **143**, 329–47.

Alexandrov, V. Ya. & Wolfenson, L. G. (1956). Reversible contractions of connective tissue cells caused by diverse agents. *Zurnal obschei Biologii* **17** N 2, 142–53 (in Russian).

Alfred, L. J., Globerson, A., Berwald, J. & Prehn, R. T. (1964). Differential toxicity response of normal and neoplastic cells *in vitro* to 3,4-benzopyrene and 3-methylcholanthrene. *Br. J. Cancer* **18**, 159–64.

Algire, G. H., Chalkley, H. W., Legallais, F. Y. & Park, H. D. (1945). Vascular reactions of normal and malignant tissues *in vivo*. I. Vascular reactions of mice to wounds and to normal and neoplastic transplants. *J. Natl. Cancer Inst.* **6**, 73–85.

Ali, I. U. & Hynes, R. O. (1978*a*). Effects of LETS glycoprotein on cell motility. *Cell* **14**, 439–46.

Ali, I. U. & Hynes, R. O. (1978*b*). Role of disulfide bonds in the attachment and function of large, external, transformation-sensitive glycoprotein at the cell surface. *Biochim. Biophys. Acta* **510**, 140–50.

Ali, I. U., Mautner, V., Lanza, R. & Hynes, R. O. (1977). Restoration of normal morphology, adhesion and cytoskeleton in transformed cells by addition of a transformation-sensitive surface protein. *Cell* **11**, 115–26.

Allen, R. W. & Moskowitz, M. (1978*a*). Arrest of cell growth in the G_1 phase of the cell cycle by serine deprivation. *Exp. Cell Res.* **116**, 127–37.

Allen, R. W. & Moskowitz, M. (1978*b*). Regulation of the rate of protein synthesis in BHK 21 cells by exogenous serine. *Exp. Cell Res.* **116**, 139–52.

Allison, A. C. (1973). The role of microfilaments and microtubules in cell movement, endocytosis and exocytosis. In *Locomotion of Tissue Cells*, ed. M. Abercrombie. Ciba Foundation Symposium, vol. 14, pp. 109–43. Amsterdam: Associated Scientific Publishers.

Altanerova, V. (1975). Transformation of BHK 21/13 cells by chemical carcinogens and mutagens. *Neoplasma* **22**, 599–606.

Altenburg, B. C., Somers, K. & Steiner, S. (1976*a*). Altered microfilament structure in cells transformed with a temperature-sensitive transformation mutant of murine sarcoma virus. *Cancer Res.* **36**, 251–7.

Altenburg, B. C., Via, D. P. & Steiner, S. H. (1976*b*). Modification of the phenotype of murine sarcoma virus-transformed cells by sodium butyrate. Effect on morphology and cytoskeletal elements. *Exp. Cell Res.* **102**, 223–31.

Ambros, V. R., Chen, L. B. & Buchanan, J. M. (1975). Surface ruffles as markers for studies of cell transformation by rous sarcoma virus. *Proc. Natl. Acad. Sci. USA* **72**, 3144–48.

Ambrose, E. J. (1976). The role of cell surface dynamics. In *Cellular Interactions, Progress in Differentiation Research*, ed. N. Müller-Berat, pp. 253–60. Amsterdam: North-Holland.

Ambrose, E. J., Batzdorf, U., Osborn, J. S. & Stuart, P. R. (1970). Sub-surface structures in normal and malignant cells. *Nature, Lond.* **227**, 397–8.

Ambrose, E. J. & Easty, D. M. (1976). Time-lapse filming of cellular interactions within living tissues. III. The role of cell shape. *Differentiation* **6**, 61–70.

Ambrose, E. J. & Ellison, M. (1968). Studies of specific properties of tumour cell membranes using stereoscan microscopy. *Eur. J. Cancer* **4**, 459–62.

Anderson, J. L. & Martin, R. G. (1976). SV-40 transformation of mouse brain cells: critical role of gene A in maintenance of the transformed phenotype. *J. Cell. Physiol.* **88**, 65–76.

Anderson, P. J. (1976). Chicken muscle and fibroblast actin structure. *Biochem. J.* **159**, 185–7.

Anderson, W. F., Evans, V. J., Price, F. M. & Dunn, T. B. (1966). Neoplastic transformation in cells explanted from the kidneys of 3-day-old C3H mice. *J. Natl. Cancer Inst.* **36**, 953–63.

Andrianov, L. A., Belitsky, G. A., Ivanova, O. J., Khesina, A. Y., Khitrovo, S. S., Shabad, L. M. & Vasiliev, J. M. (1967). Metabolic degradation of 3,4-benzopyrene in the cultures of normal and neoplastic fibroblasts. *Br. J. Cancer* **21**, 566–75.

Andrianov, L. A., Belitsky, G. A., Vasiliev, J. M., Domnina, L. V., Ivanova, O. Y. Olshevskaja, L. V., Panov, M. A., Slavnaja, I. L. & Khesina, A. Y. (1971). *Action of Carcinogenic Hydrocarbons on Cells*. Moscow: Medizina (in Russian).

Antoniades, H. N. & Scher, C. D. (1977). Radio-immunoassay of a human serum growth factor for Balb/c-3T3 cells: derivation from platelets. *Proc. Natl. Acad. Sci. USA* **74**, 1973–7.

Armato, U., Draghi, E. & Andreis, P. G. (1977). Effects of purine cyclic nucleotides on the growth of neonatal rat hepatocytes in primary tissue culture. *Exp. Cell Res.* **105**, 337–47.

Aremelin, H. A. (1973). Pituitary extracts and steroid hormones in the control of 3T3 cells growth. *Proc. Natl. Acad. Sci. USA* **70**, 2702–6.

Armelin, M. C. S. & Armelin, H. A. (1978). Steroid hormones mediate reversible phenotypic transition between transformed and untransformed states in mouse fibroblasts. *Proc. Natl. Acad. Sci. USA* **75**, 2805–9.

Arronet, N. J. (1971). *Muscular and Cellular Contractile Models.* Leningrad: Nauka (in Russian).

Ash, J. F., Louvard, D. & Singer, S. J. (1977). Antibody-induced lingages of plasma membrane proteins to intracellular actomyosin-containing filaments in cultured fibroblasts. *Proc. Natl. Acad. Sci. USA* **74**, 5584–8.

Ash, J. F. & Singer, S. J. (1976). Concanavalin-A-induced transmembrane linkage of concanavalin A surface receptors to intracellular myosin-containing filaments. *Proc. Natl. Acad. Sci. USA* **73**, 4575–9.

Ashwell, G. & Morell, A. G. (1974). The role of surface carbohydrates in the hepatic recognition and transport of circulating glycoproteins. *Adv. Enzymol.* **41**, 99–128.

Aubery, M., Bourrillin, R. & O'Neill, C. H. (1975). Stimulation of the proliferation of normal BHK 21 cultured fibroblasts by plant lectins. *Exp. Cell Res.* **93**, 47–54.

Aubin, J. E., Carlsen, S. A. & Ling, V. (1975). Colchicine permeation is required for inhibition of concanavalin A capping in Chinese hamster ovary cells. *Proc. Natl. Acad. Sci. USA* **72**, 4516–20.

Auerbach, R., Kubai, L. & Sidky, G. (1976). Angiogenesis induction by tumors, embryonic tissues and lymphocytes. *Cancer Res.* **36**, 3435–40.

Auerbach, R., Morrissey, L. W. & Sidky, Y. A. (1978). Regional differences in the incidence and growth of mouse tumors following intradermal or subcutaneous inoculation. *Cancer Res.* **38**, 1739–44.

Augenlicht, L. H. & Baserga, R. (1974). Changes in the G_0 state of WI-38 fibroblasts at different times after confluence. *Exp. Cell Res.* **89**, 255–62.

Augenlicht, L. H. & Lipkin, M. (1977). Serum stimulation of human fibroblasts: effect on non-histones of nuclear ribonucleoprotein and chromatin. *J. Cell Physiol.* **92**, 129–36.

Averdunk, R. & Lauf, P. K. (1975). Effect of mitogens on sodium, potassium transport, [^{3}H]ouabain binding and adenosine triphosphatase activity in lymphocytes. *Exp. Cell Res.* **93**, 331–42.

Avery, R. J. & Levy, J. A. (1978). Relationship of endogenous murine xenotropic type C virus production to spontaneous transformation of cultured cells. *J. Gen. Virol.* **39**, 427–35.

Axelrod, D., Ravdin, P., Koppel, D. E., Schlessinger, J., Webb, W. W., Elson, E. L. & Podleski, T. R. (1976). Lateral motion of fluorescently labelled acetylcholine receptors in membranes of developing muscle fibers. *Proc. Natl. Acad. Sci. USA.* **73**, 4594–8.

Ayad, S. R. & Foster, S. J. (1977). Adenylate cyclase activity from normal, malignant and somatic cell hybrid, effect of PGE_1, NaF and guanyl nucleotides. *Exp. Cell Res.* **109**, 87–94.

Azarnia, R. & Loewenstein, W. R. (1976). Intercellular communication and tissue growth. VII. A cancer cell strain with retarded formation of permeable membrane junction and reduced exchange of a 330-dalton molecule. *J. Membrane Biol.* **30**, 175–86.

Azarnia, R. & Loewenstein, W. R. (1977). Intercellular communication and tissue growth: VIII. A genetic analysis of junctional communication and cancerous growth. *J. Membrane Biol.* **34**, 1–28.

Bader, J. P. (1976). Sodium: a regulator of glucose uptake in virus-transformed and nontransformed cells. *J. Cell. Physiol.* **89**, 677–82.

Bader, J. P., Sege, R. & Brown, N. R. (1978). Sodium concentrations affect metabolite uptake and cellular metabolism. *J. Cell. Physiol.* **95**, 179–88.

Badley, R. A., Lloyd, C. W., Woods, A., Carruthers, L., Allcock, C. & Rees, D. A. (1978). Mechanisms of cellular adhesion. III. Preparation and preliminary characterisation of adhesions. *Exp. Cell Res.* **117**, 231–44.

Baker, M. E. (1976). Colchicine inhibits mitogenesis in CI300 neuroblastoma cells that have been arrested in G_0. *Nature, Lond.* **262**, 785–86.

Baker, M. E. (1977). Density dependent regulation of colchicine inhibition of mitogenesis in 3T3 cells arrested in G_0. *Biochem. Biophys. Res. Commun.* **77**, 738–45.

Balk, S. D. (1971). Calcium as a regulator of the proliferation of normal but not of transformed chicken fibroblasts in a plasma-containing medium. *Proc. Natl. Acad. Sci. USA* **68**, 271–5.

Balk, S. D., Whitfield, J. F., Youdale, T. & Braun, A. C. (1973). Roles of calcium, serum, plasma and folic acid in the control of proliferation of normal and Rous sarcoma virus-infected chicken fibroblasts. *Proc. Natl. Acad. Sci. USA* **70**, 675–9.

Baltimore, D. (1976). Viruses, polymerases, and cancer. *Science* **192**, 632–6.

Banks, J. R., Bhavanada, V. R. & Davidson, E. A. (1977). Chemical and biological properties of B16 murine melanoma cells grown in defined medium containing bovine serum albumin. *Cancer Res.* **37**, 4336–45.

Bard, J. & Elsdale, T. (1971). Specific growth regulation in early subcultures of human diploid fibroblasts. In *Growth Control in Cell Cultures*, ed. G. E. W. Wolstenholme & J. Knight, pp. 187–97. Edinburgh: Churchill Livingstone.

Barker, B. E. & Sanford, K. K. (1970). Cytologic manifestation of neoplastic transformation *in vitro*. *J. Natl. Cancer Inst.* **44**, 39–63.

Barnes, D. W. & Colowick, S. P. (1977). Stimulation of sugar uptake and thymidine incorporation in mouse 3T3 cells by calcium phosphate and other extracellular particles. *Proc. Natl. Acad. Sci. USA.* **74**, 5593–7.

Barrett, J. C., Crawford, B. D., Grady, D. L., Hester, L. D., Jones, P. A., Benedict, W. F. & Ts'o, P. O. P. (1977). Temporal acquisition of enhanced fibrinolytic activity by Syrian hamster embryo cells following treatment with benzo(a)pyrene. *Cancer Res.* **37**, 3815–23.

Barrett, J. C. & Ts'o, P. O. P. (1978*a*). Relationship between somatic mutation and neoplastic transformation. *Proc. Natl. Acad. Sci. USA* **75**, 3297–3301.

Barrett, J. C. & Ts'o, P. O. P. (1978*b*). Evidence for the progressive nature of neoplastic transformation *in vitro*. *Proc. Natl. Acad. Sci. USA.* **75**, 3761–65.

Barski, G. & Belehradek, J. (1965). Etude microcinematographique du mecanisme d'invasion cancereuse en cultures de tissu normal associé aux cellules malignes. *Exp. Cell Res.* **37**, 464–80.

Bartholomew, J. C., Neff, N. T. & Ross, P. A. (1976*a*). Stimulation of WI-38 cell cycle transit: effect of serum concentration and cell density. *J. Cell. Physiol.* **89**, 251–8.

Bartholomew, J. C., Yokota, H. & Ross, P. A. (1976*b*). Effect of serum on the growth of Balb 3T3 A3I mouse fibroblasts and an SV-40-transformed derivative. *J. Cell. Physiol.* **88**, 277–86.

Baserga, R. (1976). *Multiplication and Division in Mammalian Cells*. New York: Margel Dekker.

Baserga, R. (1978). Resting cells and G_1 phase of the cell cycle. *J. Cell Physiol.* **95**, 377–86.

Basilico, C. (1978). Selective production of cell cycle specific ts mutants. *J. Cell. Physiol.* **95**, 367–76.

Baylin, S. B., Gann, D. S. & Hsu, S. H. (1976). Clonal origin of inherited medullary thyroid carcinoma and pheochromocytoma. *Science* **193**, 321–3.

Begg, D. A., Rodewald, R. & Rebhun, L. J. (1978). The visualization of actin filament polarity in thin sections. Evidence for the uniform polarity of membrane-associated filaments. *J. Cell Biol.* **79**, 846–52.

Belitsky, G. A., Vasiliev, J. M., Ivanova, O. Y., Lavrova, N. A., Prigozhina, E. L., Samoilina, N. L., Stavrovskaya, A. A., Khesina, A. Ya. & Shabad, L. M. (1970). Metabolism of benzo(a)pyrene by cells of different mammals *in vitro* and toxic effect of polycyclic hydrocarbons on the cells. *Voprosi Oncologii* **16**, 53–58 (in Russian).

Bell, P. B. Jr. (1977). Locomotory behavior, contact inhibition, and pattern formation of 3T3 and polyoma virus-transformed 3T3 cells in culture. *J. Cell Biol.* **74**, 963–82.

Belyaeva, N. N. & Vasiliev, J. M. (1971*a*). DNA synthesis in the stromal cells of spontaneous mouse tumors. *Byulleten experimentalnoj Biologii i Meditsini* **72**, N 7, 77–80 (in Russian).

Belyaeva, N. N. & Vasiliev, J. M. (1971*b*). Proliferation of parenchyma and stroma in the precancerous hyperplastic nodules of murine mammary glands. *Byulleten experimentalnoj Biologii i Meditsini*. **72**, N 10, 75–7 (in Russian).

Benditt, E. P. & Benditt, J. M. (1973). Evidence for a monoclonal origin of human atherosclerotic plaques. *Proc. Natl. Acad. Sci. USA* **70**, 1753–6.

Benedict, W. F., Rucker, N., Faust, J. & Kouri, R. E. (1975). Malignant transformation of mouse cells by cigarette smoke condensate. *Cancer Res.* **35**, 857–60.

Bennett, G. S., Fellini, S. A. & Holtzer, H. (1978). 100Å filaments in different cell types: distribution and antigenic specificity. *J. Cell Biol.* **79**, 261a.

Berenblum, I. (1941). The cocarcinogenic action of croton resin. *Cancer Res.* **1**, 44–8.

Berenblum, I. (1974). *Carcinogenesis as a Biological Problem*. Amsterdam: North-Holland.

Berenblum, I. (1978). Established principles and unresolved problems in carcinogenesis. *J. Natl. Cancer Inst.* **60**, 723–6.

Berenblum, I. & Shubik, P. (1949). The persistence of latent tumour cells induced in the mouse's skin by a single application of 9,10-dimethyl-1,2-benzanthracene. *Br. J. Cancer* **3**, 384–6.

Berlin, R. D. & Oliver, J. M. (1978). Analogous ultrastructure and surface properties during capping and phagocytosis in leukocytes. *J. Cell Biol.* **77**, 789–804.

Bershadsky, A. D. & Gelfand, V. I. (1970). Ageing and malignization of cell strains. *Tsitologiya* **12**, 423–35 (in Russian).

Bershadsky, A. D., Gelfand, V. I., Guelstein, V. I., Vasiliev, J. M. & Gelfand, I. M. (1976). Serum dependence of expression of the transformed phenotype: experiments with subline of mouse L fibroblasts adapted to growth in serum-free medium. *Int. J. Cancer* **18**, 83–92.

Bershadsky, A. D., Gelfand, V. I., Svitkina, T. M. & Tint, I. S. (1978*b*). Microtubules in mouse embryo fibroblasts extracted with Triton X-100. *Cell Biol. Int. Rep.* **2**, 425–32.

Bershadsky, A. D. & Guelstein, V. I. (1973). Characteristics of cell aggregation on the surface of glutaraldehyde-fixed monolayer. *Ontogenez* **4**, 472–80 (in Russian).

Bershadsky, A. D. & Guelstein, V. I. (1976). Reversible inhibition of cell aggregation by cytochalasin B and colcemid. *Byulleten experimentalnoj Biologii i Meditsini* **81**, 49–51 (in Russian).

Bershadsky, A. D. & Lustig, T. M. (1975). Aggregation of normal and transformed cells attached to substratum. *Tsitologiya* **17**, 639–46 (in Russian).

Bershadsky, A. D., Tint, I. S., Gelfand, V. I., Rosenblat, V. A., Vasiliev, J. M. &

Gelfand, I. M. (1978*a*). Microtubular system in cultured mouse epithelial cells. *Cell Biol. Int. Rep.* **2**, 345–51.

Bershadsky, A. D., Vasiliev, J. M., Stavrovskaya, A. A., Stromskaya, T. P. (1979). Cell rounding in variant L sublines induced by a moderate lowering of the temperature. *Tsitologia* **21**, 703–10 (in Russian).

Berwald, Y. & Sachs, L. (1963). In-vitro cell transformation with chemical carcinogens. *Nature, Lond.* **200**, 1182–4.

Beug, H. & Graf, T. (1977). Isolation of clonal strains of chicken embryo fibroblasts. *Exp. Cell Res.* **107**, 417–28.

Beyer, C. F. & Bowers, W. E. (1975). Periodate and concanavalin A induce blast transformation of rat lymphocytes by an indirect mechanism. *Proc. Natl. Acad. Sci. USA* **72**, 3590–3.

Bhisey, A. N. & Freed, J. J. (1971). Ameboid movement induced in cultured macrophages by colchicine or vinblastine. *Exp. Cell Res.* **64**, 419–29.

Biquard, J. M. (1974). Agglutination by concanavalin A of normal chick embryo fibroblasts treated by 5-bromodeoxyuridine (BrdU). *J. Cell. Physiol.* **84**, 459–62.

Birdwell, C. R. & Gospodarowicz, D. (1977). Factors from 3T3 cells stimulate proliferation of cultured vascular endothelial cells. *Nature, Lond.* **268**, 528–30.

Birdwell, C. R., Gospodarowicz, D. & Nicolson, G. L. (1978). Identification, localization, and role of fibronectin in cultured bovine endothelial cells. *Proc. Natl. Acad. Sci. USA* **75**, 3273–7.

Bischoff, F. & Bryson, G. (1964). Carcinogenesis through solid state surface. *Progr. Exp. Tumor Res.* **5**, 85–133.

Bissel, M. J., Hatié, C. & Rubin, H. (1972). Patterns of glucose metabolism in normal and virus-transformed chick cells in tissue culture. *J. Natl. Cancer Inst.* **49**, 555–65.

Bloch-Stacher, N. & Sachs, L. (1977). Identification of a chromosome that controls malignancy in Chinese hamster cells. *J. Cell. Physiol.* **93**, 205–12.

Blose, S. H., Shelanski, M. L. & Chacko, S. (1977). Localization of bovine brain filament antibody on intermediate (100Å) filaments in guinea pig vascular endothelial cells and chick cardiac muscle cells. *Proc. Natl. Acad. Sci. USA* **74**, 662–5.

Blumberg, P. M., Driedger, E. & Rossow, P. W. (1976). Effect of a phorbol ester on a transformation-sensitive surface protein of chick fibroblasts. *Nature, Lond.* **264**, 446–7.

Blumberg, P. M. & Robbins, P. W. (1975). Effect of proteases on activation of resting chick embryo fibroblasts and on cell surface proteins. *Cell* **6**, 137–47.

Boettiger, D. (1974). Reversion and induction of Rous sarcoma virus expression in virus-transformed baby hamster kidney cells. *Virology* **62**, 522–9.

Boitsova, L. Y. & Potapova, T. V. (1974). The capacity to formation of intercellular contacts in normal mouse fibroblasts and L cells grown in a mixed culture with normal fibroblasts. *Byulleten experimentalnoi Biologii i Meditsini* **78**, N 8, 84–7 (in Russian).

Boitsova, L. Y., Vasiliev, J. M. & Potapova, T. V. (1975). Low-resistance intercellular junctions in the cultures of normal and transformed fibroblasts. *Tsitologiya* **7**, 279–88 (in Russian).

Bolen, J. B. & Smith, G. L. (1977). Effects of withdrawal of a mitogenic stimulus on progression of fibroblasts into S phase: differences between serum and purified multiplication-stimulating activity. *J. Cell. Physiol.* **91**, 441–8.

Boon, T. & Kellermann, O. (1977). Rejection by syngeneic mice of cell variants obtained by mutagenesis of a malignant teratocarcinoma cell line. *Proc. Natl. Acad. Sci. USA* **74**, 272–5.

Boone, C. W. (1975). Malignant hemangioendotheliomas produced by subcutaneous inoculation of Balb/3T3 cells attached to glass beads. *Science* **188**, 68–70.

Boone, C. W. (1976). Comment on comparing the membrane transport properties of 'normal' and 'transformed' Balb/3T3 cells. *J. Cell. Physiol.* **89**, 757–8.

Boone, C. W. & Jacobs, J. B. (1976). Sarcomas routinely produced from putatively non-tumorigenic Balb/3T3 and C3H/10T 1/2 cells by subcutaneous inoculation attached to plastic platelets. *J. Supramol. Struc.* **5**, 131–7.

Boone, C. W., Takeichi, N., Paranjpe, M. & Gilden, R. (1976). Vasoformative sarcomas arising from Balb/3T3 cells attached to solid substrates. *Cancer Res.* **36**, 1626–33.

Borek, C. & Fenoglio, C. M. (1976). Scanning electron microscopy of surface features of hamster cells transformed *in vitro* by X-irradiation. *Cancer Res.* **36**, 1325–34.

Borek, C., Hall, E. J. & Rossi, H. H. (1978). Malignant transformation in cultured hamster embryo cells produced by X-rays, 430-keV monoenergetic neutrons, and heavy ions. *Cancer Res.* **38**, 2997–3005.

Borenfreund, E., Higgins, P. G., Steinglass, M. & Bendich, A. (1975). Properties and malignant transformation of established rat liver parenchymal cells in culture. *J. Natl. Cancer Inst.* **55**, 375–84.

Bornstein, P. & Ash, J. F. (1977). Cell surface-associated structural proteins in connective tissue cells. *Proc. Natl. Acad. Sci. USA* **74**, 2480–4.

Bose, S. K. & Zlotnick, B. J. (1973). Growth and density-dependent inhibition of deoxyglucose transport in Balb 3T3 cells and its absence in cells transformed by murine sarcoma virus. *Proc. Natl. Acad. Sci. USA.* **70**, 2374–8.

Bosmann, H. B., Lockwood, T. & Morgan, H. R. (1974). Surface biochemical changes accompanying primary infection with Rous sarcoma virus. II. Proteolytic and glycosidase activity and sublethal autolysis. *Exp. Cell Res.* **83**, 25–30.

Botchan, M., Topp, W. & Sambrook, J. (1976). The arrangement of Simian virus 40 sequences in the DNA of transformed cells. *Cell* **9**, 269–87.

Bouck, N., Beales, N., Shenk, T., Berg, P. & Di Mayorca, G. (1978). New region of the Simian virus 40 genome required for efficient viral transformation. *Proc. Natl. Acad. Sci. USA* **75**, 2473–7.

Bouck, N. & Di Mayorca, G. (1976). Somatic mutation as the basis for malignant transformation of BHK cells by chemical carcinogens. *Nature, Lond.* **264**, 722–7.

Bourguignon, L. Y. & Singer, S. J. (1977). Transmembrane interactions and the mechanism of capping of surface receptors by their specific ligands. *Proc. Natl. Acad. Sci. USA* **74**, 5031–5.

Bourne, H. R. & Rozengurt, E. (1976). An 18000 molecular weight polypeptide induces early events and stimulates DNA synthesis in cultured cells. *Proc. Natl. Acad. Sci. USA* **3**, 4555–9.

Bowen-Pope, D. E. & Rubin, H. (1977). Magnesium and calcium effects on uptake of hexoses and uridine by chick embryo fibroblasts. *Proc. Natl. Acad. Sci. USA* **74**, 1585–9.

Bowles, D. J. & Kauss, H. (1976). Isolation of a lectin from liver plasma membrane and its binding to cellular membrane receptors *in vitro*. *FEBS Lett.* **66**, 16–19.

Boyde, A., Bailey, E. & Vesely, P. (1974). SEM studies on the surface of various rat cell lines treated with cytochalasin B, colcemide and vinblastine. In *Scanning Electron Microscopy/1974 (Part III)*, ed. O. Johari, pp. 597–603. Chicago: IIT Research Institute.

Boyde, A., Weiss, R. A. & Vesely, P. (1972). Scanning electron microscopy of cells in culture. *Exp. Cell Res.* **71**, 313–24.

Boynton, A. L. & Whitfield, J. F. (1976). Different calcium requirements for proliferation of conditionally and unconditionally tumorigenic mouse cells. *Proc. Natl. Acad. Sci. USA* **73**, 1651–4.

Boynton, A. L. & Whitfield, J. F. (1978). Calcium requirements for the proliferation of cells infected with a temperature-sensitive mutant of Rous sarcoma virus. *Cancer Res.* **38**, 1237–40.

Boynton, A. L., Whitfield, J. F. & Isaacs, R. J. (1976). Calcium-dependent stimulation of BALB/c 3T3 mouse cell DNA synthesis by a tumor-promoting phorbol ester (PMA). *J. Cell. Physiol.* **87**, 25–32.

Boynton, A. L., Whitfield, J. F., Isaacs, R. J. & Morton, H. J. (1974). Control of 3T3 cell proliferation by calcium. *In vitro* **10**, 12–17.

Boynton, A. L., Whitfield, J. F., Isaacs, R. J. & Tremblay, R. G. (1977*a*). Different extracellular calcium requirements for proliferation of nonneoplastic, preneoplastic and neoplastic mouse cells. *Cancer Res.* **37**, 2657–61.

Boynton, A. L., Whitfield, J. F., Isaacs, R. J. & Tremblay, R. (1977*b*). The control of human WI-38 cell proliferation by extracellular calcium and its elimination by SV-40 virus-induced proliferative transformation. *J. Cell. Physiol.* **92**, 241–8.

Boynton, A. L., Whitfield, J. F., Isaacs, R. J. & Tremblay, R. G. (1978). An examination of the roles of cyclic nucleotides in the initiation of cell proliferation. *Life Sciences* **2**, 703–10.

Bradley, M. O., Kohn, K. W., Sharkey, N. A. & Ewig, R. A. (1977). Differential cytotoxicity between transformed and normal human cells with combinations of aminonucleoside and hydroxyurea. *Cancer Res.* **37**, 2126–31.

Bradley, W. E. C. & Culp, L. A. (1974). Stimulation of 2-deoxyglucose uptake in growth inhibited Balb/c 3T3 and revertant SV-40-transformed 3T3 cells. *Exp. Cell Res.* **84**, 335–50.

Bradley, W. E. C. & Culp, L. A. (1977). Contact-inhibited revertant cell lines isolated from Simian virus-40-transformed cells. *J. Virol.* **21**, 1228–31.

Bragina, E. E. (1975). Electron microscope study of interaction between substratum and neoplastic fibroblasts of L strain. *Tsitilogiya* **17**, 569–80 (in Russian).

Bragina, E. E., Vasiliev, J. M. & Gelfand, I. M. (1976). Formation of bundles of microfilaments during spreading of fibroblasts on the substrate. *Exp. Cell Res.* **97**, 241–8.

Bramwell, M. E. & Harris, H. (1978). An abnormal membrane glycoprotein associated with malignancy in a wide range of different tumours. *Proc. R. Soc. Lond., Ser.* B **210**, 87–106.

Brand, K. G. (1975). Foreign body induced sarcomas. In *Cancer: a Comprehensive Treatise*, ed. F. F. Becker, vol. 1, pp. 485–511. New York: Plenum Press.

Brand, K. G. (1976). Diversity and complexity of carcinogenic processes: conceptual inferences from foreign body tumorgenesis. *J. Natl. Cancer Inst.* **57**, 973–76.

Brand, K. G., Buoen, L. C. & Brand, I. (1967). Carcinogenesis from polymer implants: new aspects from chromosomal and transplantation studies during premalignancy. *J. Natl. Cancer Inst.* **39**, 663–79.

Brand, K. G., Buoen, L. C., Johnson, K. H. & Brand, I. (1975). Etiological factors, stages, and the role of the foreign body in foreign body tumorigenesis: a review. *Cancer Res.* **35**, 279–86.

Branton, D. (1969). Membrane structure. *Annu. Rev. Plant Physiol.* **20**, 209–38.

Branton, P. E. & Landry-Magnan, J. (1978). Plasma membrane protein kinase activity in normal and Rous sarcoma virus-transformed chick embryo fibroblasts. *Biochim. Biophys. acta* **508**, 246–59.

Braun, A. C. (1974). *The Biology of Cancer*. Reading, Mass.: Addison-Wesley.

Braun, A. C. (1975). Plant Tumors. In *Cancer: a Comprehensive Treatise*, ed. F. F. Becker, vol. 4, pp. 411–27. New York: Plenum Press.

Braun, A. C. (1976). Differentiation and dedifferentiation. In *Cancer: a Comprehensive Treatise*, ed. F. F. Becker, vol. 3, pp. 3–20. New York: Plenum Press.

Braun, A. C. (1978). Plant tumors. *Biochim. Biophys. Acta* **516**, 167–91.

Braun, A. C. & Wood, H. N. (1976). Suppression of the neoplastic state with the acquisition of specialized functions in cells, tissues and organs of crown gall teratomas of tobacco. *Proc. Natl. Acad. Sci. USA* **73**, 496–500.

Bray, D. (1973). Model for membrane movements in the neural growth cone. *Nature, Lond.* **244**, 93–6.

Bray, D. & Thomas, C. (1975). The actin content of fibroblasts. *Biochem. J.* **147**, 221–8.

Bray, D. & Thomas, C. (1976). Unpolymerized actin in tissue cells. In *Cell Motility*, ed. R. Goldman, T. Pollard, J. Rosenbaum. Cold Spring Harbor Conferences on Cell Proliferation, vol. 3, pp. 461–73. New York: Cold Spring Harbor Laboratory.

Brem, S. S., Gullino, P. M. & Medina, D. (1977). Angiogenesis: a marker for neoplastic transformation of mammary papillary hyperplasia. *Science* **125**, 880–1.

Bretscher, M. S. (1976). Directed lipid flow in cell membranes. *Nature, Lond.* **260**, 21–3.

Bretscher, M. S. & Raff, M. C. (1975). Mammalian plasma membranes. *Nature, Lond.* **258**, 43–9.

Bretscher, A. & Weber, K. (1978). Tropomyosin from bovine brain contains two polypeptide chains of slightly different molecular weights. *FEBS Lett.* **85**, 145–8.

Bretton, R., Wicker, R. & Bernhard, W. (1972). Ultrastructural localization of concanavalin A receptors in normal and SV-40-transformed hamster and rat cells. *Int. J. Cancer* **10**, 397–410.

Bridges, B. A. (1976). Short term screening tests for carcinogens. *Nature, Lond.* **261**, 195–200.

Brinkley, B. R., Fuller, G. M. & Highfield, D. P. (1975). Cytoplasmic microtubules in normal and transformed cells in culture: analysis by tubulin antibody immuno-fluorescence. *Proc. Natl. Acad. Sci. USA* **72**, 4981–5.

Brinster, R. L. (1976). Participation of teratocarcinoma cells in mouse embryo development. *Cancer Res.* **36**, 3412–14.

Brondz, B. D. & Rocklin, O. W. (1978). *Molecular and Cellular Basis of Immunological Recognition*. Moscow: Nauka (in Russian).

Bronty-Boyé, D. & Little, J. B. (1977). Enhancement of X-ray induced transformation in C3H/10T 1/2 cells by interferon. *Cancer Res.* **37**, 2714–16.

Brookes, P. (1977). Mutagenicity of polycyclic aromatic hydrocarbons. *Mutation Res.* **39**, 257–84.

Brooks, R. F. (1976). Regulation of the fibroblast cell cycle by serum. *Nature* **260**, 248–50.

Brown, M. & Kiehn, D. (1977). Protease effects on specific growth properties of normal and transformed baby hamster kidney cells. *Proc. Natl. Acad. Sci. USA* **74**, 2874–8.

Brown, S. S. & Revel, J. P. (1976). Reversibility of cell surface label rearrangement. *J. Cell Biol.* **68**, 629–41.

Brunk, U., Ericsson, J. L. E., Pontén, J. & Westermark, B. (1971). Specialization of cell surfaces in contact inhibited human glia-like cells *in vitro. Exp. Cell Res.* **67**, 407–15.

Brunk, U., Schellens, J. & Westermark, B. (1976). Influence of epidermal growth factor (EGF) on ruffling activity, pinocytosis and proliferation of cultivated human glia cells. *Exp. Cell Res.* **103**, 295–302.

Buck, C. A., Glick, M. C. & Warren, L. (1971). Glycopeptides from the surface of control and virus transformed cells. *Science* **172**, 169–71.

Buck, C. A. & Warren, L. (1976). The repair of the surface structure of animal cells. *J. Cell. Physiol.* **89**, 189–200.

Buckley, I. K., Gordon, W. E., Raju, T. R. & Irving, D. O. (1978). The distribution of F-actin in several types of non-muscle cell. *J. Cell Biol.* **79**, 262a.

Buckley, I. K. & Porter, K. (1967). Cytoplasmic fibrils in living cultured cells: a light and electron microscope study. *Protoplasma* **64**, 349–80.

Buckley, I. K. & Porter, K. R. (1975). Electron microscopy of critical point dried whole cultured cells. *J. Micros.* **104**, 107–20.

Buehring, G. C. & Williams, R. R. (1976). Growth rates of normal and abnormal human mammary epithelia in cell culture. *Cancer Res.* **36**, 3742–7.

Burger, M. M. (1970). Proteolytic enzymes initiating cell division and escape from contact inhibition of growth. *Nature, Lond.* **227**, 170–1.

Burger, M. M. (1971). The significance of surface structure changes for growth control under crowded conditions. In *Growth Control in Cell Cultures*. A Ciba Foundation Symposium, ed. G. E. W. Wolstenholme & J. Knight, pp. 45–63. Edinburgh: Churchill Livingstone.

Burger, M. M. (1973). Surface changes in transformed cells detected by lectins. *Fed. Proc.* **32**, 91–101.

Burger, M. M., Bombik, B. M. & Noonan, K. (1972). Cell surface alterations in transformed tissue culture cells and their possible significance in growth control. *J. Invest. Dermatol.* **59**, 24–6.

Bürk, R. R. (1976). Induction of cell proliferation by a migration factor released from a transformed cell line. *Exp. Cell Res.* **101**, 293–8.

Burks, J. K. & Peck, W. A. (1977). Bone cells: a serum-free medium supports proliferation in primary culture. *Science* **199**, 542–4.

Burns, F. J. & Tannock, I. F. (1970). On the existence of a G_0-phase in the cell cycle. *Cell Tissue Kinet.* **3**, 321–4.

Burstin, S. J., Renger, H. C. & Basilico, C. (1974). Cyclic AMP levels in temperature sensitive SV-40 transformed cell lines. *J. Cell. Physiol.* **84**, 69–74.

Bush, H. & Shodell, M. (1976). Cell cycle in transformed cells growing under serum-free conditions. *J. Cell. Physiol.* **90**, 573–84.

Byyny, R. L., Orth, D. N., Cohen, S. & Doyne, E. S. (1974). Epidermal growth factor: effects of androgens and adrenergic agents. *Endocrinology* **95**, 776–82.

Carlsson, L., Nyström, L. E., Sundkvist, J., Markey, F. & Lindberg, U. (1977). Actin polymerizability is influenced by profilin, a low molecular weight protein in non-muscle cells. *J. Mol. Biol.* **115**, 465–83.

Carney, D. H. & Cunningham, D. D. (1977). Initiation of chick cell division by trypsin action at the cell surface. *Nature, Lond.* **268**, 602–6.

Carney, D. H. & Cunningham, D. D. (1978). Demonstration of a surface receptor for thrombin on mouse embryo fibroblasts: involvement in initiation of cell division. *J. Cell Biol.* **79**, 44a.

Carney, D. H., Glenn, K. & Cunningham, D. D. (1978). Conditions which affect initiation of animal cell division by trypsin and thrombin. *J. Cell. Physiol.* **95**, 13–22.

Carpenter, G. & Cohen, S. (1976a). ^{125}I-labelled human epidermal growth factor. Binding, internalization and degradation in human fibroblasts. *J. Cell Biol.* **71**, 159–71.

Carpenter, G. & Cohen, S. (1976b). Human epidermal growth factor and the proliferation of human fibroblasts. *J. Cell. Physiol.* **88**, 227–38.

Carpenter, G., Lembach, K. J., Morrison, M. M. & Cohen, S. (1975). Characterization of the binding of ^{125}I-labelled epidermal growth factor to human fibroblasts. *J. Biol. Chem.* **250**, 4297–304.

Carrino, D. & Gershman, H. (1977). Division of BALB/c mouse 3T3 and Simian virus-40-transformed 3T3 cells in cellular aggregates. *Proc. Natl. Acad. Sci. USA* **74**, 3874–8.

Carter, S. B. (1967a). Haptotaxis and the mechanism of cell motility. *Nature, Lond.* **213**, 256–60.

Carter, S. B. (1967b). Effects of cytochalasins on mammalian cells. *Nature, Lond.* **213**, 261–4.

Carter, W. G. & Hakomori, S. (1978). A protease-resistant, transformation-sensitive membrane glycoprotein and an intermediate filament-forming protein of hamster embryo fibroblasts. *J. Biol. Chem.* **253**, 2867–74.

Caspar, D. L. D., Goodenough, D. A., Makowski, L. & Phillips, W. C. (1977). Gap junction structures. I. Correlated electron microscopy and X-ray diffraction. *J. Cell Biol.* **74**, 605–28.

Casto, B. C., Janosko, N. & Di Paolo, J. A. (1977). Development of a focus assay model for transformation of hamster cells *in vitro* by chemical carcinogens. *Cancer Res.* **37**, 3508–15.

Castor, L. R. N. (1975). The role of protein synthesis and degradation in growth control of diploid and transformed cells. *J. Cell Biol.* **63**, 58a.

Ceccarini, C. & Eagle, H. (1971). pH as a determinant of cell growth and contact-inhibition. *Proc. Natl. Acad. Sci. USA* **68**, 229–33.

Celis, J. E., Small, J. V., Anderson, P. & Celis, A. (1978). Microfilament bundles in cultured cells. Correlation with anchorage independence and tumorigenicity in nude mice. *Exp. Cell Res.* **114**, 335–48.

Cereijido, M., Robbins, E. S., Dolan, W. J., Rotunno, C. A. & Sabatini, D. D. (1978). Polarized monolayers formed by epithelial cells on a permeable and translucent support. *J. Cell Biol.* **77**, 853–80.

Chello, P. L. & Bertino, J. R. (1973). Dependence of 5-methylterahydrofolate utilization by L5178Y murine leukemia cells *in vitro* on the presence of hydroxy-cobalamin and transcobalamin II. *Cancer Res.* **33**, 1898–1904.

Chen, L. B. & Buchanan, J. M. (1975a). Mitogenic activity of blood components. I. Thrombin and prothrombin. *Proc. Natl. Acad. Sci. USA* **72**, 131–5.

Chen, L. B. & Buchanan, J. M. (1975b). Plasminogen-independent fibrinolysis by proteases produced by transformed chick embryo fibroblasts. *Proc. Natl. Acad. Sci. USA* **72**, 1132–6.

Chen, L. B., Callimore, P. H. & McDougall, J. K. (1976). Correlation between tumor induction and the large external transformation sensitive protein on the cell surface. *Proc. Natl. Acad. Sci. USA* **73**, 3570–4.

Chen, L. B., Gudor, R. C., Sun, T. T., Chen, A. B. & Mosesson, M. W. (1977a). Control of a cell surface major glycoprotein by epidermal growth factor. *Science* **197**, 776–8.

Chen, L. B., Moser, F. G., Chen, A. B. & Mosesson, M. W. (1977b). Distribution of cell surface LETS protein in co-cultures of normal and transformed cells. *Exp. Cell Res.* **108**, 375–83.

Chen, L. B., Teng, N. N. H. & Buchanan, J. H. (1976). Mitogenicity of thrombin and surface alterations on mouse splenocytes. *Exp. Cell Res.* **101**, 41–6.

Chen, W.-T. (1978). Induction of spreading during fibroblast movement. *J. Cell Biol.* **79**, 83a.

Cherny, A. P., Vasiliev, J. M. & Gelfand, I. M. (1975). Spreading of normal and transformed fibroblasts in dense cultures. *Exp. Cell Res.* **90**, 317–27.

Chertkov, J. L. & Friedenstein, A. Y. (1977). *Cellular Basis of Haematopoiesis. Haematopoetic Precursor Cells.* Moscow: Medizina (in Russian).

Chibber, B. A., Niles, R. M., Prehn, L. & Sorof, S. (1975). High extracellular fibrinolytic activity of tumors and control normal tissues. *Biochem. Biophys. Res. Commun.* **65**, 806–12.

Cho, H. Y., Cutchins, E. C., Rhim, J. S. & Huebner, R. J. (1976). Revertants of human cells transformed by murine sarcoma virus. *Science* **194**, 951–3.

Ciechanover, A. & Hershko, A. (1976). Early effects of serum on phospholipid metabolism in untransformed and oncogenic virus-transformed cultured fibroblasts. *Biochem. Biophys. Res. Commun.* **73**, 85–91.

Clark, J. I. & Albertini, D. F. (1976). Filaments, microtubules and colchicine receptors in capped ovarian granulosa cells. In *Cell Motility*, ed. R. Goldman, T. Pollard & J. Rosenbaum. Cold Spring Harbor Conferences on Cell Proliferation, vol. 3, pp. 323–31. New York: Cold Spring Harbor Laboratory.

Clark, J. M. & Pateman, J. A. (1978). Isolation method affects transformed cell line karyotype. *Nature, Lond.* **272**, 262–3.

Clarke, G. D. & Stoker, M. G. P. (1971). Conditions affecting the response of cultured cells to serum. In *Growth Control in Cell Cultures*, ed. G. E. W. Wolstenholme & J. Knight, pp. 17–28. Edinburgh: Churchill Livingstone.

Clarke, G. D., Stoker, M. G. P., Ludlow, A. & Thornton, M. (1970). Requirement of serum for DNA synthesis in BHK-21 cells: effects of density, suspension and virus transformation. *Nature, Lond.* **227**, 798–801.

Clifton, K. H. & Sridharan, B. N. (1975). Endocrine factors and tumor growth. In *Cancer: a Comprehensive Treatise*, ed. F. F. Becker, vol. 3, pp. 249–286. New York: Plenum Press.

Cloyd, M. W. & Bigner, D. D. (1977). Surface morphology of normal and neoplastic rat cells. *Amer. J. Pathol.* **83**, 29–53.

Coffin, J. M. (1976). Genes responsible for transformation by avian RNA tumor viruses. *Cancer Res.* **36**, 4282–8.

Cohen, S. (1962). Isolation of a mouse submaxillary gland protein accelerating eruption and eyelid opening in the newborn animal. *J. Biol. Chem.* **237**, 1555–62.

Cohen, S. & Carpenter, G. (1975). Human epidermal growth factor; isolation and chemical and biological properties. *Proc. Natl. Acad. Sci. USA* **72**, 1317–21.

Colburn, N. H., Vor der Bruegge, W. F., Bates, J. R., Gray, R. H., Rossen, J. D., Kelsey, W. H. & Shimada, T. (1978). Correlation of anchorage-independent growth with tumorigenicity of chemically transformed mouse epidermal cells. *Cancer Res.* **38**, 624–34.

Colby, C. & Romano, H. (1976). Phosphorylation but not transport of sugars is enhanced in virus-transformed mouse 3T3 cells. *J. Cell. Physiol.* **89**, 701–10.

Collard, J. G. & Temmink, J. H. M. (1976). Surface morphology and agglutinability with concanavalin A in normal and transformed murine fibroblasts. *J. Cell Biol.* **68**, 101–12.

Collett, M. S. & Erikson, R. L. (1978). Protein kinase activity associated with the avian sarcoma virus *src* gene product. *Proc. Natl. Acad. Sci. USA* **75**, 2921–4.

Collett, M. S., Erikson, E., Purchio, A. F., Brugge, J. S. & Erikson, R. L. (1979). A normal cell protein similar in structure and function to the avian sarcoma virus transforming gene product. *Proc. Natl. Acad. Sci. USA* **76**, 3159–63.

Collins, J. M. (1977). Deoxyribonucleic acid structure in human diploid fibroblasts stimulated to proliferate. *J. Biol. Chem.* **252**, 141–7.

Coman, D. R. (1944). Decreased mutual adhesiveness, a property of cells from squamous cell carcinomas. *Cancer Res.* **4**, 625–9.

Comoglio, M., Tarone, G. & Bertini, M. (1978). Immunochemical purification of probe-labeled plasma membrane proteins: an approach to the molecular anatomy of the cell surface. *J. Supramol. Struct.* **8**, 39–49.

Condeelis, J. S., Taylor, D. L., Moore, P. L. & Allen, R. D. (1976). The mechano-chemical basis of amoeboid movement. II. Cytoplasmic filament stability at low divalent cation concentrations. *Exp. Cell Res.* **101**, 134–42.

Cone, C. D. & Tongier, M., Jr. (1973). Contact inhibition of division involvement of the electrical transmembrane potential. *J. Cell. Physiol.* **82**, 373–86.

Connolly, J. A., Kalnins, V. I., Cleveland, D. W. & Kirschner, M. W. (1977). Immunofluorescent staining of cytoplasmatic and spindle microtubules in mouse fibroblasts with antibody to τ protein. *Proc. Natl. Acad. Sci. USA* **74**, 2437–40.

Cornell, R. (1969). Spontaneous neoplastic transformation *in vitro:* ultrastructure of transformed cell strains and tumors produced by injection of cell strains. *J. Natl. Cancer Inst.* **43**, 891–906.

Corsaro, C. M. & Migeon, B. R. (1977). Comparison of contact-mediated communication in normal and transformed human cells in culture. *Proc. Natl. Acad. Sci. USA* **74**, 4476–80.

Corwin, L. M., Humphrey, L. P. & Shloss, J. (1977). Effect of lipids on the expression of cell transformation. *Exp. Cell Res.* **108**, 341–7.

Costlow, M. & Baserga, R. (1973). Changes in membrane transport function in G_0 and G_1 cells. *J. Cell Physiol.* **82**, 411–20.

Couzin, D. (1978). Plating efficiency measurements and the experimental control of ageing of adult human skin fibroblasts *in vitro. Exp. Cell Res.* **116**, 115–26.

Crawford, L. V., Cole, C. N., Smith, A. E., Paucha, E., Tegtmeyer, P., Rundell, K. & Berg, P. (1978). Organization and expression of early genes of Simian virus 40. *Proc. Natl. Acad. Sci. USA* **75**, 117–21.

Cristofalo, V. J. (1977). Senescence in cell culture: an accumulation of errors or terminal differentiation? In *Senescence, Dominant or Recessive in Somatic Cell Crosses?* ed. W. W. Nichols & D. G. Murphy, pp. 13–21. New York: Plenum Press.

Cristofalo, V. J. & Sharf, B. B. (1973). Cellular senescence and DNA synthesis: thymidine incorporation as a measure of population age in human diploid cells. *Exp. Cell Res.* **76**, 419–27.

Croce, C. M. (1977). Assignment of the integration site for Simian virus-40 to chromosome 17 in GM 54 VA, a human cell line transformed by Simian virus 40. *Proc. Natl. Acad. Sci. USA* **74**, 315–18.

Croce, C. M. & Koprowski, H. (1974). Somatic cell hybrids between mouse peritoneal macrophages and SV-40 transformed human cells. I. Positive control of the transformed phenotype by the human chromosome 7 carrying the SV-40 genome. *J. Exp. Med.* **140**, 1221–9.

Culp, L. A. (1976). Electrophoretic analysis of substrate-attached proteins from normal and virus-transformed cells. *Biochemistry* **15**, 4094–5104.

Culp, L. A. & Bensusan, H. (1978). Search for collagen in substrate adhesion site of two murine cell lines. *Nature, Lond.* **273**, 680–2.

Culp, L. A. & Black, P. H. (1972). Contact-inhibited revertant cell lines isolated from Simian virus 40-transformed cells. III. Concanvalin A-selected revertant cells. *J. Virology* **9**, 611–20.

Culp, L. A., Grimes, W. J. & Black, P. H. (1971). Contact-inhibited revertant cell lines isolated from SV-40-transformed cells. I. Biologic, virologic and chemical properties. *J. Cell Biol.* **50**, 682–90.

Cunningham, D. D. & Ho, T. S. (1975). Effects of added proteases on Concanavalin A-specific agglutinability and proliferation of quiescent fibroblasts. In *Proteases and Biological Control*, ed. E. Reich, D. B. Bifkin & E. Shaw. Cold Spring Harbor Conferences on Cell Proliferation 2, pp. 795–806. New York: Cold Spring Harbor Laboratory.

Cunningham, D. D. & Pardee, A. B. (1969). Transport changes rapidly initiated by serum addition to 'contact inhibited' 3T3 cells. *Proc. Natl. Acad. Sci. USA* **64**, 1049–56.

Curtis, A. S. G. (1964). The mechanism of adhesion of cells on glass. A study by interference–reflection microscopy. *J. Cell Biol.* **20**, 199–215.

Curtis, A. S. G. (1967). *The Cell Surface: its Molecular Role in Morphogenesis.* London: Logos Press.

Curtis, A. S. G. & Varde, M. (1964). Control of cell behaviour: topological factors. *J. Natl. Cancer Inst.* **33**, 15–26.

Dalen, H. & Todd, P. W. (1971). Surface morphology of trypsinized human cells *in vitro. Exp. Cell Res.* **66**, 353–61.

D'Amore, P. A. & Shepro, D. (1978). Calcium flux and ornithine decarboxylase activity in cultured endothelial cells. *Life Sciences* **22**, 571–6.

Danø, K. & Reich, E. (1978). Serine enzymes released by cultured neoplastic cells. *J. Exp. Med.* **147**, 745–57.

Das, M. & Fox, C. F. (1978). Molecular mechanism of mitogen action: processing of receptor induced by epidermal growth factor. *Proc. Natl. Acad. Sci. USA* **72**, 2644–8.

Das, M., Miyakawa, T., Fox, C. F., Pruss, R. M., Aharonov, A. & Herschman, H. R. (1977). Specific radiolabeling of a cell surface receptor for epidermal growth factor. *Proc. Natl. Acad. Sci. USA* **74**, 2790–4.

Davison, P. F., Hong, B.-S. & Cooke, P. (1977). Classes of distinguishable 10 nm cytoplasmic filaments. *Exp. Cell Res.* **109**, 471–4.

de Brabander, M., de Mey, J., van de Veire, R., Aerts, F. & Geuens, G. (1977). Microtubules in mammalian cell shape and surface modulation: an alternative hypothesis. *Cell Biol. Int. Reports.* **1**, 453–61.

Defendi, V. (1966). Transformation *in vitro* of mammalian cells by polyoma and simian 40 viruses. In *Progress in Experimental tumor research*, ed. F. Homburger, pp. 125–88. Basel: Karger.

Defendi, V. & Stoker, M. G. P. (1973). General polyploid produced by cytochalasin B. *Nature New Biol.* **242**, 24–6.

De Larco, J. E. & Todaro, G. J. (1978a). Growth factors from murine sarcoma virus-transformed cells. *Proc. Natl. Acad. Sci. USA* **75**, 4001–5.

De Larco, J. E. & Todaro, G. (1978b). Epithelioid and fibroblastic rat kidney cell clones: epidermal growth factor (EGF) receptors and the effect of mouse sarcoma virus transformation. *J. Cell. Physiol.* **94**, 335–42.

DeMars, R. (1977). Comments about the paper by Norwood *et al.* and an idea relating somatic cell mutations to cellular senescence. In *Senescence: Dominant or Recessive in Somatic Cell Crosses?* ed. W. W. Nichols & D. G. Murphy, pp. 39–55. New York: Plenum Press.

De Mello, W. C. (ed.) (1977). *Intercellular communication.* New York: Plenum Press.

De Mey, J., Joniau, M., de Brabander, M., Moens, W. & Geuens, G. (1978). Evidence for unaltered structure and in-vitro assembly of microtubules in transformed cells. *Proc. Natl. Acad. Sci. USA* **75**, 1339–43.

Deng, C. T., Boettiger, D., Macpherson, I. & Varmus, H. E. (1974). The persistence and expression of virus-specific DNA in revertants of Rous sarcoma virus-transformed BHK-21 cells. *Virology* **62**, 512–21.

De Ome, K. B. (1965). Formal discussion of multiple factors in mammary tumori-genesis. *Cancer Res.* **25**, 1348–51.

De Ome, K. B., Bern, H. A., Elias, J. J., Nandi, S. & Faulkin, L. J., Jr. (1958). Studies on the growth potential of hyperplastic nodules of the mammary gland of the C₃H/He CRGL mouse. In *International Symposium on Mammary Cancer*, ed. L. Severei, pp. 595–604. Perugia: Division of Cancer Research.

De Ome, K. B. & Medina, D. (1969). A new approach to mammary tumorigenesis in rodents. *Cancer* **24**, 1255–0.

De Petris, S. (1975). Concanavalin A receptors, immunoglobulins, and antigen of the lymphocyte surface. Interactions with concanavalin A and with cytoplasmic structures. *J. Cell Biol.* **65**, 123–46.

De Petris, S. (1977). Distribution and mobility of plasma membrane components on lymphocytes. In *Dynamic Aspects of Cell Surface Organization*, ed. G. Poste & G. L. Nicolson, Cell surface reviews, vol. 3, pp. 643–728. Amsterdam: North-Holland.

De Petris, S. (1978*a*). Preferential distribution of surface immunoflobulins on microvilli. *Nature, Lond.* **272**, 66–8.

De Petris, S. (1978*b*). Non-uniform distribution of concanavalin A receptors and surface antigens on uropod-forming thymocytes. *J. Cell. Biol.* **79**, 235–51.

De Petris, S. (1978*c*). Immunoelectron microscopy and immunofluorescence in membrane biology. In *Methods in Membrane Biology*, ed. E. D. Korn, vol. 9, pp. 1–201. New York: Plenum Press.

De Petris, S. & Raff, M. C. (1973). Fluidity of the plasma membrane and its implications for cell movement. In *Locomotion of Tissue Cells*. ed. M. Abercrombie. Ciba Foundation Symposium 14, pp. 27–41 Amsterdam: Associated Scientific Publishers.

De Pierre, J. W. & Ernster, L. (1978). The metabolism of polycyclic hydrocarbons and its relationship to cancer. *Biochim. Biophys. Acta* **473**, 148–86.

Dexter, T. M., Allen, T. D. & Lajtha, L. G. (1977). Conditions controlling the proliferation of haemopoietic stem cells *in vitro*. *J. Cell. Physiol.* **91**, 335–44.

Diamond, I., Legg, A., Schneider, J. A. & Rozengurt, E. (1978). Glycolysis in quiescent cultures of 3T3 cells. Stimulation by serum, epidermal growth factor and insulin in intact cells and persistence of the stimulation after cell homogenization. *J. Biol. Chem.* **253**, 866–71.

Diamond, L. (1971). Metabolism of polycyclic hydrocarbons in mammalian cell cultures. *Int. J. Cancer* **8**, 451–62.

Diamond, L., Defendi, V. & Brookes, P. (1967). The interaction of 7,12-dimethylbenz(a)anthracene with cells sensitive and resistant to toxicity induced by this carcinogen. *Cancer Res.* **27**, 890–7.

Diamond, L., Defendi, V. & Brookes, P. (1968). The development of resistance to carcinogen-induced cytotoxicity in hamster embryo cultures. *Exp. Cell Res.* **52**, 180–4.

Diglio, C. A. & Dougherty, R. M. (1977). Control of transformed focus morphology in chicken cell cultures infected with Rous sarcoma virus. *J. Gen. Virol.* **36**, 413–27.

Di Mayorca, G., Callender, J., Marin, G. & Giorgano, R. (1969). Temperature-sensitive mutants of polyoma virus. *Virology* **38**, 126–33.

Dinowitz, M. (1977). A continuous line of Rous sarcoma virus-transformed chick embryo cells. *J. Nat. Cancer Inst.* **58**, 307–12.

Di Paolo, J. A., Donovan, P. J. & Nelson, R. L. (1969). Quantitative studies of in-vitro transformation by chemical carcinogens. *J. Nat. Cancer Inst.* **42**, 867–74.

Di Paolo, J. A., Nelson, R. L. & Donovan, P. J. (1972*a*). In-vitro transformation of Syrian hamster embryo cells by diverse chemical carcinogens. *Nature, Lond.* **235**, 278–80.

Di Paolo, J. A., Takano, K. & Popescu, N. C. (1972*b*). Quantitation of chemically induced neoplastic transformation of BALB/3T3 cloned cell lines. *Cancer Res.* **32**, 2686–95.

Di Pasquale, A. (1975*a*). Locomotory activity of epithelial cells in culture. *Exp. Cell Res.* **94**, 191–215.

Di Pasquale, A. (1975*b*). Locomotion of epithelial cells: factors involved in extension of the leading edge. *Exp. Cell Res.* **95**, 425–39.

Di Pasquale, A. & Bell, P. B., Jr. (1974). The upper cell surface: its inability to support active cell movement in culture. *J. Cell Biol.* **62**, 198–214.

Di Pasquale, A. & Bell, P. B., Jr. (1975). Comments on reported observations of cells spreading on the upper surfaces of other cells in culture. *J. Cell Biol.*, **66**, 216–18.

Di Pasquale, A. M., McGuire, J., Moellmann, G. & Wasserman, S. J. (1976). Microtubule assembly in cultivated Greene melanoma cells is stimulated by dibutyryl adenosine 3′:5′-cyclic monophosphate or cholera toxin. *J. Cell Biol.* **71**, 735–48.

Domnina, L. V., Ivanova, O. Y., Margolis, L. B., Olshevskaya, L. V., Rovensky, Y. A., Vasiliev, J. M. & Gelfand, I. M. (1972). Defective formation of the lamellar cytoplasm by neoplastic fibroblasts. *Proc. Natl. Acad. Sci. USA* **69**, 248–52.

Domnina, L. V., Pletyushkina, O. Y., Vasiliev, J. M. & Gelfand, I. M. (1977). Effect of antitubulins on the redistribution of crosslinked receptors on the surface of fibroblasts and epithelial cells. *Proc. Natl. Acad. Sci. USA* **74**, 2865–8.

Doroszewsky, J., Skierski, J. & Przadka, L. (1977). Interaction of neoplastic cells with glass surface under flow conditions. *Exp. Cell Res.* **104**, 335–43.

Dubrow, R., Pardee, A. B. & Pollack, R. (1978). 2-aminoisobutyric acid and 3-*O*-methyl-D-glucose transport in 3T3, SV-40-transformed 3T3 and revertant cell lines. *J. Cell. Physiol.* **95**, 203–12.

Duc-Nguyen, H., Rosenblum, E. N. & Zeigel, R. F. (1966). Persistent infection of a rat kidney cell line with Rauscher murine leukemia virus. *J. Bacteriol.* **92**, 1133–40.

Dulak, N. C. & Shing, Y. W. (1976). Large scale purification and further characterization of a rat liver cell conditioned medium multiplication stimulating activity. *J. Cell. Physiol.* **90**, 127–38.

Dulak, N. C. & Temin, H. M. (1973*a*). A partially purified polypeptide fraction from rat liver cell conditioned medium with multiplication-stimulating activity for embryo fibroblasts. *J. Cell. Physiol.* **81**, 153–60.

Dulak, N. C. & Temin, H. M. (1973*b*). Multiplication-stimulating activity for chicken embryo fibroblasts from rat liver cell conditioned medium: a family of small polypeptides. *J. Cell. Physiol.* **81**, 161–70.

Dulbecco, R. (1970). Topoinhibition and serum requirement of transformed and untransformed cells. *Nature, Lond.* **227**, 802–6.

Dulbecco, R. (1973). Cell transformation by viruses and the role of viruses in cancer. *J. Gen. Microbiol.* **79**, 7–17.

Dulbecco, R. (1976). From the molecular biology of oncogenic DNA viruses to cancer. *Science* **192**, 437–40.

Dulbecco, R. & Eckhart, W. (1970). Temperature-dependent properties of cells transformed by a thermosensitive mutant of polyoma virus. *Proc. Natl. Acad. Sci. USA* **67**, 1775–81.

Dulbecco, R. & Elkington, J. (1975). Induction of growth in resting fibroblastic cell cultures by Ca²⁺. *Proc. Natl. Acad. Sci. USA* **72**, 1584–88.

Dulbecco, R. & Stoker, M. G. P. (1970). Conditions determining initiation of DNA synthesis in 3T3 cells. *Proc. Natl. Acad. Sci. USA* **66**, 204–10.

Dunkel, V. C., Wolff, J. S., III, & Pienta, R. J. (1977). In-vitro transformation as a presumptive test for detecting chemical carcinogens. *Cancer Bull.* **29**, 167–74.

Dunn, G. A. & Ebendal, T. (1978). Contact guidance on oriented collagen gels. *Exp. Cell Res.* **111**, 475–79.

Dunn, G. A. & Heath, J. P. (1976). A new hypothesis of contact guidance in tissue cells. *Exp. Cell Res.* **101**, 1–14.

Dunn, T. B. (1945). Morphology and histogenesis of mammary tumors. In *A Symposium on Mammary Tumors in Mice*, ed. F. R. Moulton, pp. 13–38. Washington: American Association for the Advancement of Science.

Dürnberger, H., Heuberger, B., Schwartz, P., Wasner, G. & Kratochwill, K. (1978). Mesenchyme-mediated effect of testosterone on embryonic mammary epithelium. *Cancer Res.* **38**, 4066–70.

Dykes, D. J., Griswold, D. P., Jr. & Schabel, F. M., Jr. (1976). Growth support of small B16 melanoma implants with nitrosourea-sterilized fractions of the same tumor. *Cancer Res.* **36**, 2031–4.

Eagle, H. (1955). Nutrition needs of mammalian cells in tissue culture. *Science* **122**, 501–4.

Eagle, H. (1973). The effect of environmental pH on the growth of normal and malignant cells. *J. Cell. Physiol.* **82**, 1–8.

Eagle, H., Foley, G. E., Koprowski, H., Lazarus, H., Levine, E. M. & Adams, R. A. (1970). Growth characteristics of virus-transformed cells. Maximum population density, inhibition by normal cells, serum requirement, growth in soft agar, and xenogeneic transplantability. *J. Exp. Med.* **131**, 863–79.

Eagle, H. & Levine, E. M. (1967). Growth regulatory effects of cellular interaction. *Nature, Lond.* **213**, 1102–6.

Eagle, H., Levine, E. M. & Koprowski, H. (1968). Species specificity in growth regulatory effects of cellular interactions. *Nature, Lond.* **220**, 266–9.

Eagle, H. & Piez, K. (1962). The population-dependent requirement by cultured mammalian cells for metabolites which they can synthesize. *J. Exp. Med.*, **116**, 29–43.

Earle, W. R. (1943). Production of malignancy *in vitro*. IV. The mouse fibroblast cultures and changes seen in the living cells. *J. Nat. Cancer Inst.* **4**, 165–212. Reprinted in *J. Nat. Cancer Inst.* **59**, 715–47, (1978).

Eastment, C. T. & Sirbaski, D. A. (1978). Platelet derived growth factor(s) for a hormone-responsive rat mammary tumor cell line. *J. Cell. Physiol.* **97**, 17–28.

Easty, G. C. & Mercer, E. H. (1960). An electron microscope study of the surfaces of normal and malignant cells in culture. *Cancer Res.* **20**, 1608–13.

Eckhart, W., Dulbecco, R. & Burger, M. M. (1971). Temperature-dependent surface changes in cells infected or transformed by a thermosensitive mutant of polyoma virus. *Proc. Natl. Acad. Sci. USA* **68**, 283–6.

Eckhart, W. & Weber, M. J. (1974). Uptake of 2-deoxyglucose by BALB/3T3 cells: changes after polyoma infection. *Virology* **61**, 223–8.

Edds, K. T. (1977). Microfilament bundles. I. Formation with uniform polarity. *Exp. Cell Res.* **108**, 452–6.

Edelman, G. M. (1976). Surface modulation in cell recognition and cell growth. *Science* **192**, 218–26.

Edelman, G. M., Rutishauser, U. & Millette, C. F. (1971). Cell fractionation and arrangement on fibers, beads, and surfaces. *Proc. Natl. Acad. Sci. USA* **68**, 2153–7.

Edelman, G. M. & Yahara, I. (1976). Temperature-sensitive changes in surface modulating assemblies of fibroblasts transformed by mutants of Rous sarcoma virus. *Proc. Natl. Acad. Sci. USA* **73**, 2047–51.

Edidin, M. & Wei, T.-Y. (1977*a*). Diffusion rates of cell surface antigens of mouse-human heterokaryons. I. Analysis of the population. *J. Cell Biol.* **75**, 475–82.

Edidin, M. & Wei, T. (1977*b*). Diffusion rates of cell surface antigens of mouse-human heterokaryons. II. Effect of membrane potential on lateral diffusion. *J. Cell Biol.* **75**, 483–9.

Edidin, M. & Weiss, A. (1972). Antigen cap formation in cultured fibroblasts: a reflection of membrane fluidity and cell motility. *Proc. Natl. Acad. Sci. USA* **69**, 2456–9.

Edidin, M., Zagyansky, Y. & Lardner, T. J. (1976). Measurement of membrane protein lateral diffusion in single cells. *Science* **191**, 466–8.

Edström, A., Kanje, M. & Walum, E. (1976). Density-dependent inhibition of 2-deoxy-D-glucose uptake into glioma and neuroblastoma cells in culture. *Exp. Cell Res.* **97**, 6–14.

Ehrmann, R. L. & Gey, G. O. (1956). The growth of cells on a transparent gel of reconstituted rat-tail collagen. *J. Nat. Cancer Inst.* **16**, 1375–1403.

Ellem, K. A. O. & Gierthy, J. F. (1977). Mechanism of regulation of fibroblastic cell replication. IV. An analysis of the serum dependence of cell replication based on Michaelis–Menten Kinetics. *J. Cell. Physiol.* **92**, 381–400.

El Mishad, A. M., McCormick, K. J., McCormick, N. K. & Trentin, J. J. (1975). Potentiation of hamster tumors by normal cells or charcoal. *Cancer Res.* **35**, 2098–103.

Elsdale, T. & Bard, J. (1972). Cellular interactions in mass cultures of human diploid fibroblasts. *Nature, Lond.* **236**, 152–55.

Elsdale, T. & Bard, J. (1974). Cellular interactions in the morphogenesis of epithelial-mesenchymal systems. *J. Cell Biol.* **63**, 343.

Engvall, E., Ruoslahti, E. & Miller, E. J. (1978). Affinity of fibronectin to collagens of different genetic types and to fibrinogen. *J. Exp. Med.* **147**, 1584–95.

Enlander, D., Tobey, R. A. & Scott, T. (1975). Cell cycle-dependent surface changes in Chinese hamster cells grown in suspension culture. *Exp. Cell Res.* **95**, 396–404.

Ephrussi, B. (1935). *Phenomênes d'intégration dans les cultures des tissus.* Paris: Herman.

Epifanova, O. I. (1977). Mechanisms underlying the differential sensitivity of proliferating and resting cells to external factors. *Int. Rev. Cytol.* Suppl. **5**, 303–35.

Epifanova, O. I. & Terskikh, V. (1969). On resting periods in the cell life cycle. *Cell Tissue Kinet.* **2**, 75–93.

Epstein, M. L. & Gilula, N. B. (1977). A study of communication specificity between cells in culture. *J. Cell Biol.* **75**, 769–87.

Erickson, C. A. & Trinkaus, J. P. (1976). Microvilli and blebs as sources of reserve surface membrane during cell spreading. *Exp. Cell Res.* **99**, 375–84.

Evans, R. B., Morhenn, V., Jones, A. L. & Tomkins, G. M. (1974). Concomitant effects of insulin on surface membrane conformation and polysome profiles of serum-starved BALB/C 3T3 fibroblasts. *J. Cell Biol.* **61**, 95–106.

Farber, E., Parker, S. & Gruenstein, M. (1976). The resistance of putative premalignant liver cell populations, hyperplastic nodules to the acute cytotoxic effects of some hepatocarcinogens. *Cancer Res.* **36**, 3897–87.

Farrow, G. M., Utz, D. C., Rife, C. C. & Greene, L. F. (1977). Clinical observations on sixty-nine cases of in-situ carcinomy of the urinary bladder. *Cancer Res.* **37**, 2794–8.

Farquahar, M. G. & Palade, G. E. (1963). Junctional complexes in various epithelia. *J. Cell Biol.* **17**, 375–412.

Fastaia, J. & Dumont, A. E. (1976). Pathogenesis of ascites in mice with peritoneal carcinomatosis. *J. Nat. Cancer Inst.* **56**, 547–50.

Fellini, S. A., Bennett, G. S. & Holtzer, H. (1978). Selective binding of antibody against gizzard 10-nm filaments to different cell types in myogenic cultures. *Am. J. Anat.* **153**, 451–7.

Fentiman, J. S. & Taylor-Papadimitriou, J. (1977). Cultured human breast cancer cells lose selectivity in direct intercellular communication. *Nature, Lond.* **269**, 156–8.

Fentiman, I., Taylor-Papadimitriou, J. & Stoker, M. (1976). Selective contact-dependent cell communication. *Nature, Lond.* **264**, 760–2.

Fernandez-Pol, J. A., Bono, V. H., Jr. & Johnson, G. S. (1977). Control of growth by picolinic acid. Differential response of normal and transformed cells. *Proc. Natl. Acad. Sci. USA* **74**, 2889–93.

Fernandez-Pol, J. A. & Johnson, G. S. (1977). Selective toxicity induced by picolinic acid in Simian virus 40–transformed cells in tissue culture. *Cancer Res.* **37**, 4276–9.

Fialkow, P. J. (1972). Use of genetic markers of study cellular origin and development of tumors in human females. *Adv. Cancer Res.* **15**, 191–226.

Fialkow, P. J. (1974). Origin and development of human tumors studied with cell markers. *New Engl. J. Med.* **291**, 26–35.

Fialkow, P. J. (1976). Clonal origin of human tumors. *Biochim. Biophys. Acta* **458**, 283–321.

Fialkow, P. J. (1977). Clonal origin and stem cell evolution of human tumors. *In Genetics of Human Cancer*, ed. J. J. Mulvihill, R. W. Miller & J. F. Fraumeni, Jr., pp. 439–54. New York: Raven Press.

Fidler, I. J. (1974). Immune stimulation-inhibition of experimental cancer metastasis. *Cancer Res.* **34**, 491–8.

Fidler, I. J. (1975a). Biological behavior of malignant melanoma cells correlated to their survival *in vivo. Cancer Res.* **35**, 218–24.

Fidler, I. J. (1975b). Mechanisms of cancer invasion and metastasis. In *Cancer: a Comprehensive Treatise*, ed. F. F. Becker, vol. 4, pp. 101–31. New York: Plenum Press.

Fidler, I. J. (1976). Patterns of tumor cell arrest and development. In *Fundamental Aspects of Metastasis*, ed. L. Weiss, pp. 275–89. Amsterdam: North Holland.

Fidler, I. J. (1977). Metastasis results from pre-existing variant cells within a malignant tumor. *Science* **197**, 893–95.

Fidler, I. J. (1978). Tumor heterogenecity and the biology of cancer invasion and metastasis. *Cancer Res.* **38**, 2651–60.

Fidler, I. J. & Kripke, M. L. (1977). Metastasis results from pre-existing variant cells within a malignant tumor. *Science* **197**, 893–95.

Fidler, I. J. & Nicolson, G. L. (1976). Organ selectivity for survival and growth of B16 melanoma variant tumor lines. *J. Nat. Cancer Inst.* **57**, 1199–1202.

Filbach, E., Reuben, R. C., Rifkind, R. A. & Marks, P. A. (1977). Effect of hexamethylene bisacetamide on the commitment to differentiation of urine erythrolenkemie cells. *Cancer Res.* **37**, 440–4.

Fine, R. E. & Taylor, L. (1976). Decreased actin and tubulin synthesis in 3T3 cells after transformation by SV-40 virus. *Exp. Cell Res.* **102**, 162–8.

Fischer, A. (1946). *Biology of Tissue Cells.* Cambridge: Cambridge University Press.

Fisher, H. W. & Yeh, J. (1967). Contact inhibition in colony formation. *Science* **155**, 581–2.

Fishinger, P. J., Nomura, S., Peebles, P. T., Haapala, D. K. & Bassin, R. H. (1972). Reversion of murine sarcoma virus transformed mouse cells: variants without a rescuable sarcoma virus. *Science* **176**, 1033–5.

Fishman, P. H. & Brady, R. O. (1976). Biosynthesis and function of gangliosides. *Science* **194**, 906–15.

Fishman, W. H. (1976). Activation of developmental genes in neoplastic transformation. *Cancer Res.* **36**, 3423–8.

Fleischer, M. & Wohlfarth-Bottermann, K. E. (1975). Correlation between tension force generation fibrillogenesis and ultrastructure of cytoplasmic actomyosin during isometric and isotonic contractions of protoplasmic strands. *Cytobiologie* **10**, 339–65.

Fleischman, E. W. & Prigogina, E. L. (1977). Karyotype peculiarities of malignant lymphomas. *Hum. Genet.* **35**, 269–79.

Fodge, D. W. & Rubin, H. (1975). Stimulation of lactic acid production in chick embryo fibroblasts by serum and high pH in the absence of external glucose. *J. Cell. Physiol.* **86**, 453–8.

Fogh, J. (1971). Longevity of strains from individual foci of human amnion cells transformed by simian virus 40. *J. Nat. Cancer Inst.* **47**, 733–8.

Fogh, J. (ed.) (1975). *Human Tumor Cells* in vitro. New York. Plenum Press.

Fogh, J., Fogh, J. M. & Orfeo, T. (1977). One hundred and twenty-seven cultured human tumor cell lines producing tumors in nude mice. *J. Nat. Cancer Inst.* **59**, 221–5.

Folkman, J. (1974). Tumor angiogenesis factor. *Cancer Res.* **34**, 2109–13.

Folkman, J. (1975). Tumor Angiogenesis. In *Cancer: a Comprehensive Treatise*, ed. F. F. Becker, vol. 3, pp. 355–88. New York: Plenum Press.

Folkman, J. & Greenspan, H. P. (1975). Influence of geometry on control of cell growth. *Biochim. Biophys. Acta* **417**, 211–36.

Folkman, J. & Hochberg, M. (1973). Self-regulation of growth in three dimensions. *J. Exp. Med.* **138**, 745–53.

Folkman, J., Merler, E., Abernathy, C. & Williams, G. (1971). Isolation of a tumor factor responsible for angiogenesis. *J. Exp. Med.* **133**, 257–88.

Folkman, J. & Moscona, A. (1978). Role of cell shape in growth control. *Nature, Lond.* **273**, 345–9.

Fonte, V. G., Anderson, K. L., Wolosewick, J. J. & Porter, K. R. (1978). The effects of dihydrocytochalasin B on the cytoskeleton of NRK cells and embryonic chick epithelial cells in culture. *J. Cell Biol.* **79**, 74a.

Foulds, L. (1954). Tumor progression: a review. *Cancer Res.* **14**, 327–39.

Foulds, L. (1956). The histological analysis of mammary tumors of mice. *J. Nat. Cancer Inst.* **17**, 701–801.

Foulds, L. (1969). *Neoplastic Development*, vol. 1. London: Academic Press.

Foulds, L. (1975). *Neoplastic Development*, vol. 2. London: Academic Press.

Fournier, R. E. & Pardee, A. B. (1975). Cell cycle studies of mononucleate and cytochalasin-B-induced binucleate fibroblasts. *Proc. Natl. Acad. Sci. USA* **72**, 869–73.

Fowler, V. & Branton, D. (1977). Lateral mobility of human erythrocyte integral membrane proteins. *Nature, Lond.* **268**, 23–6.

Fox, C. H., Dvorak, J. A. & Sanford, K. K. (1976). Cytometric analysis of neoplastic transformation of vertebrate cell populations. *Cancer Res.* **36**, 1556–61.

Fox, C. H., Caspersson, T., Kudynowski, J., Sanford, K. K. & Tarone, R. E. (1977). Morphometric analysis of neoplastic transformation in rodent fibroblast cell lines. *Cancer Res.* **37**, 892–7.

Fox, T. O. Sheppard, J. R. & Burger, M. M. (1971). Cyclic membrane changes in animal cells: transformed cells permanently display a surface architecture detected in normal cells only during mitosis. *Proc. Natl. Acad. Sci. USA* **68**, 244–7.

Fradkin, A., Janoff, A., Lane, B. P. & Kuschner, M. (1975). In-vitro transformation of BHK 21 cells grown in the presence of calcium chromate. *Cancer Res.* **35**, 1058–63.

Franke, W. W., Grund, C., Fink, A., Weber, K., Jockusch, B. M., Zentgraf, M. & Osborn, M. (1978a). Location of actin in the microfilament bundles associated with the junctional specialisations between sertoli cells and spermatids. *Biologie Cellulaire* **31**, 7–14.

Franke, W. W., Grund, C., Osborn, M. & Weber, K. (1978b). The intermediate-sized filaments in rat kangaroo PtK$_2$ cells. I. Morphology *in situ*. *Cytobiologie, Eur. J. Cell Biol.* **17**, 365–91.

Frankel, A. E., Haapala, D. K., Neubauer, R. L. & Fischinger, P. J. (1976). Elimination of the sarcoma genome from murine sarcoma virus transformed cat cells. *Science* **191**, 1264–6.

Frankel, F. R. (1976). Organization and energy-dependent growth of microtubules in cells. *Proc. Natl. Acad. Sci. USA* **73**, 2798–802.

Frankfurt, O. S. (1975). *Cell Cycle in Tumours*. Moscow: Medizina (in Russian).

Franks, L. M., Chesterman, F. C. & Rowlatt, C. (1970). The structure of tumours derived from mouse cells after 'spontaneous' transformation *in vitro*. *Brit. J. Cancer* **24**, 843–8.

Franks, L. M. & Cooper, T. W. (1972). The origin of human embryo lung cells in culture: a comment on cell differentiation, in vitro growth and neoplasia. *Int. J. Cancer* **9**, 19–29.

Frazier, W. A. (1976). The role of cell surface components in the morphogenesis of the cellular slime molds. *Trends Biochem. Sci.* **1**, 130–3.

Freed, J. J. & Lebowitz, M. M. (1970). The association of class of saltatory movements with microtubules in cultured cells. *J. Cell Biol.* **45**, 334–53.

Freedman, M. H., Raff, M. C. & Gomperts, B. (1976). Induction of increased calcium intake in mouse T lymphocytes by concanavalin A and its modulation by cyclic nucleotides. *Nature, Lond.* **255**, 378–83.

Freedman, V. H. & Shin, S. (1977). Isolation of human diploid cell variants with enhanced colony-forming efficiency in semi-solid medium after a single-step chemical mutagenesis. *J. Nat. Cancer Inst.* **58**, 1873–5.

Frelin, C. & Padieu, P. (1976). Pleiotypic response of rat heart cells in culture to serum stimulation. *Biochimie* **58**, 953–9.

Friedell, G. H. (1976). Carcinoma, carcinoma *in situ* and 'Early Lesions' of the uterine cervix and the urinary bladder: introduction and definitions. *Cancer Res.* **36**, 2482–4.

Friedenstein, A. J. (1973). Determined and inducible osteogenic precursor cells. In *Hard Tissue Growth, Repair and Remineralization* ed. R. F. Sognnaes. Ciba Foundation Symposium 11, pp. 169–86. Amsterdam: Associated Scientific Publishers.

Friedenstein, A. J. (1976). Precursor cells of mechanocytes. *Int. Rev. Cytol.* **47**, 327–59.

Friedenstein, A. J., Rapoport, R. I. & Luria, E. A. (1967). Histiotypic structures arising in organ cultures of human diploid strains. *Dokl. Acad. Nauk. SSSR*, **176**, 452–55 (in Russian).

Friedewald, W. F. & Rous, P. (1944). The initiating and promoting elements in tumor production. An analysis of the effects of tar, benzpyrene, and methylcholanthrene on rabbit skin. *J. Exp. Med.* **80**, 101–26.

Friend, C., Scher, W., Holland, J. G. & Sato, T. (1971). Hemoglobin synthesis in murine virus-induced leukemia cells *in vitro*: stimulation of erythroid differentiation by dimethyl sulfoxide. *Proc. Natl. Acad. Sci. USA* **68**, 378–82.

Friend, D. S. & Gilula, N. B. (1972). Variations in tight and gap junctions in mammalian tissues. *J. Cell Biol.* **53**, 758–76.

Froelich, J. & Rachmeler, M. (1972). Effect of adenosine 3',5'-cyclic monophosphate on cell proliferation. *J. Cell Biol.* **55**, 19–32.

Froelich, J. E. & Anastassiades, T. P. (1974). Role of pH in fibroblast proliferation. *J. Cell. Physiol.* **84**, 253–60.

Froelich, J. E. & Anastassiades, T. P. (1975). Possible limitation of growth in human fibroblast cultures by diffusion. *J. Cell. Physiol.* **86**, 567–80.

Frye, L. D. & Edidin, M. (1970). The rapid intermixing of all surface antigens after formation of mouse–human heterokaryons. *J. Cell Sci.* **7**, 319–35.

Furcht, L. T., Mosher, D. F., Wendelschafer-Crabb, G. (1978). Differences in the ultrastructural organization of fibronectin (LETS protein) on transformed and normal human cells. *Fed. Proc.* **37**, 909.

Furcht, L. T. & Wendelschafer-Crabb, G. (1978). Trypsin-induced co-ordinate alterations in cell shape, cytoskeleton, and intrinsic membrane structure of contact-inhibited cells. *Exp. Cell Res.* **114**, 1–14.

Furshpan, E. J. & Potter, D. D. (1968). Low-resistance junctions between cells in embryos and tissue culture. *Curr. Top. Dev. Biol.* **3**, 95–127.

Furth, J. (1953). Conditioned and autonomous neoplasms: a review. *Cancer Res.* **13**, 477–92.

Gabbiani, G., Chaponnier, C., Zumbe, A. & Vassalli, P. (1977). Actin and tubulin co-cap with surface immunoglubins in mouse B lymphocytes. *Nature, Lond.* **269**, 697–8.

Gaffney, B. J. (1975). Fatty acid chain flexibility in the membranes of normal and transformed fibroblasts. *Proc. Natl. Acad. Sci. USA* **72**, 664–8.

Gahmberg, C. G. & Hakomori, S. (1973). Altered growth behavior of malignant cells associated with changes in externally labelled glycoprotein and glycolipid. *Proc. Natl. Acad. Sci. USA* **70**, 3329–33.

Gail, M. H. & Boone, C. W. (1971*a*). Effect of colcemid on fibroblast motility. *Exp. Cell Res.* **65**, 221–7.

Gail, M. H. & Boone, C. W. (1971*b*). Density inhibition of motility in 3T3 fibroblasts and their SV-40 transformants. *Exp. Cell Res.* **64**, 156–62.

Gail, M. H., Boone, C. W. & Thompson, C. S. (1973). A calcium requirement for fibroblast motility and proliferation. *Exp. Cell Res.* **79**, 386–90.

Galasko, C. S. B. (1976). Mechanisms of bone destruction in the development of skeletal metastases. *Nature, Lond.* **263**, 507–8.

Gallagher, M., Detwiler, T. C. & Stracher, A. (1976). Two forms of platelet actin that differ from skeletal muscle actin. In *Cell Motility*, ed. R. Goldman, T. Pollard & J. Rosenbaum. Cold Spring Harbor Conferences on Cell Proliferation, vol. 3, pp. 475–85. New York: Cold Spring Harbor Laboratory.

Gammon, M. T. & Isselbacher, K. J. (1976). Neoplastic potentials and regulation of uptake of nutrients. I. A glutamine independent variant of polyoma BHK with a very high neoplastic potential. *J. Cell. Physiol.* **89**, 759–64.

Gardner, W. U. (1953). Hormonal aspects of experimental tumorigenesis. *Adv. Cancer Res.* **1**, 173–232.

Garrido, J., Burglen, M.-J., Samolyk, D., Wicker, R. & Bernhard, W. (1974). Ultrastructural comparison between the distribution of concanavalin A and wheat germ agglutinin cell surface receptors of normal and transformed hamster and rat cell lines. *Cancer Res.* **34**, 230–43.

Garrod, D. R. & Steinberg, M. S. (1975). Cell locomotion within a contact-inhibited monolayer of chick embryonic liver parenchyma cells. *J. Cell Sci.* **18**, 405–25.

Garschin, W. G. (1928). Über die Bedeutung der atypischen Epithelwucherungen. *Z. Krebsforsch.* **27**, 569–78.

Garschin, W. G. (1939). *Inflammatory proliferation of epithelium; its biological significance and relation to the cancer problem.* Moscow: Medgiz (in Russian).

Gartler, S. M. (1974). Utilization of mosaic systems in the study of the origin and progression of tumors. In *Chromosomes and Cancer*, ed. J. German, pp. 314–34. New York: J. Wiley & Sons.

Gartner, T. K. & Podleski, T. R. (1975). Evidence that a membrane bound lectin mediates fusion of L_6 myoblasts. *Biochem. Biophys. Res. Commun.* **67**, 972–8.

Gasic, G. J., Gasic, T. B., Galanti, N., Johnson, T. & Murphy, S. (1973). Platelet-tumor cell interaction in mice. The role of platelets in the spread of malignant disease. *Int. J. Cancer* **11**, 704–18.

Gay, S., Martin, G. R., Müller, P. K., Timpl, R. & Kühn, K. (1976). Simultaneous synthesis of types I and III collagen by fibroblasts in culture. *Proc. Natl. Acad. Sci. USA* **73**, 4037–40.

Gearhart, J. D. & Mintz, B. (1974). Contact-mediated myogenesis and increased acetylcholinesterase activity in primary cultures of mouse teratocarcinoma cells. *Proc. Natl. Acad. Sci. USA* **71**, 1734–8.

Gearhart, J. D. & Mintz, B. (1975). Creatine kinase, myokinase, and acetylcholinesterase activities in muscle-forming primary cultures of mouse teratocarcinoma cells. *Cell* **6**, 61–6.

Gelboin, H. V., Okuda, T., Selkirk, J., Nemoto, N., Yang, S. K., Wiebel, F. J., Whitlock, J. P., Jr, Rapp, H. J. & Bast, R. C., Jr. (1976). Benzo(a)pyrene metabolism: enzymatic and liquid chromatographic analysis and application to human liver, lymphocytes and monocytes. In *Screening Tests in Chemical Carcinogenesis*, ed. R. Montesano, H. Bartsch & L. Tomatis, pp. 225–247. Lyon: International Agency for Research on Cancer.

Gelboin, H. V. & Wiebel, F. J. (1971). Studies on the mechanism of aryl hydrocarbon hydroxylase induction and its role in cytotoxicity and tumorigenicity. *Ann. N.Y. Acad. Sci.* **179**, 529–47.

Gelboin, H. V., Wiebel, F. J. & Kinoshita, N. (1974). Aryl hydrocarbon hydroxylase regulation and role in polycyclic hydrocarbon. In *Chemical Carcinogenesis*, ed. P.O.P. Ts'o & J. D. Di Paolo, pp. 309–51. New York: Marcel Dekker.

Gelfand, V. I. (1973). Binding of concanavalin A by normal and neoplastic cells. *Byulleten experimentalnoj Biologii i Medicini* 75, No. 4, 82–4 (in Russian, translated into English).

Gelfand, V. I. (1974). Reversion of neoplastic transformation under the action of serum-free medium. *Dokl. Akad. Nauk SSR* **215**, 460–3 (in Russian).

Gelfant, S. (1977). A new concept of tissue and tumor cell proliferation. *Cancer Res.* **37**, 3845–62.

Gelfant, S. & Smith, J. G., Jr. (1972). Aging, non-cycling cells: an explanation. *Science* **178**, 375–81.

Gershman, H. & Rosen, J. J. (1978). Cell adhesion and cell surface topography in aggregates of 3T3 and SV-40 virus-transformed 3T3 cells. Visualization of interior cells by scanning electron microscopy. *J. Cell Biol.* **76**, 639–51.

Gharrett, A. J., Malkinson, A. M. & Sheppard, J. R. (1976). Cyclic AMP-dependent protein kinases from normal and SV-40-transformed 3T3 cells. *Nature, Lond.* **264**, 673–5.

Ghosh, N. K., Deutsch, S. I., Griffin, M. J. & Cox, R. P. (1975). Regulation of growth and morphological modulation of Hela$_{65}$ cells in monolayer culture by dibutyryl cyclic AMP, butyrate and their analogs. *J. Cell. Physiol.* **86**, 663–72.

Giaever, I. & Ward, E. (1978). Cell adhesion to substrates containing adsorbed or attached IgC. *Proc. Natl. Acad. Sci. USA* **75**, 1366–8.

Gidwitz, S., Weber, M. J. & Storm, D. R. (1976). Solubilization of adenylate cyclase from normal and Rous sarcoma-transformed chicken embryo fibroblasts. *J. Biol. Chem.* **251**, 7950–1.

Gilula, N. B. (1974). Junctions between cells. In *Cell Communication*, ed. R. P. Cox, pp. 1–29. New York: J. Wiley & Sons.

Gilula, N. B. & Satir, P. (1972). The ciliary necklace: a ciliary membrane specialization. *J. Cell Biol.* **53**, 494–509.

Gimbrone, M. A., Jr. & Gullino, P. M. (1976*a*). Angiogenic capacity of preneoplastic lesions of the murine mammary gland as a marker of neoplastic transformation. *Cancer Res.* **36**, 2611–20.

Gimbrone, M. A., Jr. & Gullino, P. M. (1976*b*). Neovascularization induced by intra-ocular xenografts of normal, preneoplastic, and neoplastic mouse Mammary tissues. *J. Nat. Cancer Inst.* **56**, 305–18.

Ginsburg, E., Salomon, D., Sreevalsan, T. & Freese, E. (1973). Growth inhibition and morphological changes caused by lipophilic acid in mammalian cells. *Proc. Natl. Acad. Sci. USA* **70**, 2457–61.

Gionti, E. & Lawrence, D. A. (1977). Cyclic AMP levels and 2-deoxyglucose uptake in cells transformed by temperature-sensitive class T mutants of Rous sarcoma virus. *Virology* **79**, 244–8.

Gipson, J. & Anderson, R. A. (1978). Comparison of intermediate filaments from three bovine tissues. *J. Cell. Biol.* **79**, pt. 2, 260a.

Glaser, R., Mumaw, V., Farrugia, R. & Munger, B. (1977). Scanning electron microscopy of the surfaces of hamster embryo cells transformed by herpes simplex virus. *Cancer Res.* **37**, 4420–2.

Glenn, K. C. & Cunningham, D. D. (1978). Cleavage of cell surface proteins and initiation of cell division by thrombin. *J. Cell Biol.* **79**, 41a.

Glimelius, B., Norling, B., Westermark, B. & Wasteson, Å. (1978). Composition and distribution of glycosaminoglycans in cultures of human normal and malignant glial cells. *Biochem. J.* **172**, 443–56.

Glynn, R. D., Thrash, C. R. & Cunningham, D. D. (1973). Maximal concanavalin A specific agglutinability without loss of density-dependent growth control. *Proc. Natl. Acad. Sci. USA* **70**, 2676–7.

Goldenberg, G. J. & Stein, W. D. (1978). Stimulation of uridine uptake in 3T3 cells is associated with increased ATP affinity of uridine-phosphorylating system. *Nature, Lond.* **274**, 475–7.

Goldfine, I. D. & Smith, G. J. (1976). Binding of insulin to isolated nuclei. *Proc. Natl. Acad. Sci. USA* **73**, 1427–31.

Goldman, R. (1976). The effect of cytochalasin B and colchicine on concanavalin A-induced vacuolation in mouse peritoneal macrophages. *Exp. Cell Res.* **99**, 385–94.

Goldman, R. (1977). Lectin-mediated attachment and ingestion of yeast cells and erythrocytes by hamster fibroblasts. *Exp. Cell Res.* **104**, 325–34.

Goldman, R., Sharon, N. & Lotan, R. (1976). A differential response elicited in macrophages on interaction with lectins. *Exp. Cell Res.* **99**, 408–22.

Goldman, R. D., Berg, G., Bushnell, A., Chang, C.-M., Dickerman, L., Hopkins, N., Miller, M. L., Pollack, R. & Wang, E. (1973). Fibrillar systems in cell motility. In *Locomotion of Tissue Cells*, ed. M. Abercrombie. Ciba Foundation Symposium, vol. 14, pp. 83–103. Amsterdam: Associated Scientific Publishers.

Goldman, R. D. & Follett, E. A. C. (1970). Birefringent filamentous organelle in BHK-21 cells and its possible role in cell spreading and motility, *Science* **169**, 286–8.

Goldman, R. D. & Knipe, D. (1973). Functions of cytoplasmic fibers in non-muscle cell motility. *Cold Spring Harbour Symp. Quant. Biol.* **37**, 523–34.

Goldman, R. D., Schloss, J. A. & Starger, J. M. (1976a). Organization changes of actin-like microfilaments during animal cell movement. In *Cell Motility*, ed. R. Goldman, T. Pollard & J. Rosenbaum. Cold Spring Harbor Conferences on Cell Proliferation, vol. 3, pp. 217–45. New York: Cold Spring Harbor Laboratory.

Goldman, R. D., Yerna, M. & Schloss, J. A. (1976b). Localization and organization of microfilaments and related proteins in normal and virus-transformed cells. *J. Supramol. Struc.* **5**, 155–83.

Gonda, M. A., Aaronson, S. A., Ellmore, N., Zeve, V. H. & Nagashima, K. (1976). Ultrastructural studies of surface features of human normal and tumor cells in tissue culture by scanning and transmission electron microscopy. *J. Nat. Cancer Inst.* **56**, 245–64.

Goodenough, D. A. (1976). In-vitro formation of gap junction vesicles. *J. Cell Biol.* **68**, 220–31.

Goodenough, D. A. & Revel, J. P. (1970). A fine structural analysis of intercellular junctions in the mouse liver. *J. Cell Biol.* **45**, 272–90.

Gordon, W. E., III, Bushell, A. & Burridge, K. (1978). Characterization of the intermediate (10 nm) filaments of cultured cells using an auto-immune rabbit antiserum. *Cell* **13**, 249–61.

Gorozhanskaya, E. C. & Shapot, V. S. (1964). Pecualiarities of glucose consumption by the ascites cancer cells. *Dokl. Akad. Nauk USSR* **155**, 947–50 (in Russian).

Gospodarowicz, D., Bialecki, H. & Greenburg, G. (1978a). Purification of the fibroblast growth factor activity from bovine brain. *J. Biol. Chem.* **253**, 3736–43.

Gospodarowicz, D., Brown, K. D., Birdwell, C. R. & Zetter, B. R. (1978*b*). Control of proliferation of human vascular endothelial cells. *J. Cell Biol.* **77**, 774–88.

Gospodarowicz, D., Lindstrom, J. & Benirschke, K. (1975). Fibroblast growth factor: its localization, purification, mode of action, and physiological significance. *Adv. Metab. Disord.* **8**, 301–35.

Gospodarowicz, D. & Moran, J. S. (1974*a*). Stimulation of division of sparse and confluent 3T3 cell population by a fibroblast growth factor, dexamethazone and insulin. *Proc. Natl. Acad. Sci. USA* **71**, 4584–8.

Gospodarowicz, D. & Moran, J. (1974*b*). Effect of a fibroblast growth factor, insulin, dexamethasone and serum on the morphology of BALB/c 3T3 cells. *Proc. Natl. Acad. Sci. USA* **71**, 4648–52.

Gospodarowicz, D. & Moran, J. S. (1976). Growth factors in mammalian cell culture. *Annu. Rev. Biochem.* **45**, 531–58.

Gospodarowicz, D., Moran, J. S. & Braun, D. L. (1977). Control of proliferation of bovine vascular endothelial cells. *J. Cell. Physiol.* **91**, 377–86.

Goto, M., Kataoka, Y., Kimura, T., Goto, K. & Sato, H. (1973). Decrease of saturation density of cells of hamster cell lines after treatment with dextran sulfate. *Exp. Cell Res.* **82**, 367–74.

Grady, S. R. & McGuire, E. J. (1976). Intercellular adhesive selectivity. III. Species selectivity of embryonic liver intercellular adhesion. *J. Cell Biol.* **71**, 96–106.

Gray, P. N., Cullum, M. E. & Griffin, M. J. (1976). Population density and regulation of cell division in 3T3 cells. I. Inorganic phosphate levels, uptake and release. *J. Cell. Physiol.* **89**, 225–34.

Green, H. (1977). Terminal differentiation of cultured human epidermal cells. *Cell* **11**, 405–16.

Green, H. & Kehinde, O. (1975). An established pre-adipose cell line and its differentiation in culture. II. Factors affecting the adipose conversion. *Cell* **5**, 19–27.

Green, H. & Kehinde, O. (1976). Spontaneous heritable changes leading to increased adipose conversion in 3T3 cells. *Cell* **7**, 105–13.

Green, H. & Thomas, J. (1978). Pattern formation by cultured human epidermal cells: development of curved ridges resembling dermatoglyphs. *Science* **200**, 1385–8.

Greenberg, D. B., Barsh, G. S., Ho, T. S. & Cunningham, D. D. (1976). Serum-stimulated phosphate uptake and initiation of fibroblast proliferation. *J. Cell. Physiol.* **90**, 193–210.

Greenberg, D. B. & Cunningham, D. D. (1973). Does hyaluronidase initiate DNA synthesis. *J. Cell. Physiol.* **82**, 511–12.

Greenblatt, M. & Shubik, P. (1968). Tumor angiogenesis: transfilter diffusion studies in the hamster by the transparent chamber technique. *J. Natl. Cancer Inst.* **41**, 111–24.

Gregory, S. H. & Bose, S. K. (1977). Density-dependent changes in hexose transport, glycolytic enzyme levels, and glycolytic rates, in uninfected and murine sarcoma virus-transformed rat kidney cells. *Exp. Cell Res.* **110**, 387–97.

Griffith, L. M. & Pollard, T. D. (1978). Evidence for actin filament-microtubule interaction mediated by microtubule-associated proteins. *J. Cell Biol.* **78**, 958–65.

Grinnell, F. (1976). Cell spreading factor. Occurrence and specificity of action. *Exp. Cell Res.* **102**, 51–62.

Grinnell, F. (1978). Cellular adhesion and extracellular substrata. *Int. Rev. Cytol.* **53**, 65–144.

Grinnell, F. & Hays, D. G. (1978*a*). Cell adhesion and spreading factor. Similarity to cold insoluble globulin in human serum. *Exp. Cell Res.* **115**, 221–9.

Grinnell, F. & Hays, D. G. (1978*b*). Induction of cell spreading by substratum-adsorbed ligands directed against the cell surface. *Exp. Cell Res.* **116**, 275–84.

Groneberg, J., Sutter, D., Soboll, H. & Doerfler, W. (1978). Morphological revertants of adenovirus type 12-transformed hamster cells. *J. Gen. Virol.* **40**, 635–45.

Grove, G. L. & Cristofalo, V. J. (1976). Characterization of the cell cycle of cultured human diploid cells: effects of aging and hydrocortisone. *J. Cell. Physiol.* **90**, 415–22.

Guelstein, V. I., Ivanova, O. Y., Margolis, L. B., Vasiliev, J. M. & Gelfand, I. M. (1973). Contact inhibition of movement in cultures of transformed cells. *Proc. Natl. Acad. Sci. USA* **70**, 2011–14.

Guelstein, V. I. & Klempner, L. B. (1973). Kinetics of cellular proliferation of slowly growing highly differentiated mouse hepatoma. *Voprosi Onkologii* **19**, N 12, 41–6 (in Russian).

Guelstein, V. I. & Stavrovskaya, A. A. (1972). The effect of colcemid on mitosis, migration and DNA synthesis of mouse fibroblasts *in vitro. Byulleten experimentalnoj Biologii i Meditsini* **74**, N 6, 94–6 (in Russian).

Guidotti, G. (1977). The structure of intrinsic membrane proteins. *J. Supramol. Struc.* **7**, 489–97.

Guillouzo, A., Oudea, P., Le Guilly, Y., Oudea, M. C., Lenoir, P. & Bourel, M. (1972). An ultrastructural study of primary cultures of adult human liver tissue. *Exp. Mol. Pathol.* **16**, 1–15.

Gullino, P. M. (1966). The internal milieu of tumors. In *Progress in experimental Tumor Research* ed. F. Homburger, vol. 8, pp. 1–25. Basel: S. Karger.

Gullino, P. M. (1975). Extracellular compartments of solid tumors. In *Cancer: a Comprehensive Treatise*, ed. F. F. Becker, vol. 3, pp. 327–54. New York: Plenum Press.

Gullino, P. M., Clark, S. H. & Grantham, F. H. (1964). The intestinal fluid of solid tumors. *Cancer Res.* **24**, 780–98.

Gurney, T., Jr. (1969). Local stimulation of growth in primary cultures of chick embryo fibroblasts. *Proc. Natl. Acad. Sci. USA* **62**, 906–11.

Hadden, J. W., Hadden, E. M., Sadlik, J. R. & Coffey, R. G. (1976). Effects of concanavalin A and succinylated derivative on lymphocyte proliferation and cyclic nucleotide levels. *Proc. Natl. Acad. Sci. USA* **73**, 1717–21.

Haddow, A. (1938). Cellular inhibition and the origin of cancer. *Acta Unio. Int. Contra Cancrum* **3**, 342–53.

Haddox, M. K., Furcht, L. T., Gentry, S. R., Moser, M. E., Stephenson, J. H. & Goldberg, N. D. (1976). Periodate-induced increase in cyclic GMP in mouse and guinea pig splenic cells in association with mitogenesis. *Nature, Lond.* **262**, 146–8.

Haigler, H., Ash, J. F., Singer, S. J. & Cohen, S. (1978). Visualization by fluorescence of the binding and internalization of epidermal growth factor in human carcinoma cells A-431. *Proc. Natl. Acad. Sci. USA* **75**, 3317–21.

Haji-Karim, M. & Carlsson, J. (1978). Proliferation and viability in cellular spheroids of human origin. *Cancer* **38**, 1457–64.

Hakomori, S. (1975). Structures and organization of cell surface glycolipids; dependency on the cells growth and malignant transformation. *Biochem. Biophys. Res. Commun.* **12**, 55–89.

Hale, A. H. & Weber, M. J. (1975). Hydrolase and serum treatment of normal chick embryo cells: effect on hexose transport. *Cell* **5**, 245–52.

Hale, A. H., Winkelhake, J. L. & Weber, M. J. (1975). Cell surface changes and Rous sarcoma virus gene expression in synchronized cells. *J. Cell Biol.* **64**, 398–407.

Hallowes, R. C., Rudland, P. S., Hawkins, R. A., Lewis, D. J., Bennett, D. & Durbin, H. (1977). Comparison of the effects of hormones on DNA synthesis in cell cultures of nonneoplastic and neoplastic mammary epithelium from rats. *Cancer Res.* **37**, 2492–504.

Hard, G. C. & Borland, R. (1975). In-vitro culture of cells isolated from dimethylnitrosamine-induced renal mesenchymal tumors of the rat. I. Quantitative morphology. *J. Natl. Cancer Inst.* **54**, 1085–95.

Hard, G. C. & Borland, R. (1977). Morphological character of transforming renal cell cultures derived from Wistar rats given dimethylnitrosamine. *J. Natl. Cancer Inst.* **58**, 1377–90.

Harel, L. & Jullien, M. (1976). Evaluation of proximity inhibition of DNA synthesis in 3T3 cells (Comments on the paper of M. C. Canagaratna & P. A. Riley). *J. Cell. Physiol.* **88**, 253–4.

Harley, C. B. & Goldstein, S. (1978). Cultured human fibroblasts: distribution of cell generations and a critical limit. *J. Cell. Physiol.* **97**, 509–16.

Harrington, W. N., Godman, G. C. & Wilner, G. D. (1978). A selective inhibitor of cell proliferation from serum. *Fed. Proc.* **37**, 402.

Harris, A. K. (1973a). Location of cellular adhesions to solid substrata. *Dev. Biol.* **35**, 97–114.

Harris, A. K. (1973b). Behavior of cultured cells on substrata of variable adhesiveness. *Exp. Cell Res.* **77**, 285–97.

Harris, A. K. (1973c). Cell surface movements related to cell locomotion. In *Locomotion in Tissue Cells*, ed. M. Abercrombie. Ciba Foundation Symposium, vol. 14. pp. 3–20. Amsterdam: Associated Scientific Publishers.

Harris, A. (1974). Contact inhibition of cell locomotion. In *Cell Communication*, ed. R. P. Cox, pp. 147–85. New York: J. Wiley & Sons.

Harris, A. K. (1976a). Recycling of dissolved plasma membrane components as an explanation of the capping phenomenon. *Nature, Lond.* **263**, 781–3.

Harris, A. K. (1976b). Is cell sorting caused by differences in the work of intercellular adhesion? A critique of the Steinberg hypothesis. *J. Theor. Biol.* **61**, 276–85.

Harris, A. K. & Dunn, G. (1972). Centripetal transport of attached particles on both surfaces of moving fibroblasts. *Exp. Cell Res.* **73**, 519–23.

Harris, M. (1974). Research on tissue culture at the National Cancer Institute, with special reference to the contributions of Dr Virginia J. Evans. *J. Natl. Cancer Inst.* **53**, 1465–9.

Hartwig, J. H. & Stossel, T. P. (1977). Actin-binding protein as an actin gelling factor: specificity and mechanism. *J. Cell Biol.* **72**, 253a.

Hartwig, J. H. & Stossel, T. P. (1978). Cytochalasin B dissolves actin gels by breaking actin filaments. *J. Cell Biol.* **79**, 271a.

Haselbacher, G. K. & Humbel, R. E. (1976). Stimulation of ornithine decarboxylase activity in chick fibroblasts by non-suppressible insulin-like activity (NSILA), insulin and serum. *J. Cell. Physiol.* **88**, 239–46.

Hassell, J. & Engelhardt, D. L. (1977). Factors regulating the multiplication of animal cells in culture. *Exp. Cell Res.* **107**, 159–67.

Hatanaka, M. (1976). Saturable and non-saturable process of sugar uptake: effect of oncogenic transformation in transport and uptake of nutrients. *J. Cell. Physiol.* **89**, 745–50.

Hatanaka, M. & Hanafusa, H. (1970). Analysis of a functional change in membrane in the process of cell transformation by Rous sarcoma virus: alteration in the characteristics of sugar transport. *Virology* **41**, 647–52.

Hatanaka, M., Huebner, R. J. & Gilden, R. V. (1969). Alterations in the characteristics of sugar uptake by mouse cells transformed by murine sarcoma viruses. *J. Natl. Cancer Inst.* **43**, 1091–6.

Hatcher, V. B., Oberman, M. S., Wertheim, M. S., Rhee, C. Y., Tsien, G. & Burk, P. G. (1977). The relationship between surface protease activity and the rate of cell proliteration in normal and transformed cells. *Biochem. Biophys. Res. Commun.* **76**, 602–8.

Hatcher, V. B., Wertheim, M. S., Rhee, C., Tsien, G. & Burk, P. G. (1976). Relationship between cell surface protease activity and doubling time in various normal and transformed cells. *Biochim. Biophys. Acta* **451**, 499–510.

Hatten, M. E., Scandella, C. J., Horwitz, A. F. & Burger, M. M. (1978). Similarities in the membrane fluidity of 3T3 and SV101-3T3 cells and its relation to concanavalin A- and wheat germ agglutinin-induced agglutination. *J. Biol. Chem.* **253**, 1972–7.

Haudenschild, C. C., Zahniser, D., Folkman, J. & Klagsbrum, M. (1976). Human vascular endothelial cells in culture. Lack of response to serum growth factors. *Exp. Cell Res.* **98**, 175–83.

Hausman, R. E. & Moscona, A. A. (1976). Isolation of retina-specific cell-aggregating factor from membranes of embryonic neural retina tissue. *Proc. Natl. Acad. Sci. USA* **73**, 3594–8.

Hayflick, L. (1973). Biology of human aging. *Am. J. Med. Sci.* 433–45.

Hayflick, L. & Moorhead, P. S. (1961). The serial cultivation of human diploid cell strains. *Exp. Cell Res.* **25**, 585.

Heath, J. P. & Dunn, G. A. (1978). Cell to substratum contacts of chick fibroblasts and their relation to the microfilament system. A correlated interference–reflexion and high-voltage electron-microscope study. *J. Cell Sci.* **29**, 197–212.

Heaysman, J. E. M. (1970). Non-reciprocal contract inhibition. *Experientia* **26**, 1344.

Heaysman, J. E. M. & Pegrum, S. M. (1973). Early contacts between fibroblasts. An ultrastructural study. *Exp. Cell Res.* **78**, 71–8.

Hedman, K., Vaheri, A. & Wartiovaara, J. (1978). External fibronectin of cultured human fibroblasts is predominantly a matrix protein. *J. Cell Biol.* **76**, 748–60.

Heggeness, M. H., Simon, M. & Singer, S. J. (1978). Association of mitochondria with microtubules in cultured cells. *Proc. Natl. Acad. Sci. USA* **75**, 3863–66.

Heggeness, M. H., Wang, K. & Singer, S. J. (1977). Intracellular distributions of mechanochemical proteins in cultured fibroblasts. *Proc. Natl. Acad. Sci. USA* **74**, 3883–7.

Heidelberger, C. (1973). Chemical oncogenesis in culture. *Adv. Cancer Res.* **18**, 317–66.

Heidelberger, C. & Boshell, P. F. (1975). Chemical oncogenesis in culture. *Gann Monogr. Cancer Res.* **17**, 39–58.

Heidrick, M. L. & Ryan, W. L. (1971). Adenosine 3′,5′-cyclic monophosphate and contact inhibition. *Cancer Res.* **31**, 1313–15.

Heilbrunn, L. V. (1956). *The Dynamics of Living Protoplasm.* New York: Academic Press.

Heldin, C. H., Wasteson, A. & Westermark, B. (1977). Partial purification and characterization of platelet factors stimulating the multiplication of normal human glial cells. *Exp. Cell Res.* **109**, 429–37.

Henderson, J. F. & Le Page, G. A. (1959). The nutrition of tumors: A review. *Cancer Res.* **19**, 887–902.

Hendil, K. B. (1977). Intracellular protein degradation in growing, in density-inhibited, and in serum-restricted fibroblast cultures. *J. Cell. Physiol.* **92**, 353–64.

Henneberry, R. C. & Fishman, P. H. (1976). Morphological and biochemical differentiation in HeLa cells. *Exp. Cell Res.* **103**, 55–62.

Herman, M. M. & Van den Berg, S. R. (1978). Neoplastic neuroepithelial differentiation in an experimental transplantable teratoma. In *Cell Differentiation and Neoplasia*, ed. G. F. Saunders, pp. 93–109. New York: Raven Press.

Hermens, A. F. & Barendsen, G. W. (1978). The proliferative status and clonogenic capacity of tumour cells in a transplantable rhabdomyosarcoma of the rat before and after irradiation with 800 rad of X-rays. *Cell Tissue Kinet.* **11**, 83–100.

Hershko, A., Mamow, P., Shields, R. & Tomkins, G. M. (1971). Pleiotropic response. *Nature New Biol.* **232**, 206–11.

Herzog, W. & Weber, K. (1977). In-vitro assembly of pure tubulin into microtubules in the absence of microtubule-associated proteins and glycerol. *Proc. Natl. Acad. Sci. USA* **74**, 1860–4.

Hesketh, T. R., Smith, G. A., Houslay, M. D., Warren, G. B. & Metcalfe, G. C. (1977). Is an early calcium flux necessary to stimulate lymphocytes? *Nature, Lond.* **267**, 490–4.

Hill, B. T. (1978). The management of human 'solid' tumours: some observations on the irrelevance of traditional cell cycle kinetics and the value of certain recent concepts. *Cell Biol. Int. Rep.* **2**, 215–30.

Hirano, A. & Kurimura, T. (1974). Virally transformed cells and cytochalasin B. I. The effect of cytochalasin B on cytokinesis, karyokinesis and DNA synthesis in cell. *Exp. Cell Res.* **89**, 111–20.

Hitchcock, S. E. (1977). Regulation of motility in non-muscle cells. *J. Cell Biol.* **74**, 1–15.

Hitotsumachi, S., Rabinowitz, Z. & Sachs, L. (1971). Chromosomal control of reversion in transformed cells. *Nature, Lond.* **231**, 511–14.

Hodges, G. M., Livingston, D. C. & Franks, L. M. (1973). The localization of trypsin in cultured mammalian cells. *J. Cell Sci.* **12**, 887–902.

Hodges, G. M. & Muir, G. M. (1972). A scanning electron-microscope study of the surface features of mammalian cells *in vitro. J. Cell Sci.* **11**, 233–47.

Hoffman, R. M. & Erbe, R. W. (1976). High in-vivo rates of methionine biosynthesis in transformed human and malignant rat cells auxotrophic for methionine. *Proc. Natl. Acad. Sci. USA* **73**, 1523–7.

Hoffman-Berling, H. (1954). Adenosintriphosphat als Betriebstoff von Zellbewegungen. *Biochim. Biophys. Acta* **14**, 182–94.

Hollenberg, M. D. & Cuatrecasas, P. (1973). Epidermal growth factor: receptors in human fibroblasts and modulation of action by cholera toxin. *Proc. Natl. Acad. Sci. USA* **70**, 2964–8.

Holley, R. W. (1972). A unifying hypothesis concerning the nature of malignant growth. *Proc. Natl. Acad. Sci. USA* **69**, 2840–1.

Holley, R. W. (1975). Control of growth of mammalian cells in cell culture. *Nature, Lond.* **258**, 487–90.

Holley, R. W., Armour, R. & Baldwin, J. H. (1978*a*). Density-dependent regulation of growth of BSC-1 cells in cell culture: control of growth by low molecular weight nutrients. *Proc. Natl. Acad. Sci. USA* **75**, 339–41.

Holley, R. W., Armour, R. & Baldwin, J. H. (1978*b*). Density-dependent regulation of growth of BSC-1 cells in cell culture: growth inhibitors formed by the cells. *Proc. Natl. Acad. Sci. USA* **75**, 1864–6.

Holley, R. W., Armour, R., Baldwin, J. H., Brown, K. D. & Yeh, Y.-C. (1977). Density-dependent regulation of growth of BSC-1 cells in cell culture. Control of growth by serum factors. *Proc. Natl. Acad. Sci. USA* **74**, 5046–50.

Holley, R. W., Baldwin, J. H., Kiernan, J. A. & Messmer, T. O. (1976). Control of growth of benzo(a)pyrene transformed 3T3 cells. *Proc. Natl. Acad. Sci. USA* **73**, 3229–32.

Holley, R. W. & Kiernan, J. A. (1968). 'Contact inhibition' of cell division in 3T3 cells. *Proc. Natl. Acad. Sci. USA* **60**, 300–4.

Holley, R. W. & Kiernan, J. A. (1971). Studies of serum factors required by 3T3 and SV3T3 cells. In *Growth Control in Cell Cultures*, ed. G. E. W. Wolstenholme & J. Knight, pp. 3–10. Edinburgh: Churchill Livingstone.

Holley, R. W. & Kiernan, J. A. (1974*a*). Control of the initiation of DNA synthesis in 3T3 cells: Serum factors. *Proc. Natl. Acad. Sci. USA* **71**, 2908–11.

Holley, R. W. & Kiernan, J. A. (1974*b*). Control of the initiation of DNA synthesis in 3T3 cells: low-molecular-weight nutrients. *Proc. Natl. Acad. Sci. USA* **71**, 2942–5.

Holliday, R. (1975). Growth and death of diploid and transformed human fibroblasts. *Fed. Proc.* **34**, 51–5.

Holtfreter, J. (1943). A study of the mechanics of gastrulation; Part I. *J. Exp. Zool.* **94**, 261–318.

Holtfreter, J. (1944). A study of the mechanics of gastrulation. Part. II. *J. Exp. Zool.* **95**, 171–212.

Holtzer, H., Biehl, J., Yeon, G., Meganathan, R. & Kaji, A. (1975). Effect of oncogenic virus on muscle differentiation. *Proc. Natl. Acad. Sci. USA* **72**, 4051–5.

Houck, J. C. & Attallah, A. M. (1975). Chalones (specific and endogenous mitotic inhibitors) and cancer. In *Cancer: a Comprehensive Treatise*, ed. F. F. Becker, vol. 3, pp. 287–326. New York: Plenum Press.

Houck, J. C. & Cheng, R. F. (1973). Isolation, purification, and chemical characterization of the serum mitogen for diploid human fibroblasts. *J. Cell. Physiol.* **81**, 257–70.

Houck, J. C., Kanagalingam, K., Hunt, C., Attallah, A. & Chung, A. (1977). Lymphocyte and fibroblast chalones: some chemical properties. *Science* **196**, 67–8.

Howard, E. F., Scott, D. F. & Bennett, C. (1976). Stimulation of thymidine uptake and cell proliferation in mouse embryo fibroblasts by conditioned medium from mammary cells in culture. *Cancer Res.* **36**, 4543–51.

Howard, E. F., Scott, D. F. & Manter, J. O. (1977). Cyclic nucleotide levels in mouse mammary epithelial cells during growth arrest and growth initiation in culture. *J. Natl. Cancer Inst.* **59**, 145–9.

Hsie, A. & Puck, T. (1971). Morphological transformation of Chinese hamster ovary cells by dibutyryl adenosine cyclic $3':5'$-monophosphate and testosterone. *Proc. Natl. Acad. Sci. USA* **68**, 358–61.

Hsu, T. C. & Cooper, J. E. K. (1974). On diploid cell lines. *J. Natl. Cancer Inst.* **53**, 1431–6.

Huberman, E., Mager, R. & Sachs, L. (1976). Mutagenesis and transformation of normal cells by chemical carcinogens. *Nature, Lond.* **264**, 360–1.

Huberman, E. & Sachs, L. (1974). Cell-mediated mutagenesis of mammalian cells with chemical carcinogens. *Int. J. Cancer* **13**, 326–33.

Huberman, E. & Sachs, L. (1976). Mutability of different genetic loci in mammalian cells by metabolically activated carcinogenic polycyclic hydrocarbons. *Proc. Natl. Acad. Sci. USA* **73**, 188–92.

Huberman, E., Sachs, L., Yang, S. K. & Gelboin, H. V. (1976). Identification of mutagenic metabolites of benzo(a)pyrene in mammalian cells. *Proc. Natl. Acad. Sci. USA* **73**, 607–11.

Hudspeth, A. J. (1975). Establishment of tight junctions between epithelial cells. *Proc. Natl. Acad. Sci. USA* **72**, 2711–13.

Huggins, C. & Yang, W. C. (1962). Induction and extinction of mammary cancer. *Science* **137**, 257–62.

Hull, B. E. & Staehelin, L. A. (1976). Functional significance of the variations in the geometrical organization of tight junction networks. *J. Cell Biol.* **68**, 688–704.

Hume, D. A. & Weidemann, M. J. (1978). On the stimulation of rat thymocyte 3-*O*-methyl-glucose transport by mitogenic stimuli. *J. Cell. Physiol.* **96**, 303–8.

Hutchings, S. E. & Sato, G. H. (1978). Growth and maintenance of HeLa cells in serum-free medium supplemented with hormones. *Proc. Natl. Acad. Sci. USA* **75**, 901–4.

Huxley, H. E. (1976). Introductory remarks: the relevance of studies on muscle to problems of cell motility. In *Cell Motility*, ed. R. Goldman, T. Pollard & J. Rosenbaum. Cold Spring Harbor Conferences on Cell Proliferation, vol. 3, pp. 115–26. New York: Cold Spring Harbor Laboratory.

Hynes, R. O. (1976). Cell surface proteins and malignant transformation. *Biochim. Biophys. Acta* **458**, 73–107.

Hynes, R. O. & Destree, A. T. (1978). 10 nm filaments in normal and transformed cells. *Cell* **13**, 151–63.

Hynes, R. O., Destree, A. T., Mautner, V. M. & Ali, I. U. (1977). Synthesis, secretion, and attachment of LETS glycoprotein in normal and transformed cells. *J. Supramol. Struc.* **7**, 397–408.

Iannaccone, P. M., Gardner, R. L. & Harris, H. (1978). The cellular origin of chemically induced tumours. *J. Cell Sci.* **29**, 249–69.

Illmensee, K. & Minz, B. (1976). Totipotency and normal differentiation of single teratocarcinoma cells cloned by injection into blastocysts. *Proc. Natl. Acad. Sci. USA* **73**, 544–8.

Inbar, M., Yuli, I. & Raz, A. (1977). Contact-mediated changes in the fluidity of membrane lipids in normal and malignant transformed mammalian fibroblasts. *Exp. Cell Res.* **105**, 325–35.

Indo, K. (1977). Biological analysis of fetal MRC rat lung epithelial cells treated with 3-methyl-cholanthrene in culture: premalignant and malignant stages. *J. Natl. Cancer Inst.* **58**, 351–60.

Indo, K. & Wilson, R. B. (1977). Fetal rat keratinizing epidermal cells in culture: effects of long-term treatment by benzo(a)pyrene on their growth characteristics. *J. Natl. Cancer Inst.* **59**, 867–80.

Ingram, V. M. (1969). A side view of moving fibroblasts. *Nature, Lond.* **222**, 641–4.

Irlin, I. S., Ivanova, O. Y., Starikova, V. B. & Vasiliev, J. M. (1968). The sensitivity of various hamster cell lines to the toxic action of carcinogenic hydrocarbons. *Voprosi Onkologii* **14**, N 1, 42–8 (in Russian).

Irlin, I. S. & Vasiliev, J. M. (1968). The increase of cellular resistance to the toxic effect of a chemical carcinogen by virus SV-40. *Voprosi Onkologii* **14**, N 6, 80–3 (in Russian).

Isaka, T., Yoshida, M., Owada, M. & Toyoshima, K. (1975). Alterations in membrane polypeptides of chick embryo fibroblasts induced by transformation with avian sarcoma viruses. *Virology* **65**, 226–37.

Iscove, N. N. (1978). Regulation of proliferation and maturation at early and late stages of erythroid differentiation. In *Cell Differentiation and Neoplasia*, ed. G. F. Saunders, pp. 195–209. New York: Raven Press.

Iscove, N. N. & Melchers, F. (1978). Complete replacement of serum by albumin, transferrin, and soybean lipid in cultures of lipopolysaccharide-reactive B lymphocytes. *J. Exp. Med.* **147**, 923–33.

Isenberg, G., Rathke, P. C., Hülsmann, N., Franke, W. W. & Wohlfarth-Bottermann, K. E. (1976). Cytoplasmic actomyosin fibrils in tissue culture cells. Direct proof of contractility by visualization of ATP-induced contraction in fibrils isolated by laser microbeam dissection. *Cell Tissue Res.* **166**, 427–43.

Ishii, Y., Elliott, J. A., Mishra, N. K. & Lieberman, M. W. (1977). Quantitative studies of transformation by chemical carcinogens and ultraviolet radiation using a subclone of BHK_{21} clone 13 Syrian hamster cells. *Cancer Res.* **37**, 2023–9.

Ishikawa, H., Bischoff, R. & Holtzer, H. (1969). Formation of arrowhead complexes with heavy meromyosin in a variety of cell types. *J. Cell Biol.* **43**, 312–28.

Ishiwata, I., Nozawa, S., Inoue, T. & Okumura, H. (1977). Development and characterization of established cell lines from primary and metastatic regions of human endometrial adenocarcinoma. *Cancer Res.* **37**, 1777–85.

Ivanova, O. Y. & Margolis, L. B. (1973). The use of phospholipid membranes for preparation of cell cultures of given shape. *Nature, Lond.* **242**, 200–1.

Ivanova, O. Y., Margolis, L. B., Vasiliev, J. M. & Gelfand, I. M. (1976). Effect of colcemid on the spreading of fibroblasts in culture. *Exp. Cell Res.* **101**, 207–19.

Izzard, C. S. & Izzard, S. L. (1975). Calcium regulation of the contractile state of isolated mammalian fibroblast cytoplasm. *J. Cell Sci.* **18**, 241–56.

Izzard, C. S. & Lochner, L. R. (1976). Cell-to-substrate contacts in living fibroblasts: an interference–reflexion study with an evaluation of the technique. *J. Cell Sci.* **21**, 129–59.

Jack, R., Rajewsky, K. & Kapp, J.-F. (1975). Adhesion to substrate as a developmental process in murine teratoma embryoid bodies. *Exp. Cell Res.* **96**, 180–8.

Jacob, F. (1977). Mouse teratocarcinoma and embryonic antigens. *Immunol. Rev.* **33**, 3–32.

Jacobs, S. & Cuatrecasas, P. (1976). The mobile receptor hypothesis and 'cooperativity' of hormone binding. Application to insulin. *Biochim. Biophys. Acta* **433**, 482–95.

Jacobson, K., Wu, E. & Poste, G. (1976). Measurement of the translational mobility of concanavalin A in glycerol-saline solutions and on the cell surface by fluorescence recovery after photobleaching. *Biochim. Biophys. Acta* **433**, 215–22.

Janchill, J. L. & Todaro, G. J. (1970). Stimulation of cell growth *in vitro* by serum with and without growth factor. *Exp. Cell Res.* **59**, 137–46.

Jandl, J. H., Files, N. M., Barnett, S. B., Macdonald, R. A. (1965). Proliferative response of the spleen and liver to hemolysis. *J. Exp. Med.* **122**, 299–326.

Jasswoin, G. (1938). Beiträge zur vergleichenden Histologie des Blutes und des Bindegewebes. VIII. Vergleichende Studien über einige Zellformen des lockeren Bindegewebes der Säugetiere. *Z. Mikr.-anat. Forsch.* **15**, 107–56.

Jasswoin, G. (1930). Beiträge zur vergleichenden Histologie des Blutes und des Bindegewebes. IX. Experimentell-morphologische Studien über einige Zellformen des lockeren Bindegewebes des Säugetiere. *Z. Mikr.-anat. Forsch.* **19**, 513–36.

Jimenez de Asua, L., Clingan, D. & Rudland, P. S. (1975). Initiation of cell proliferation in cultured mouse fibroblasts by prostaglandin $F_{2\alpha}$. *Proc. Natl. Acad. Sci. USA* **72**, 2724–8.

Jimenez de Asua, L., O'Farrell, M. K., Clingan, D. & Rudland, P. S. (1977). Temporal sequence of hormonal interactions during the prereplicative phase of quiescent cultured 3T3 fibroblasts. *Proc. Natl. Acad. Sci. USA* **74**, 3845–9.

Jimenez de Asua, L., Rozengurt, E. & Dulbecco, R. (1974). Kinetics of early changes in phosphate and uridine transport and cyclic AMP levels stimulated by serum in density-inhibited 3T3 cells. *Proc. Natl. Acad. Sci. USA* **71**, 96–8.

Jimenez de Asua, L., Surian, E. C., Flawia, M. M. & Torres, H. N. (1973). Effect of insulin on the growth pattern and adenylate cyclase activity by BHK fibroblasts. *Proc. Natl. Acad. Sci. USA* **70**, 1388–92.

Johnson, G., Friedman, R. & Pastan, I. (1971). Restoration of several morphological characteristics of normal fibroblasts in sarcoma cells treated with adenosine 3′:5′-cyclic monophosphate and its derivatives. *Proc. Natl. Acad. Sci. USA* **68**, 425–9.

Johnson, G. S., Morgan, W. D. & Pastan, I. (1972). Cyclic AMP increases the adhesion of fibroblasts to substratum. *Nature New Biol.* **236**, 247–9.

Johnson, K. H., Ghobrial, H. K. G., Buoen, L. C., Brand, I. & Brand, K. G. (1973). Non-fibroblastic origin of foreign body sarcomas implicated by histological and electron microscopic studies. *Cancer Res.* **33**, 3139–54.

Johnson, L. F., Abelson, H. T., Green, H. & Penman, S. (1974). Changes in RNA in relation to growth of the fibroblast: I. Amounts of mRNA, rRNA, and tRNA in resting and growing cells. *Cell* **1**, 95–100.

Johnson, L. F., Levis, R., Abelson, H. T., Green, H. & Penman, S. (1976). Changes in RNA in relation to growth of the fibroblast. IV. Alterations in the production and processing of mRNA and rRNA in resting and growing cells. *J. Cell Biol.* **71**, 933–8.

Johnson, L. F., Williams, J. G., Abelson, H. T., Green, H. & Penman, S. (1975). Changes in RNA in relation to growth of the fibroblast. III. Post-transcriptional regulation of mRNA formation in resting and growing cells. *Cell* **4**, 69–75.

Jones, P. A., Benedict, W. F., Baker, H. S., Mondal, S., Rapp, U. & Heidelberger, C. (1976*a*). Oncogenic transformation of C3H/10T 1/2 clone 8 mouse embryo cells by halogenated pyrimidine nucleosides. *Cancer Res.* **36**, 101–7.

Jones, P., Benedict, W., Strickland, S. & Reich, E. (1975). Fibrin overlay methods for the detection of single transformed cells and colonies of transformed cells. *Cell* **5**, 323–9.

Jones, P. A., Laug, W. E., Gardner, A., Nye, C. A., Fink, L. M. & Benedict, W. F. (1976*b*). In-vitro correlates of transformation in C3H/10T 1/2 clone 8 mouse cells. *Cancer Res.* **36**, 2863–7.

Jones, P. A., Rhim, J. S., Isaacs, H., Jr. & McAllister, R. M. (1975). The relationship between tumorigenicity, growth in agar and fibrinolytic activity in a line of human osteosarcoma cells. *Int. J. Cancer* **16**, 616–21.

Juliano, R. L. & Gagalang, E. (1977). The adhesion of Chinese hamster cells. I. Effects of temperature, metabolic inhibitors and proteolytic dissection of cell surface macromolecules. *J. Cell. Physiol.* **92**, 209–20.

Jullien, M. & Harel, L. (1976). Stimulation by serum of the phosphorylation reactions in density-inhibited 3T3 cells. *Exp. Cell Res.* **97**, 23–30.

Kahan, B. D., Rutzky, L. P., Kahan, A. V., Oyasu, R., Wiseman, F. & LeGrue, S. (1977). Cell surface changes associated with malignant transformation of bladder epithelium *in vitro. Cancer Res.* **37**, 2866–71.

Kahn, C. R. (1976). Membrane receptors for hormones and neurotransmitters. *J. Cell Biol.* **70**, 261–86.

Kakunaga, T. (1975). The role of cell division in the malignant transformation of mouse cells treated with 3-methylcholantrene. *Cancer Res.* **36**, 1637–42.

Kamely, D. & Rudland, P. S. (1976). Induction of DNA synthesis and cell division in human diploid skin fibroblasts by fibroblast growth factor. *Exp. Cell Res.* **97**, 120–6.

Kamine, J. & Rubin, H. (1976). Magnesium required for serum-stimulation of growth in cultures of chick embryo fibroblasts. *Nature, Lond.* **263**, 143–4.

Kamine, J. & Rubin, H. (1977). Co-ordinate control of collagen synthesis and cell growth in chick embryo fibroblasts and the effect of viral transformation on collagen synthesis. *J. Cell. Physiol.* **92**, 1–12.

Kane, R. E. (1976). Actin polymerization and interaction with other proteins in temperature-induced gelation of sea urchin egg extracts. *J. Cell Biol.* **71**, 704–14.

Kano-Tanaka, K., Tanaka, T., Emura, M. & Hanaichi, T. (1976). Malignant transformation and viral replication of rat bone and muscle cells after in-vitro infection with rat-adapted murine sarcoma virus (Moloney). *Cancer Res.* **36**, 3924–35.

Kapeller, M., Gal-oz, R., Grover, N. B. & Doljanski, F. (1973). Natural shedding of carbohydrate-containing macromolecules from cell surfaces. *Exp. Cell Res.* **79**, 152–8.

Kapeller, M., Plesser, Y. M., Kapeller, N. & Doljanski, F. (1976). Turnover and shedding of cell-surface constituents in normal and neoplastic chicken cells. In

Progress in Differentiation Research, ed. N. Müller-Berat, pp. 397–408. Amsterdam: North-Holland.

Kaplowitz, P. B. & Moscona, A. A. (1976). Lectin-mediated stimulation of DNA synthesis in cultures of embryonic neural retina cells. *Exp. Cell Res.* **100**, 177–89.

Kapp, L. N. & Klevecz, R. R. (1976). The cell cycle of low passage and high passage human diploid fibroblasts. *Exp. Cell Res.* **101**, 154–8.

Karasaki, S., Simard, A., de Lamirand, G. (1977). Surface morphology and nucleoside phosphatase activity of rat liver epithelial cells during oncogenic transformation *in vitro*. *Cancer Res.* **37**, 3516–25.

Karnovsky, M. J., Unanaue, E. R. & Leventhal, M. (1972). Ligand-induced movement of lymphocyte membrane macromolecules. II. Mapping of surface moieties. *J. Exp. Med.* **136**, 907–30.

Karp, R. D., Johnson, K. H., Buoen, L. C., Chobrial, H. K. G., Brand, I. & Brand, K. G. (1973). Tumorigenesis by milipore filters in mice: histology and ultrastructure of tissue reactions as related to pore size. *J. Natl. Cancer Inst.* **51**, 1275–85.

Katsuta, H. & Takaoka, T. (1973). Cultivation of cells in protein and lipid-free synthetic media. In *Methods in Cell Biology*, ed. D. M. Prescott, vol. 6, pp. 1–42. New York: Academic Press.

Kellermayer, M., Jobst, K. & Szücs, Gy. (1978). Inhibition of internuclear transport of SV-40-induced T antigen in heterokaryons. *Cell Biol. Int. Rep.* **2**, 19–24.

Kelly, F. & Sambrook, J. (1973). Differential effect of cytochalasin B on normal and transformed mouse cell. *Nature New Biol.* **242**, 217–19.

Kelly, L. S., Brown, B. A. & Dobson, E. L. (1962). Cell division and phagocytic activity in liver reticuloendothelial cells. *Proc. Natl. Acad. Sci. USA* **110**, 555–9.

Keski-Oja, J. (1976). Polymerization of a major surface-associated glycoprotein fibronectin, in cultured fibroblasts. *FEBS Lett.* **71**, 325–9.

Key, D. J. & Todaro, G. J. (1974). Xeroderma pigmentosum cell susceptibility to SV-40 virus transformation: lack of effect of low dosage ultraviolet radiation in enhancing virus-induced transformation. *J. Invest. Dermatol.* **62**, 7–10.

Killen, P. D., Striker, G. E. & Byers, P. H. (1978). Synthesis of basal lamina collagen by human glomerular epithelial cells *in vitro*. *Fed. Proc.* **37**, 642.

Killion, J. J. & Kollmorgen, G. M. (1975). Induced agglutinability of 3T3 mouse fibroblasts. *Nature, Lond.* **254**, 247–8.

Kimura, G. & Itagaki, A. (1975). Initiation and maintenance of cell transformation by Simian virus 40: a viral genetic property. *Proc. Natl. Acad. Sci. USA* **72**, 673–7.

Kimura, G., Itagaki, A. & Summers, J. (1975). Rat cell line 3YI and its virogenic polyoma- and SV-40-transformed derivatives. *Int. J. Cancer* **15**, 694–706.

Kirkland, D. J. (1976). Chemical transformation of Chinese hamster cells. I. A comparison of some properties of transformed cells. *Br. J. Cancer* **34**, 134–44.

Kish, A. L., Keller, R. O., Crissman, H. & Paxton, L. (1973). Dimethyl sulfoxide-induced reversion of several features of polyoma transformed baby hamster kidney cells (BHK-21). *J. Cell Biol.* **57**, 38–53.

Klein, G. (1953). Conversion of solid into ascites tumors. *Nature, Lond.* **171**, 398–9.

Klein, G. & Klein, E. (1956). Conversion of solid neoplasms into ascites tumors. *Ann. N.Y. Acad. Sci.* **63**, 640–65.

Kleinsmith, L. J. & Pierce, G. B., Jr. (1964). Multipotentiality of single embryonal carcinoma cell. *Cancer Res.* **24**, 1544–52.

Klempner, L. B. & Guelstein, Y. I. (1976). Kinetics of cellular proliferation and growth of transplantable mouse hepatomas. *Voprosi Onkologii* **22**, N 12, 61–8 (in Russian).

Kletzien, R. F. & Perdue, J. F. (1974a). Sugar transport in chick embryo fibroblasts. I. A functional change in the plasma membrane associated with the rate of cell growth. *J. Biol. Chem.* **249**, 3366–74.

Kletzien, R. F. & Perdue, J. F. (1974*b*). Sugar transport in chick embryo fibroblasts. III. Evidence for post-transcriptional and post-translational regulation of transport following serum addition. *J. Biol. Chem.* **249**, 3383–7.

Knecht, M. E. & Lipkin, G. (1977). Biochemical studies of a protein which restores contact inhibition of growth to malignant melanocytes. *Exp. Cell Res.* **108**, 15–22.

Knighton, D., Ausprunk, D., Tapper, D. & Folkman, J. (1977). Avascular and vascular phases of tumour growth in the chick embryo. *Br. J. Cancer* **35**, 347–56.

Knutton, S., Sumner, M. C. B. & Pasternak, C. A. (1975). Role of microvilli in surface changes of synchronized P815Y mastocytoma cells. *J. Cell. Biol.* **66**, 568–76.

Kohler, N. & Lipton, A. (1974). Platelets as a source of fibroblast growth-promoting activity. *Exp. Cell Res.* **87**, 287–301.

Kolb, H., Schudt, C., Kolb-Bachofen, V. & Kolb, H.-A. (1978). Cellular recognition by rat liver cells of neuraminidase-treated erythrocytes. Demonstration and analysis of cell contacts. *Exp. Cell Res.* **113**, 319–25.

Kopelovich, L. (1977). Phenotypic markers in human skin fibroblasts as possible diagnostic indices of hereditary adenomatosis of the colon and rectum. *Cancer* **40**, 2534–41.

Kopelovich, L., Conlon, S. & Pollack, R. (1977). Defective organization of actin in cultured skin fibroblasts from patients with inherited adenocarcinoma. *Proc. Natl. Acad. Sci. USA* **74**, 3019–22.

Kopelovich, L., Pfeffer, L. & Lipkin, M. (1976). Recent studies on the identification of proliferative abnormalities and of the oncogenic potential of cutaneous cells in individuals at increased risk of colon cancer. *Semin. Oncol.* **3**, 369–71.

Koprowsky, H. & Croce, C. M. (1977). Tumorigenicity of Simian virus-40-transformed human cells and mouse-human hybrids in nude mice. *Proc. Natl. Acad. Sci. USA* **74**, 1142–6.

Koprowski, H., Ponten, J. A., Jensen, F., Ravidin, R. G., Moorhead, P. S. & Saksela, E. (1962). Transformation of cultures of human tissue infected with Simian virus SV-40. *J. Cell. Physiol.* **59**, 281–92.

Korchak, H. M. & Weissmann, G. (1978). Changes in membrane potential of human granulocytes antecede the metabolic responses to surface stimulation. *Proc. Natl. Acad. Sci. USA* **75**, 3818–22.

Koren, P., Shohami, E., Bibi, O. & Stein, W. D. (1978). Uridine transport properties of mammalian cell membranes are not directly involved with growth control or oncogenesis. *FEBS Lett.* **86**, 71–5.

Korn, E. D. (1978). Biochemistry of actomyosin-dependent motility (a review). *Proc. Natl. Acad. Sci. USA* **75**, 588–99.

Kouri, R. E., Kurtz, S. A., Price, P. J. & Benedict, W. F. (1975). 1-β-D-arabinofuranosylcytosine-induced malignant transformation of hamster and rat cells in culture. *Cancer Res.* **35**, 2413–19.

Kucherlapati, R., Hwang, S. P., Shimisu, N., McDougall, J. K. & Botchan, M. R. (1978). Another chromosomal assignment for a simian virus 40 integration site in human cells. *Proc. Natl. Acad. Sci. USA* **75**, 4460–4.

Kuettner, K. E., Pauli, B. U. & Soble, L. (1978). Morphological studies on the resistance of cartilage to invasion by osteosarcoma cells *in vitro* and *in vivo*. *Cancer Res.* **38**, 277–87.

Kuri-Harcuch, W., Wise, L. S. & Green, H. (1978). Interruption of the adipose conversion of 3T3 cells by biotin deficiency: differentiation without triglyceride accumulation. *Cell* **14**, 53–9.

Kuroki, T. (1975). Contributions of tissue culture to the study of chemical carcinogenesis, a review. *Gann Mongr. Cancer Res.* **17**, 69–85.

Kuroki, T., Miyashita, S. Y. & Yuasa, Y. (1975). Development of a 3T3-like line from an embryo culture of an inbred strain of Syrian golden hamster. *Cancer Res.* **35**, 1819–25.

Kuwata, T., Oda, T., Sekiya, S. & Morinaga, N. (1976). Characteristics of a human cell line successively transformed by Rous sarcoma virus and Simian virus 40. *J. Natl. Cancer Inst.* **56**, 919–26.

Laerum, O. D. & Rajewsky, M. F. (1975). Neoplastic transformation of fetal rat brain cells in culture after exposure to ethylnitrosourea in vivo. *J. Natl. Cancer Inst.* **55**, 1177–87.

Laerum, O. D., Rajewsky, M. F., Schachner, M., Stavrou, D., Haglid, K. G. & Haugen, Å. (1977). Phenotypic properties of neoplastic cell lines developed from fetal rat brain cells in culture after exposure to ethylnitrosourea in vivo. *Z. Krebsforsch.* **89**, 273–95.

Laishes, B. A., Roberts, E. & Farber, E. (1978). In-vitro measurement of carcinogen-resistant liver cells during hepatocarcinogenesis. *Int. J. Cancer* **21**, 186–93.

Lajtha, L. G. (1963). Differential sensitivity of the cell life cycle. *J. Cell. Comp. Physiol.* **62** (Suppl. 1), 141–56.

Lajtha, L. G., Lord, B. I., Dexter, T. M., Wright, E. G. & Allen, T. D. (1978). Interrelationship of differentiation and proliferation control in hemopoietic stem cells. In *Cell Differentiation and Neoplasia*, ed. G. F. Saunders, pp. 179–93. New York: Raven Press.

Lang, D. R. & Weber, M. J. (1978). Increased membrane transport of 2-deoxyglucose and 3-*O*-methylglucose is an early event in the transformation of chick embryo fibroblasts by Rous sarcoma virus. *J. Cell. Physiol.* **94**, 315–20.

Langenbach, R., Freed, H. J., Raveh, D. & Huberman, E. (1978). Cell specificity in metabolic activation of aflatoxin B_1 and benzo(a)pyrene to mutagens for mammalian cells. *Nature, Lond.* **276**, 277–9.

Langenbach, R., Malick, I. & Kennedy, S. (1977). Ganglioside and morphological changes in mouse embryo cells with time. *Cancer Lett.* **4**, 13–19.

Lasne, C., Gentil, A. & Chouroulinkov, I. (1974). Two-stage malignant transformation of rat fibroblast in tissue culture. *Nature, Lond.* **247**, 490–1.

Lasne, C., Gentil, A. & Chouroulinkov, I. (1977). Two-stage carcinogenesis with rat embryo cells in tissue culture. *Br. J. Cancer* **35**, 722–9.

Laurent, M., Lonchampt, M.-O., Regnault, F., Tassin, J. & Courtois, Y. (1978). Biochemical ultrastructural and immunological study of in-vitro production of collagen by bovine lens epithelial cells in culture. *Exp. Cell Res.* **115**, 127–42.

Lazarides, E. (1976*a*). Actin, α-actinin and tropomyosin interaction in the structural organization of actin filaments in non-muscle cells. *J. Cell Biol.* **68**, 202–19.

Lazarides, E. (1976*b*). Aspects of the structural organization of actin filaments in tissue culture cells. In *Cell Motility*, ed. R. Goldman, T. Pollard & J. Rosenbaum, vol. 3, pp. 347–60. New York: Cold Spring Harbor Laboratory.

Lazarides, E. (1976*c*). Two general classes of cytoplasmic actin filaments in tissue culture cells: the role of tropomyosin. *J. Supramol. Struc.* **5**, 531–63.

Lazarides, E. (1978). The distribution of desmin (100 Å) filaments in primary cultures of embryonic chick cardiac cells. *Exp. Cell Res.* **112**, 265–73.

Lazarides, E. & Burridge, K. (1975). α-Actinin: Immunofluorescent localization of a muscle structural protein in non-muscle cells. *Cell* **6**, 289–98.

Lazarides, E. & Weber, K. (1974). Actin antibody: the specific visualization of actin filaments in non-muscle cells. *Proc. Natl. Acad. Sci. USA* **71**, 2268–72.

Leblond, C. P. & Cheng, H. (1976). Identification of stem cells in the small intestine of the mouse. In *Stem Cells of Renewing Cell Population*, ed. A. B. Cairnie, P. K. Lala & D. G. Osmond, pp. 7–31. New York: Academic Press.

Lee, L. S. & Weinstein, I. B. (1978). Epidermal growth factor, like phorbol esters, induces plasminogen activator in HeLa cells. *Nature, Lond.* **274**, 696–7.

Leffert, H. L., Koch, K. S. & Rubaclava, B. (1976). Present paradoxes in the environmental control of hepatic proliferation. *Cancer Res.* **36**, 4250–6.

Leffert, H. L. & Paul, D. (1973). Serum-dependent growth of primary cultured differentiated fetal rat hepatocytes in arginine-deficient medium. *J. Cell. Physio.* **81**, 113–24.

Leffert, H. L. & Weinstein, D. B. (1976). Growth control of differentiated fetal rat hepatocytes in primary monolayer culture. IX. Specific inhibition of DNA synthesis initiation by very low density lipoprotein and possible significance to the problem of liver regeneration. *J. Cell Biol.* **70**, 20–32.

Lehman, J. M. (1974). Early chromosome changes in diploid Chinese hamster cells after infection with Simian virus 40. *Int. J. Cancer* **13**, 164–72.

Lehman, J. M. & Defendi, V. (1970). Changes in deoxyribonucleic acid synthesis regulation in Chinese hamster cells infected with Simian virus 40. *J. Virology* **6**, 738–49.

Leibovitz, A., Stinson, J. C., McCombs, W. B. III, McCoy, C. E., Mazur, K. C. & Mabry, N. D. (1976). Classification of human colorectal adenocarcinoma cell lines. *Cancer Res.* **36**, 4562–9.

Leighton, J., Siar, W. J. & Mahoney, M. J. (1962). Examination of invasion by manipulating stroma and parenchyma of carcinomas *in vitro* and *in vivo*. In *Biological Interactions in Normal and Neoplastic Growth. A Contribution to the Host–Tumor Problem*, ed. M. J. Brennan & W. L. Simpson, pp. 681–702. Boston: Little, Brown & Co.

Lembach, K. J. (1976*a*). Enhanced synthesis and extracellular accumulation of hyaluronic acid during stimulation of quiescent human fibroblasts by mouse epidermal growth factor. *J. Cell. Physiol.* **89**, 277–88.

Lembach, K. J. (1976*b*). Induction of human fibroblast proliferation by epidermal growth factor (EGF): enhancement by an EGF-binding arginine esterase and by ascorbate. *Proc. Natl. Acad. Sci. USA* **73**, 183–7.

Lernhardt, W., Andersson, J., Coutinho, A. & Melchers, F. (1978). Cloning of murine transformed cell lines in suspension culture with efficiencies near 100%. *Exp. Cell Res.* **111**, 300–16.

Lever, J. E. (1976). Regulation of amino acid and glucose transport activity expressed in isolated membranes from untransformed and SV-40-transformed mouse fibroblasts. *J. Cell. Physiol.* **89**, 779–88.

Levin, W., Wood, A. W., Chang, R. L., Slaga, T. J., Yagi, H., Jerina, D. M. & Conney, A. H. (1977). Marked differences in the tumor-initiating activity of optically pure (+) and (−) -*trans*-7,8-dihydroxy- 7,8-dihydrobenzo(a)pyrene on mouse skin. *Cancer Res.* **37**, 2721–5.

Levine, A. J. (1978). Approaches to mapping the temporal events in the cell cycle using conditional lethal mutants. *J. Cell. Physiol.* **95**, 387–92.

Levine, E. M., Becker, Y., Boone, C. W. & Eagle, H. (1965). Contact inhibition, macromolecular synthesis, and polyribosomes in cultured human fibroblasts. *Proc. Natl. Acad. Sci. USA* **53**, 350–6.

Levine, E. M., Jeng, D.-Y. & Chang, Y. (1974). Contact inhibition, polyribosomes, and cell membranes in cultured mammalian cells. *J. Cell. Physiol.* **84**, 349–64.

Levinson, A. & Levine, A. J. (1977). The group C adenovirus tumor antigens: identification in infected and transformed cells and a peptide map analysis. *Cell* **11**, 871–80.

Levinson, W., Bhatnagar, R. S. & Liu, T.-Z. (1975). Loss of ability to synthesize collagen in fibroblasts transformed by Rous sarcoma virus. *J. Natl. Cancer Inst.* **55**, 807–10.

Lewis, J., Slack, J. M. W. & Wolpert, L. (1977). Thresholds in development. *J. Theor. Biol.* **65**, 579–90.

Lewis, J. H. & Wolpert, L. (1976). The principle of non-equivalence in development. *J. Theor. Biol.* **62**, 479–90.

Lin, D. C. & Lin, S. (1978). High affinity binding of [³H]dihydrocytochalasin B to peripheral membrane proteins related to the control of cell shape in the human red cell. *J. Biol. Chem.* **253**, 1415–19.

Lin, S., Lin, D. C. & Flanagan, M. D. (1978). Specificity of the effect of cytochalasin B on transport and motile processes. *Proc. Natl. Acad. Sci. USA* **75**, 329–33.

Lindgren, A. & Westermark, B. (1976). Subdivision of the G_1 phase of human glia cells in culture. *Exp. Cell Res.* **99**, 357–62.

Lindgren, A. & Westermark, B. (1977a). Reset of the pre-replicative phase of human glia cells in culture. *Exp. Cell Res.* **106**, 89–93.

Lindgren, A. & Westermark, B. (1977b). Serum requirement and density-dependent inhibition of human malignant glioma cells in culture. *Exp. Cell Res.* **104**, 293–9.

Liotta, L. A., Kleinerman, J., Catanzaro, P. & Rynbrandt, D. (1977). Degradation of basement membrane by murine tumor cells. *J. Natl. Cancer Inst.* **58**, 1427–31.

Lipkin, G. & Knecht, M. E. (1976). Contact inhibition of growth is restored to malignant melanocytes of man and mouse by a hamster protein. *Exp. Cell Res.* **102**, 341–8.

Lipkin, G., Knecht, E. & Rosenberg, M. (1978). A protein inhibitor of normal and transformed cell growth derived from contact-inhibited cells. *Cancer Res.* **38**, 635–43.

Littlefield, J. W. (1976). *Variation, senescence and neoplasia in cultured somatic cells.* Cambridge, Mass.: Harvard University Press.

Littlefield, J. W., Choy, W. N. & Epstein, J. (1977). Research on cellular senescence: present and future directions. In *Senescence. Dominant or Recessive in Somatic Cell Crosses?* ed. W. W. Nichols & D. G. Murphy, pp. 111–20. New York: Plenum Press.

Lloyd, C. W., Smith, C. G., Woods, A. & Rees, D. A. (1977). Mechanisms of cellular adhesion. II. The interplay between adhesion, the cytoskeleton and morphology in substrate-attached cells. *Exp. Cell Res.* **110**, 427–37.

Loewenstein, W. R. (1967). On the genesis of cellular communication. *Dev. Biol.* **15**, 503–20.

Loewenstein, W. R. (1979). Junctional intercellular communication and the control of growth. *Biochim. Biophys. Acta* **560**, 1–65.

Lohmander, S., Moskalewski, S., Madsen, K., Thyberg, J. & Friberg, U. (1976). Influence of colchicine on the synthesis and secretion of proteoglycans and collagen by fetal guinea pig chondrocytes. *Exp. Cell Res.* **99**, 333–45.

Loor, F. (1976). Cell surface design. *Nature, Lond.* **264**, 272–3.

Lord, B. I. (1976). Stem cell reserve and its control. In *Stem Cells of Renewing Cell Populations,* ed. A. B. Cairnie, P. K. Lala & D. G. Osmond, pp. 165–79. New York: Academic Press.

Lu, R. & Elzinga, M. (1976). Comparison of amino acid senquences of actins from bovine brain and muscles. In *Cell Motility,* ed. R. Goldman, T. Pollard, J. Rosenbaum. Cold Spring Harbor Conferences on Cell Proliferation, vol. 3, pp. 487–92. New York: Cold Spring Harbor Laboratory.

Luckasen, J. R., White, J. G. & Kersey, J. H. (1974). Mitogenic properties of calcium ionophore, A23187. *Proc. Natl. Acad. Sci. USA* **71**, 5088–90.

Lundgren, E. & Roos, G. (1976). Cell surface changes in HeLa cells as an indication of cell cycle events. *Cancer Res.* **36**, 4044–51.

Lynch, R. G., Graff, R. J., Sirisinha, S., Simms, E. S. & Eisen, H. N. (1972). Myeloma proteins as tumor-specific transplantation antigens. *Proc. Natl. Acad. Sci. USA* **69**, 1540–4.

Lyons, L. B. & Thompson, E. B. (1976). Delayed malignancy and altered growth properties of somatic cell hybrids between rat hepatoma and mouse L-cells. *J. Cell. Physiol.* **90**, 179–92.

Lyubimov, A. V. (1978*a*). Changes in surface morphology of normal and transformed mouse fibroblasts during alteration of cell–substrate adhesion. *Byuleten experimentalnoj Biologii i Meditsini* **86**, 589–91 (in Russian).

Lyubimov, A. V. (1978*b*). The influence of agents impairing cell–substrate adhesion on protein and glycopropein pattern of transformed mouse fibroblasts in culture. *Tsitologiya* **20**, 1179–85 (in Russian).

McCann, J. & Ames, B. N. (1976). Detection of carcinogens as mutagens in the Salmonella/microsome test: assay of 300 chemicals: Discussion. *Proc. Natl. Acad. Sci. USA* **73**, 950–4.

McClain, D. A., D'Eustachio, P. & Edelman, G. M. (1977). Role of surface modulating assemblies in growth control of normal and transformed fibroblasts. *Proc. Natl. Acad. Sci. USA* **74**, 66–70.

McClain, D. A., Maness, P. F. & Edelman, G. M. (1978). Assay for early cytoplasmic effects of the *src* gene product of Rous sarcoma virus. *Proc. Natl. Acad. Sci. USA* **75**, 2750–4.

Machatkova, M. & Pospišil, Z. (1975). Biological characteristics of cell lines derived from the respiratory tract of a bovine foetus. *Folia biologica* (*Praha*) **21**, 117–21.

MacIntyre, E. & Pontén, J. (1967). Interaction between normal and transformed bovine fibroblasts in culture. 1. Cells transformed by Rous sarcoma virus. *J. Cell Sci.* **2**, 309–22.

McKeehan, W. L. & Ham, R. G. (1976). Stimulation of clonal growth of normal fibroblasts with substrata coated with basic polymers. *J. Cell Biol.* **71**, 727–34.

McKeehan, W. L., Hamilton, W. G. & Ham, R. G. (1976). Selenium is an essential trace nutrient for growth of WI-38 diploid human fibroblasts. *Proc. Natl. Acad. Sci. USA* **73**, 2023–7.

MacLean, S., Griffith, L. M. & Pollard, T. D. (1978). A direct effect of cytochalasin-B upon actin filaments. *J. Cell Biol.* **79**, 267a.

McNutt, N. S., Culp, L. A. & Black, P. H. (1971). Contact-inhibited revertant cell lines isolated from SV-40-transformed cells. II. Ultrastructural study. *J. Cell Biol.* **50**, 691–708.

McNutt, N. S., Culp, L. A. & Black, P. H. (1973). Contact-inhibited revertant cell lines isolated from SV-40-transformed cells. IV. Microfilament distribution and cell shape in untransformed, transformed and revertant BALB/c 3T3 cells. *J. Cell Biol.* **56**, 412–28.

Macpherson, I. (1965). Reversion in hamster cells transformed by Rous sarcoma virus. *Science* **148**, 1731–3.

Macpherson, I. (1971). Reversion in cell transformed by tumor viruses. *Proc. Roy. Soc. Lond., Ser. B.* **177**, 41–8.

Macpherson, I. A. & Montagnier, L. (1964). Agar suspension culture for the selective assay of cells transformed by polyoma virus. *Virology* **23**, 291–4.

McQuilkin, W. T., Evans, V. J. & Earle, W. R. (1957). The adaptation of additional lines of NCTC clone 929 (strain L) cells to chemically defined protein-free medium NCTC 109. *J. Natl. Cancer Inst.* **19**, 885–908.

Mahdavi, V. & Hynes, R. O. (1978). Effects of cocultivation with transformed cells on surface proteins of normal cells. *Biochim. Biophys. Acta* **542**, 191–208.

Maino, V. C., Green, N. M. & Crumpton, M. J. (1974). The role of calcium ions in initiating transformation of lymphocytes. *Nature, Lond.* **251**, 324–7.

Makowski, L., Caspar, D. L. D., Phillips, W. C. & Goodenough, D. A. (1977). Gap junction structures. II. Analysis of the X-ray diffraction data. *J. Cell Biol.* **74**, 629–45.

Malick, L. E. & Langenbach, R. (1976). Scanning electron microscopy of in-vitro chemically transformed mouse embryo cells. *J. Cell Biol.* **68**, 654–64.

Manes, C. (1976). Murine teratocarcinoma and the developmental approach to neoplasia. *Cancer Res.* **36**, 4206–7.

Marciani, D. J. & Okazaki, T. (1976). Interactions of concanavalin A with chick embryo fibroblasts transformed by Rous sarcoma virus. Study with an RSV mutant thermosensitive for transformation. *Biochim. Biophys. Acta* **455**, 849–64.

Mareel, M., de Ridder, L., de Brabander, M. & Vakaet, L. (1975). Characterization of spontaneous, chemical and viral transformants of a C3H/3T3-type mouse cell line by transplantation into young chick blastoderms. *J. Natl. Cancer Inst.* **54**, 923–9.

Margolis, L. B. & Bergelson, L. D. (1979). Cell-lipid interaction: induction of microvilli on the cell surface by liposomes. *Exp. Cell Res.* **119**, 145–70.

Margolis, L. B., Dyatlovitskaya, E. V. & Bergelson, L. D. (1978). Cell–lipid interactions. Cell attachment to lipid substrates. *Exp. Cell Res.* **111**, 454–7.

Margolis, L. B., Samoilov, V. I., Vasiliev, J. M. & Gelfand, I. M. (1975). Quantitative evaluation of cell orientation in culture. *J. Cell Sci.* **17**, 1–10.

Margolis, R. L. & Wilson, L. (1977). Addition of colchinine–tubulin complex to microtubule ends: the mechanism of substoichiometric colchicine poisoning. *Proc. Natl. Acad. Sci. USA* **74**, 3466–70.

Mark, J., Levan, G. & Mitelman, F. (1972). Identification by fluorescence of the G chromosome lost in human meningeomas. *Heraditas* **71**, 163–8.

Maroudas, N. G. (1972). Anchorage dependence: correlation between amount of growth and diameter of bead, for single cells grown on individual glass beads. *Exp. Cell Res.* **74**, 337–42.

Maroudas, N. G. (1973a). Chemical and mechanical requirements for fibroblast adhesion. *Nature, Lond.* **244**, 353–4.

Maroudas, N. G. (1973b). Growth of fibroblasts on linear and planar anchorages of limiting dimensions. *Exp. Cell Res.* **81**, 104–10.

Maroudas, N. G. (1977). Sulphonated polystyrene as an optimal substratum for the adhesion and spreading of mesenchymal cells in monovalent and divalent saline solution. *J. Cell Physiol.* **90**, 511–20.

Marquardt, H., Baker, S., Tierney, B., Grover, P. L. & Sims, P. (1977). The metabolic activation of 7 methylbenz(a)anthracene: the induction of malignant transformation and mutation in mammalian cells by non-K-region dihydrodiols. *Int. J. Cancer* **19**, 828–33.

Marquardt, H., Grover, P. L. & Sims, P. (1976). In-vitro malignant transformation of mouse fibroblasts by non-K-region dihydrodiols derived from 7-methylbenz(a)-anthracene, 7,12-dimethylbenz(a)anthracene and benzo(a)pyrene. *Cancer Res.* **36**, 2059–64.

Marquardt, H., Sodergren, J., Grover, P. L. & Sims, P. (1974). Malignant transform-ation *in vitro* of mouse fibroblasts by 7,12-dimethylbenz(a)anthracene and 7-hydroxymethylbenz(a)anthracene and by their K-region derivatives. *Int. J. Cancer* **13**, 304–10.

Marshak, M. I., Varshaver, N. B. & Shapiro, N. I. (1975). Induction of gene mutations and chromosomal aberrations by Simian virus-40 in cultured mammalian cells. *Mutation Res.* **30**, 383–96.

Martin, G. R. (1975). Teratocarcinomas as a model system for the study of embryo-genesis and neoplasia. *Cell* **5**, 229–43.

Martin, G. R. & Rubin, H. (1974). Effects of cell adhesion to the substratum on the growth of chick embryo fibroblasts. *Exp. Cell Res.* **85**, 319–33.

Martin, R. G. & Stein, S. (1976). Resting state in normal and Simian virus 40 transformed Chinese hamster lung cells. *Proc. Natl. Acad. Sci. USA* **73**, 1655–9.

Martinez-Palomo, A., Braislovsky, C. & Bernhard, W. (1969). Ultrastructural modification of the cell surface and intercellular contacts of some transformed cell strains. *Cancer Res.* **29**, 925–37.

Martz, E. & Steinberg, M. (1972). The role of cell–cell contact in 'contact' inhibition of cell division: a review and new evidence. *J. Cell Physiol.* **79**, 189–210.

Maslow, D. E. & Mayhew, E. (1975). Inhibition of embryonic cell aggregation by neoplastic cells. *J. Natl. Cancer Inst.* **54**, 1097–102.

Maslow, D. E., Mayhew, E. & Minowada, J. (1976). Differential inhibition of embryonic cell aggregation by cultured human cells with 'malignant' or 'normal' characteristics. *Cancer Res.* **36**, 2707–9.

Maslow, D. E. & Weiss, L. (1978). The effects of tumour cells on embryonic cell aggregation, adhesion and detachment. *J. Cell Sci.* **29**, 271–5.

Mautner, V. & Hynes, R. O. (1977). Surface distribution of LETS protein in relation to the cytoskeleton of normal and transformed cells. *J. Cell Biol.* **75**, 743–68.

Mazia, D. (1937). The release of calcium in Arbacia eggs in fertilization. *J. Cell. Comp. Physiol.* **10**, 291–304.

Medina, D. (1975). Tumor progression. In *Cancer: a Comprehensive Treatise*, ed. F. F. Becker, vol. 3, pp. 99–119. New York: Plenum Press.

Meek, R. L., Bowman, P. D. & Daniel, C. W. (1977). Establishment of mouse embryo cells *in vitro*. Relationship of DNA synthesis, senescence and malignant transformation. *Exp. Cell Res.* **107**, 277–84.

Mendelsohn, M. L. (1960). The growth fraction: a new concept applied to tumours. *Science* **132**, 1496.

Mets, T. & Verdonk, G. (1978). The theory of transition probability and the division pattern of WI-38 cells. *Cell Biol. Int. Rep.* **2**, 561–4.

Michalke, W. & Loewenstein, W. R. (1971). Communication between cells of different types. *Nature, Lond.* **232**, 121–2.

Micklem, K. J., Abra, R. M., Knutton, S., Graham, J. M. & Pasternak, C. A. (1976). The fluidity of normal and virus-transformed cell plasma membrane. *Biochem. J.* **154**, 561–6.

Middleton, C. A. (1973). The control of epithelial cell locomotion in tissue culture. In *Locomotion of Tissue Cells*, ed. M. Abercrombie. Ciba Foundation Symposium, vol. 14, pp. 251–62. Amsterdam: Associated Scientific Publishers.

Middleton, C. A. (1976). Contact-induced spreading is a new phenomenon depending on cell–cell contact. *Nature, Lond.* **259**, 311–13.

Middleton, C. A. (1977). The effects of cell–cell contact on the spreading of pigmented retina epithelial cells in culture. *Exp. Cell Res.* **109**, 349–59.

Mierzejewski, K. & Rozengurt, E. (1977). Density-dependent inhibition of fibroblast growth is overcome by pure mitogenic factors. *Nature, Lond.* **269**, 155–6.

Mierzejewski, K. & Rozengurt, E. (1978). A partially purified serum fraction synergistically enhances the mitogenic activity of epidermal growth factor and insulin in quiescent cultures of 3T3 cells. *Biophys. Biochem. Res. Commun.* **83**, 874–80.

Miettinen, A., Virtanen, J. & Linder, E. (1978). Cellular Actin and junction formation during reaggregation of epithelial cell sheets. *J. Cell Sci.* **31**, 341–53.

Milam, M., Grinnell, F. & Srere, P. A. (1973). Effect of centrifugation on cell adhesion. *Nature New Biol.* **244**, 83–4.

Miller, E. C. (1978). Some current perspectives on chemical carcinogenesis in humans and experimental animals: presidential address. *Cancer Res.* **38**, 1479–96.

Milo, G. E., Jr. & Di Paolo, J. A. (1978). Neoplastic transformation of human diploid cells *in vitro* after chemical carcinogen treatment. *Nature, Lond.* **275**, 130–2.

Mintz, B. (1978). Genetic mosaicism and in-vivo analyses of neoplasia and differentiation. In *Cell Differentiation and Neoplasia*, ed. G. F. Saunders, pp. 27–53. New York: Raven Press.

Mintz, B. & Illmensee, K. (1975). Normal genetically mosaic mice producted from malignant teratocarcinoma cells. *Proc. Natl. Acad. Sci. USA* **72**, 3585–9.

Mir-Lechaire, F. J., Barondes, S. H. (1978). Two distinct developmentally regulated lectins in chick embryo muscle. *Nature, Lond.* **272**, 256–8.

Misfeldt, D. S., Hamamoto, S. T. & Pitelka, D. R. (1976). Transepithelial transport in cell culture. *Proc. Natl. Acad. Sci. USA* **73**, 1212–16.

Mitchel, B. F. & Tupper, J. T. (1978). Normal and transformed WI-38 fibroblasts have different requirements for Ca^{2+} and Mg^{2+} during G_1 phase of the cell cycle. *J. Cell Biol.* **79**, 5a.

Mitelman, F. & Levan, G. (1976). Clustering of aberrations to specific chromosomes in human neoplasms. II. A survey of 287 neoplasms. *Hereditas* **82**, 167–74.

Mitsui, Y. & Schneider, E. L. (1976). Increased nuclear sizes in senescent human diploid fibroblast cultures. *Exp. Cell Res.* **100**, 147–52.

Mittelman, L. A., Sharovskaja, J. J. & Vasiliev, J. M. (1972). Toxic effect of 7,12-dimethylbenz-L-anthracene on neoplastic cells grown in mixed cultures with normal fibroblasts. *Int. J. Cancer* **10**, 667–74.

Moizhess, T. G. (1969). Some characteristics of phago- and pinocytosis in normal fibroblast-like cells of the mouse and L-strain cells *in vitro*. *Tsytologiya* **11**, 493–98 (in Russian).

Moizhess, T. G. & Prigogina, E. L. (1973). Localization of pretumour elements in plastic carcinogenesis. *Byulleten experimentalnoj Biologii i Meditsini* **76**, N 9, 92–4 (in Russian).

Mondal, S., Brankow, D. W. & Heidelberger, C. (1976). Two-stage chemical oncogenesis in cultures of C3H/10T 1/2 cells. *Cancer Res.* **36**, 2254–60.

Mondal, S., Embleton, M. J., Marquardt, H. & Heidelberger, C. (1971). Production of variants of decreased malignancy and antigenicity from clones transformed *in vitro* by methylcholanthrene. *Int. J. Cancer* **8**, 410–20.

Mondal, S. & Heidelberger, C. (1976). Transformation of C3H/10T 1/2 CL8 mouse embryo fibroblasts by ultraviolet irradiation and a phorbol ester. *Nature, Lond.* **260**, 710–11.

Montagnier, L. (1971). Factors controlling the multiplication of untransformed and transformed BHK_{21} cells under various environmental conditions. In *Growth Control in Cell Cultures*, ed. G. E. W. Wolstenholme & J. Knight, pp. 33–41. Edinburgh: Churchill Livingstone.

Montesano, R., Bartsch, H. & Tomatis, L. (eds.) (1976). *Screening tests in chemical carcinogenesis*. Lyon: International Agency for Research on Cancer.

Montesano, R., Drevon, C., Kuroki, T., Saint Vincent, L., Handleman, S., Sanford, K. K., DeFeo, D. & Weinstein, I. B. (1977). Test for malignant transformation of rat liver cells in culture: cytology, growth in soft agar, and production of plasminogen activator. *J. Natl. Cancer Inst.* **59**, 1651–8.

Montesano, R., Saint Vincent, L. & Tomatis, L. (1973). Malignant transformation *in vitro* of rat liver cells by dimethylnitrosamine and N-methyl-N'-nitro-N-nitrosoguanidine. *Br. J. Cancer* **28**, 215–20.

Moore, M. A. S., Kurland, J. & Broxmeyer, H. E. (1976). The granulocytic and monocytic stem cell. In *Stem Cells of Renewing Cell Populations*, ed. A. B. Cairnie, P. K. Lala & D. G. Osmond, pp. 181–8. New York: Academic Press.

Mooseker, M. S. (1976). Actin filament–membrane attachment in the microvilli of intestinal epithelial cells. In *Cell Motility*, ed. R. Goldman, T. Pollard & J. Rosenbaum. Cold Spring Harbor Conferences on Cell Proliferation, vol. 3, pp. 631–50. New York: Cold Spring Harbor Laboratory.

Mora, P. T., Chang, C., Couvillion, L., Kuster, J. M. & McFarland, V. W. (1977). Immunological selection of tumour cells which have lost SV-40 antigen expression. *Nature, Lond.* **269**, 36–40.

Morgan, H. R. (1964). The biologic properties of cells infected with Rous sarcoma virus *in vitro*. In *International Conference on Avian Tumor Viruses*, ed. J. W. Beard, National Cancer Institute Monograph, 17, pp. 395–406. Bethesda, Md.: National Cancer Institute.

Morgan, H. R. & Ganapathy, S. (1963). Comparative studies in Rous sarcoma. IV. Glucose metabolism of normal and Rous sarcoma virus-infected cells. *Proc. Soc. Exp. Biol. Med.* **113**, 312–15.

Moroney, J., Smith, A., Tomei, L. D. & Wenner, C. E. (1978). Stimulation of ^{86}Rb$^+$ and ^{32}P$_i$ movements in 3T3 cells by prostaglandins and phorbol esters. *J. Cell. Physiol.* **95**, 287–94.

Moscatelli, D. & Rubin, H. (1975). Increased hyaluronic acid production on stimulation of DNA synthesis in chick embryo fibroblasts. *Nature, Lond.* **254**, 65–6.

Moscatelli, D. & Rubin, H. (1977). Hormonal control of hyaluronic acid production in fibroblasts and its relation to nucleic acid and protein synthesis. *J. Cell. Physiol.* **91**, 79–88.

Moscona, A. A. (1965). Recombination of dissociated cells and the development of cell aggregates. In *Cells and tissues in culture*, ed. E. N. Willmer, vol. 1, pp. 489–529. New York: Academic Press.

Moses, H. L., Proper, J. A., Volkenant, M. E., Wells, D. Y. & Getz, M. J. (1978). Mechanism of growth arrest of chemically transformed cells in culture. *Cancer Res.* **38**, 2807–12.

Mostafapour, M.-K. & Green, H. (1975). Effect of withdrawal of a serum stimulus on the protein-synthesizing machinery of cultured fibroblasts. *J. Cell. Physiol.* **86**, 313–20.

Mott, D. M., Fabisch, P. H., Sani, B. P. & Sorof, S. (1974). Lack of correlation between fibrinolysis and the transformed state of cultured mammmalian cells. *Biochem. Biophys. Res. Commun.* **61**, 621–7.

Moyer, W. A. & Steinberg, M. S. (1976). Do rates of intercellular adhesion measure the cell affinities reflected in cell-sorting and tissue-spreading configurations? *Dev. Biol.* **52**, 246–62.

Murphy, D. B. & Borisy, G. G. (1975). Association of high molecular weight proteins with microtubules and their role in microtubule assembly *in vitro*. *Proc. Natl. Acad. Sci. USA* **72**, 2696–700.

Nachmias, V. T. & Ash, A. (1976). Regulation and polarity: results with myxomycete plasmodium and with human platelets. In *Cell Motility*, ed. R. Goldman, T. Pollard & J. Rosenbaum. Cold Spring Harbor Conference on Cell Proliferation, vol. 3, pp. 771–83. New York: Cold Spring Harbor Laboratory.

Naiditch, W. P. & Cunningham, D. D. (1977). Hexose uptake and control of fibroblast proliferation. *J. Cell. Physiol.* **92**, 319–32.

Nandi, S. (1978). Role of hormones in mammary neoplasia. *Cancer Res.* **38**, 4046–9.

Nath, K. & Srere, P. A. (1977). Effects of temperature, metabolic and cytoskeletal inhibitors on the rate of BHK cell adhesion to polystyrene. *J. Cell. Physiol.* **92**, 33–42.

Negendank, W. G. & Collier, C. R. (1976). Ion contents of human lymphocytes. The effect of concanavalin A and ouabain. *Exp. Cell Res.* **101**, 31–40.

Nesbitt, J. A., III, Anderson, W. B., Miller, Z. & Pastan, I. (1976). Guanylate cyclase and cyclic guanosine 3′:5′-monophosphate phosphodiesterase activities and cyclic guanosine 3′:5′-monophosphate levels in normal and transformed fibroblasts in culture. *J. Biol. Chem.* **251**, 2344–52.

Neupert, G. (1972). Strukturalterationen bei Leberzellen während der Gewebsdissociation und Monolayerformation. *Z. Mikrosk.-anat. Forsch.* **86**, 443–64.

Newbold, R. F., Wigley, C. B., Thompson, M. H. & Brookes, P. (1977). Cell-meditated mutagenesis in cultured chinese hamster cells by carcinogenic polycyclic hydrocarbons: nature and extent of the associated hydrocarbon–DNA. *Mutation Res.* **43**, 101–16.

Ng, C. E. & Inch, W. R. (1978). Comparison of the densities of clonogenic cells from EMT6 fibrosarcoma monolayer cultures, multicell spheroids, and solid tumours in ficoll density gradients. *J. Natl. Cancer Inst.* **60**, 1017–22.

Nicolson, G. L. (1971). Difference in topology in normal and tumour cell membranes shown by different surface distributions of ferritin-conjugated concanavalin A. *Nature New Biol.* **233**, 244–6.

Nicolson, G. L. (1974). The interactions of lectins with animal cell surfaces. *Int. Rev. Cytol.* **39**, 89–190.

Nicolson, G. L. (1976a). Transmembrane control of the receptors on normal and tumor cells. I. Cytoplasmic influence over cell surface components. *Biochim. Biophys. Acta* **457**, 1–40.

Nicolson, G. L. (1976b). Trans-membrane control of the receptors on normal and tumor cells. II. Surface changes associated with transformation and malignancy. *Biochim. Biophys. Acta* **458**, 1–72.

Nicolson, G. L., Brunson, K. W. & Fidler, I. J. (1978). Specificity of arrest, survival, and growth of selected metastatic variant cell lines. *Cancer Res.* **38**, 4105–11.

Nicolson, G. L. & Winkelhake, J. L. (1975). Organ specificity of blood-borne tumour metastasis determined by cell adhesion? *Nature, Lond.* **255**, 230–2.

Nilausen, K. (1978). Role of fatty acids in growth-promoting effect of serum albumin on hamster cells *in vitro*. *J. Cell. Physiol.* **96**, 1–14.

Njeuma, D. L. (1971a). Mitosis and population density in cultures of embryonic chick and mouse fibroblasts. *Exp. Cell Res.* **66**, 237–43.

Njeuma, D. L. (1971b). Non-reciprocal density-dependent mitotic inhibition in mixed cultures of embryonic chick and mouse fibroblasts. *Exp. Cell Res.* **66**, 244–50.

Noble, R. L. (1977). Hormonal control of growth and progression in tumours of Nb rats and a theory of action. *Cancer Res.* **37**, 82–94.

Nomura, S., Dunn, K. J. & Fishinger, P. J. (1973). Rapid screening assay for revertants derived from MSV-transformed cells. *Nature, Lond.* **246**, 213–14.

Noonan, K. D. (1976). Role of serum in protease-induced stimulation of 3T3 cell division past the monolayer stage. *Nature, Lond.* **259**, 573–6.

Noonan, K. D. & Burger, M. M. (1973). Induction of 3T3 cell division at the monolayer stage. *Exp. Cell Res.* **80**, 405–14.

Nowell, P. C. (1974). Chromosome changes and the clonal evolution of cancer. In *Chromosomes and Cancer*, ed. J. German, pp. 268–85. New York: J. Wiley & Sons.

Nowell, P. C. (1976). The clonal evolution of tumor cell populations. *Science* **194**, 23–8.

O'Brien, T. & Diamond, L. (1977). Ornithine decarboxylase induction and DNA synthesis in hamster embryo cell cultures treated with tumor-promoting phorbol diester, *Cancer Res.* **37**, 3895–900.

Odell, W. D. & Wolfsen, A. (1976). Ectopic hormone secretion by tumors. In *Cancer: a Comprehensive Treatise*, ed. F. F. Becker, vol. 3, pp. 81–97. New York: Plenum Press.

Ogata, S., Muramatsu, T. & Kobata, A. (1976). New structural characteristic of the large glycopeptides from transformed cells. *Nature, Lond.* **259**, 580–2.

O'Keefe, E. & Cuatrecasas, P. (1978). Cholera toxin and membrane gangliosides:

binding and adenylate cyclase activation in normal and transformed cells. *J. Membrane Biol.* **42**, 61–79.

Okuda, A. & Kimura, G. (1978). Serum stimulation of DNA synthesis in rat 3-Y1 cells. *Exp. Cell Res.* **111**, 55–62.

Olden, K. & Yamada, K. M. (1977). Mechanism of the decrease in the major cell surface protein of chick embryo fibroblasts after transformation. *Cell* **11**, 957–69.

O'Neill, F. J. (1974). Control of nuclear division in normal but not in neoplastic mouse cells. *Cancer Res.* **34**, 1070–3.

O'Neill, F. J. (1975). Selective destruction of cultured tumor cells with uncontrolled nuclear division by cytochalasin B and cytosine arabinoside. *Cancer Res.* **35**, 3111–15.

O'Neill, F. J. (1976). Loss of controlled nuclear division in BHK21 cells passed *in vitro*. *Cancer Res.* **36**, 2019–24.

O'Neill, F. J. (1978). Differential in-vitro growth properties of cells transformed by DNA and RNA tumor viruses. *Exp. Cell Res.* **117**, 393–401.

O'Neill, F. J., Miller, T. H., Hoen, J., Stradley, B. & Devlahovich, V. (1975). Differential response to cytochalasin B among cells transformed by DNA and RNA tumor viruses. *J. Natl. Cancer Inst.* **55**, 951–5.

Oppenheimer, B. S., Oppenheimer, E. T. & Stout, A. P. (1948). Sarcomas induced in rats implanting cellophane. *Proc. Soc. Exp. Biol. Med.* **67**, 33–4.

Oppenheimer, B. S., Oppenheimer, E. T., Stout, A. P., Willhite, M. & Danishefsky, I. (1958). The latent period in carcinogenesis by plastic in rats and its relation to the presarcomatous stage. *Cancer* **11**, 204–13.

Oppenheimer, S. B., Bales, B. L., Brenneman, G., Knapp, L., Lesin, E. S., Neri, A. & Pollock, E. G. (1977). Modulation of agglutinability by alteration of the surface topography in mouse ascites tumor cells. *Exp. Cell Res.* **105**, 291–300.

Orci, L., Carpentier, J.-L., Perrelet, A., Anderson, R. G. W., Goldstein, J. L. & Brown, M. S. (1978). Occurrence of low density lipoprotein receptors within large pits on the surface of human fibroblasts as demonstrated by freeze-etching. *Exp. Cell Res.* **111**, 1–13.

Orci, L., Like, A. A., Amherdt, M., Blondel, B., Kanazawa, Y., Marliss, E. B., Lambert, A. E., Wollheim, C. B. & Renold, A. E. (1973). Monolayer cell culture of neonatal rat pancreas: an ultrastructural and biochemical study of functioning endocrine cells. *J. Ultrastruct. Res.* **43**, 270–97.

Orgel, L. E. (1963). The maintenance of the accuracy of protein synthesis and its relevance to ageing. *Proc. Natl. Acad. Sci. USA* **49**, 517–21.

Osborn, M., Franke, W. W. & Weber, K. (1977). Visualization of a system of filaments 7–10 nm thick in cultured cells of an epithelioid line (PtK$_2$) by immunofluorescence microscopy. *Proc. Natl. Acad. Sci. USA* **74**, 2490–4.

Osborn, M. & Weber, K. (1976a). Cytoplasmic microtubules in tissue culture cells appear to grow from an organizing structure towards the plasma membrane. *Proc. Natl. Acad. Sci. USA* **73**, 867–71.

Osborn, M. & Weber, K. (1976b). Tubulin-specific antibody and the expression on microtubules in 3T3 cells after attachment to a substratum. Further evidence for the polar growth of cytoplasmic microtubules *in vivo*. *Exp. Cell Res.* **103**, 331–40.

Osborn, M. & Weber, K. (1977). The display of microtubules in transformed cells. *Cell* **12**, 561–71.

Oshima, R. G., Pellett, O. L., Robb, J. A. & Schneider, J. A. (1977). Transformation of human cystinotic fibroblasts by SV-40: characteristics of transformed cells with limited and unlimited growth potential. *J. Cell. Physiol.* **93**, 129–36.

Oshiro, Y. & Di Paolo, J. A. (1974). Changes in the uptake of 2-deoxy-D-glucose in BalbT3 cells chemically transformed in culture. *J. Cell. Physiol.* **83**, 193–201.

Ossowski, L., Quigley, J. P., Kellerman, G. M. & Reich, E. (1973). Fibrinolysis associated with oncogenic transformation. Requirement of plasminogen for correlated changes in cellular morphology, colony formation in agar and cell migration. *J. Exp. Med.* **138**, 1056–64.

Otten, J., Bader, J., Johnson, G. S. & Pastan, I. (1972). A mutation in a Rous sarcoma virus gene that controls adenosine 3′,5′-monophosphate levels and transformation. *J. Biol. Chem.* **247**, 1632–33.

Otten, J., Johnson, G. & Pastan, I. (1971). Cyclic AMP levels in fibroblasts: relationship to growth rate and contact inhibition of growth. *Biochem. Biophys. Res. Commun.* **44**, 1192–8.

Overton, J. (1977). Is there an instance of Steinberg's 'site frequency model'? *J. Theor. Biol.* **65**, 787–9.

Owens, R., Smith, H., Nelson-Rees, W. & Springer, E. (1976). Epithelial cell cultures from normal and cancerous tissues. *J. Cell Biol.* **56**, 843–9.

Ozanne, B. & Vogel, A. (1974). Selection of revertants of Kirsten sarcoma virus transformed non-producer BALB/3T3 cells. *J. Virol.* **14**, 239–48.

Ozato, K., Huang, L. & Ebert, J. D. (1977). Accelerated calcium ion uptake in murine thymocytes induced by concanvalin A. *J. Cell. Physiol.* **93**, 153–60.

Paranjpe, M. S. & Boone, C. W. (1975). Intermittent sphering of virus-transformed and other neoplastic cells observed by time lapse cinematography. *Exp. Cell Res.* **94**, 147–51.

Paranjpe, M. S., Boone, C. W. & Eaton, S. A. (1975). Selective growth of malignant cells by in-vitro incubation on Teflon. *Exp. Cell Res.* **93**, 508–12.

Paranjpe, M. S., Eaton, S. & Boone, C. W. (1978). Neoplasms produced from C3H/1OT 1/2 cells attached to plastic plates; saturation density, anchorage dependence and serum requirement of in-vitro lines correlated with growth aggressiveness *in vitro*. *J. Cell. Physiol.* **96**, 63–72.

Pardee, A. B. (1974). A restriction point for control of normal animal cell proliferation. *Proc. Natl. Acad. Sci. USA* **71**, 1286–90.

Pardee, A. B. & James, L. J. (1975). Selective killing of transformed baby hamster kidney (BHK) cells. *Proc. Natl. Acad. Sci. USA* **72**, 4994–8.

Parnes, J. R., Carvey, T. Q. & Isselbacher, K. J. (1976). Amino acid transport by membrane vesicles of virally transformed and non-transformed cells: effects of sodium gradient and cell density. *J. Cell. Physiol.* **89**, 789–94.

Parshad, R. & Sanford, K. K. (1977). Intermittent exposure to fluorescent light extends lifespan of human diploid fibroblasts in culture. *Nature, Lond.* **268**, 736–7.

Pastan, I. & Johnson, G. S. (1974). Cyclic AMP and the transformation of fibroblasts. *Adv. Cancer Res.* **19**, 303–29.

Pastan, I. H., Johnson, G. S. & Anderson, W. B. (1974). Role of cyclic nucleotides in growth control. *Annu. Rev. Biochem.* **44**, 491–522.

Pastan, I. & Willingham, M. (1978). Cellular transformation and the 'morphologic phenotype' of transformed cells. *Nature, Lond.* **274**, 645–50.

Paucha, E., Mellor, A., Harvey, R., Smith, A. E., Hewick, R. M. & Waterfield, M. D. (1978). Large and small tumor antigens from Simian virus 40 have identical amino termini mapping at 0.65 map units. *Proc. Natl. Acad. Sci. USA* **75**, 2165–9.

Paul, D. (1978). Growth control in HeLa cells by serum and anchorage. *Exp. Cell Res.* **114**, 434–8.

Pearlstein, E. (1976). Plasma membrane glycoprotein which mediates adhesion of fibroblasts to collagen. *Nature, Lond.* **262**, 497–9.

Pearlstein, E., Hynes, R. O., Franks, L. M. & Hemmings, V. J. (1976). Surface proteins and fibrinolytic activity of cultured mammalian cells. *Cancer Res.* **36**, 1475–80.

Pearson, T. A., Dillman, J. M., Solez, K. & Heptinstall, R. H. (1978). Clonal markers in the study of the origin and growth of human atherosclerotic lesions. *Cir. Res.* **43**, 10–18.

Perdue, J. F. (1976). Loss of the post-translational control of nutrient transport in in-vitro and in-vivo virus-transformed chicken cells. *J. Cell. Physiol.* **89**, 729–36.

Perdue, J. F., Lubenskyi, W., Kivity, E. & Susanto, I. (1978). Identification and characterization of thrombin receptors on avian cells. *J. Cell Biol.* **79**, 50a.

Perecko, J. P., Berezesky, I. K. & Grimley, P. M. (1973). Surface features of some established murine cell lines under various conditions of oncogenic virus infection. In *Scanning Electron Microscopy/1973*, Pt 3, ed. O. Johari & I. Corvin, pp. 521–28. Chicago: IIT Research Institute.

Pfeffer, L., Lipkin, M., Stutman, O. & Kopelovich, L. (1976). Growth abnormalities of cultured human skin fibroblasts derived from individuals with hereditary adenomatosis of the colon and rectum. *J. Cell. Physiol.* **89**, 29–38.

Pfeffer, L. M. & Kopelovich, L. (1977). Differential genetic susceptibility of cultured human skin fibroblasts to transformation by Kirsten murine sarcoma virus. *Cell* **10**, 313–20.

Phillips, H. M., Steinberg, M. S. & Lipton, B. H. (1977). Embryonic tissues as elasticoviscous liquids. II. Direct evidence for cell slippage in centrifuged aggregates. *Devl. Biol.* **59**, 124–34.

Pickett, P. B., Pitelka, D. R., Hamamoto, S. T. & Misfeldt, D. S. (1975). Occluding junctions and cell behavior in primary cultures of normal and neoplastic mammary gland cells. *J. Cell Biol.* **66**, 316–32.

Pierce, G. B. (1967). Teratocarcinoma: model for a developmental concept of cancer. *Curr. Top. Dev. Biol.* **2**, 223–46.

Pierce, G. B. (1974). Neoplasms, differentiations and mutations. *Am. J. Pathol.* **77**, 103–18.

Pierce, G. B. (1976). Origin of Neoplastic Stem Cells. In '*Progress in Differentation Research*', ed. N. Müller-Berat, pp. 269–73. Amsterdam: North-Holland.

Pierce, G. B. & Cox, W. F., Jr. (1978). Neoplasms as caricatures of tissue renewal. In *Cell Differentiation and Neoplasia*, ed. G. F. Saunders, pp. 57–66. New York: Raven Press.

Pinto da Silva, P. & Gilula, N. B. (1972). Gap junctions in normal and transformed fibroblasts in culture. *Exp. Cell Res.* **71**, 393–401.

Pisam, M. & Ripoche, P. (1976). Redistribution of surface macromolecules in dissociated epithelial cells. *J. Cell Biol.* **71**, 907–20.

Pitts, J. D. (1971). Appendix. Growth properties of normal and transformed cells. In *Growth Control in Cell Cultures*, ed. G. E. W. Wolstenholme & J. Knight, pp. 261–66. Edinburgh: Churchill Livingstone.

Pitts, J. D. & Burk, R. K. (1976). Specificity of junctional communication between animal cells. *Nature, Lond.* **264**, 762–4.

Pitts, J. D. & Simms, J. W. (1977). Permeability of junctions between animal cells. Intercellular transfer of nucleotides but not of macromolecules. *Exp. Cell Res.* **104**, 153–63.

Pledger, W. J., Gardner, R. M., Epstein, P. M., Thompson, W. J., Strada, S. J. & Wlodyka, L. (1979). Cell cycle traverse and macromolecular synthesis in BHK fibroblasts as affected by insulin. *Exp. Cell Res.* **118**, 389–94.

Pledger, W. I., Stiles, C. D., Antoniades, H. N. & Scher, C. D. (1977). Induction of DNA synthesis in BALB/c 3T3 cells by serum components: Re-evaluation of the commitment process. *Proc. Natl. Acad. Sci. USA* **74**, 4481–5.

Pletyushkina, O. Y., Vasiliev, J. M. & Gelfand, I. M. (1975). Neoplastic fibroblasts sensitive to the growth inhibition by homologous cells but insensitive to inhibition by parent normal cells. *Br. J. Cancer* **31**, 535–43.

Pogosianz, H. E. (1973). Comparative study of mutagenic and carcinogenic action of certain chemicals. *Neoplasma* **20**, 527–30.

Pohjanpelto, P. (1977). Proteases stimulate proliferation of human fibroblasts. *J. Cell. Physiol.* **91**, 387–92.

Pohjanpelto, P. (1978). Stimulation of DNA synthesis in human fibroblasts by thrombin. *J. Cell. Physiol.* **95**, 189–94.

Pollack, R. (1977). A strategy for the in-vitro analysis of the metastatic process. *Gann Monogr. Cancer Res.* **20**, 37–45.

Pollack, R. E., Green, H. & Todaro, G. J. (1968). Growth control in cultured cells: selection of sublines with increased sensitivity to contact inhibition and decreased tumor-producing ability *Proc. Natl. Acad. Sci. USA* **60**, 126–33.

Pollack, R., Osborn, M. & Weber, K. (1975). Patterns of organization of actin and myosin in normal and transformed cultured cells. *Proc. Natl. Acad. Sci. USA* **72**, 994–8.

Pollack, R. & Rifkin, D. (1975). Actin-containing cables within anchorage-dependent rat embryo cells are dissociated by plasmin and trypsin. *Cell* **6**, 495–506.

Pollack, R., Risser, R., Conlon, S. & Rifkin, D. (1974). Plasminogen activator production accompanies loss of anchorage regulation in transformation of primary rat embryo cells by Simian virus 40. *Proc. Natl. Acad. Sci. USA* **71**, 4792–6.

Pollack, R., Wolman, S. & Vogel, A. (1970). Reversion of virus transformed cell lines: hyperploidy accompanies retention of viral genes. *Nature, Lond.* **228**, 967–70.

Pollard, T. D. (1976). The role of actin in the temperature-dependent gelation and contraction of extracts of *Acanthamoeba*. *J. Cell Biol.* **68**, 579–601.

Pollard, T. D., Fujiwara, K., Niederman, R. & Maupin-Szamier, P. (1976). Evidence for the role of cytoplasmic actin and myosin in cellular structure and motility. In *Cell Motility*, ed. R. Goldman, T. Pollard & J. Rosenbaum. Cold Spring Harbor Conferences on Cell Proliferation, vol. 3, pp. 689–724. New York: Cold Spring Harbor Laboratory.

Pontén, J. (1964). The in-vivo growth mechanism of avian Rous sarcoma. In *International Conference on Avian Tumor viruses*, ed. J. W. Beard, National Cancer Institute Monograph 17, pp. 31–6. Bethesda Md.: National Cancer Institute.

Pontén, J. (1975). Neoplastic glia cells in culture. In *Human Tumor Cells in vitro*, ed. J. Fogh, pp. 175–206. New York: Plenum Press.

Pontén, J. (1976). The relationship between in-vitro transformation and tumor formation in-vivo. *Biochim. Biophys. Acta* **458**, 397–422.

Pontén, J. & MacIntyre, E. H. (1968). Interaction between normal and transformed bovine fibroblasts in culture. II. Cells transformed by polyoma virus. *J. Cell Sci.* **3**, 603–13.

Poo, M.-M., Poo, W.-J. H. & Lam, J. W. (1978). Lateral elecrophoresis and diffusion of concanavalin A receptors in the membrane of embryonic muscle cells. *J. Cell Biol.* **76**, 483–501.

Poo, M.-M. & Robinson, K. R. (1977). Electrophoresis of concanavalin A receptors along embryonic muscle membrane. *Nature, Lond.* **265**, 602–5.

Poole, A. R., Tiltman, K. J., Reckless, A. D. & Stoker, T. A. M. (1978). Differences in secretion of the proteinase cathepsin B at the edges of human breast carcinomas and fibroadenomas. *Nature, Lond.* **273**, 545–7.

Porter, K. R. (1976). Introduction: motility in cells. In *Cell Motility*, ed. R. Goldman, T. Pollard & J. Rosenbaum. Cold Spring Harbor Conferences on Cell Proliferation, vol. 3, pp. 1–28. New York: Cold Spring Harbor Laboratory.

Porter, K. R. & Fonte, V. G. (1973). Observations on the topography of normal and cancer cells. In *Scanning Electron Microscopy/1973*, Pt 3, ed. O. Johari & I. Corvin, pp. 683–8. Chicago: IIT Research Institute.

Porter, K., Prescott, D. & Frye, J. (1973*a*). Changes in surface morphology of Chinese hamster ovary cells during the cell cycle. *J. Cell. Biol.* **57**, 815–36.

Porter, K. R., Todaro, G. J. & Fonte, V. (1973*b*). A scanning electron microscope study of surface features of viral and spontaneous transformants of mouse BALB/3T3 cells. *J. Cell Biol.* **59**, 633–42.

Poste, G. (1975). Production of a serine-protease with macrophage migration-inhibitory factor activity by virus-transformed cells and human tumor cell lines. *Cancer Res.* **35**, 2588–66.

Poste, G., Papahedjopoulos, D. & Nicolson, G. L. (1975). Local anesthetics affect transmembrane cytoskeletal control of mobility and distribution of cell surface receptors. *Proc. Natl. Acad. Sci. USA* **72**, 4430–4.

Potten, C. S. (1976*a*). Small intestinal crypt stem cells. In *Stem Cells of Renewing Cell Populations*, ed. A. B. Cairnie, P. K. Lala & D. G. Osmond, pp. 79–84. New York: Academic Press.

Potten, C. S. (1976*b*). Identification of clonogenic cells in the epidermis and the structural arrangement of the epidermal proliferative unit (EPU). In *Stem Cells of Renewing Cell Populations*, ed. A. B. Cairnie, P. K. Lala & D. G. Osmond, pp. 91–102. New York: Academic Press.

Potter, M. & Cancro, M. (1978). Plasmacytomagenesis and the differentiation of immunoglobulin-producing cells. In *Cell Differentiation and Neoplasia*, ed. G. F. Saunders, pp. 145–61. New York: Raven Press.

Pouysségur, J. M. & Pastan, I. (1976). Mutants of Balb/c 3T3 fibroblasts defective in adhesiveness to the substratum. Evidence for alteration in cell surface proteins. *Proc. Natl. Acad. Sci. USA* **73**, 544–8.

Pouysségur, J. M., Shiu, R. P. C. & Pastan, I. (1977). Induction of two transformation-sensitive membrane polypeptides in normal fibroblasts by a block in glycoprotein synthesis or glucose deprivation. *Cell* **11**, 941–7.

Prasad, K. & Hsie, A. (1971). Morphologic differentiation of mouse neuroblastoma cells induced *in vitro* by dibutyryl adenosine 3':5' cyclic monophosphate. *Nature New Biol.* **233**, 141–2.

Prasad, K. N. & Sinha, P. K. (1976). Effect of sodium butyrate on mammalian cells in culture: a review. *In vitro* **12**, 125–32.

Prasad, K. N. & Sinha, P. K. (1978). Regulation of differentiated functions and malignancy in neuroblastoma cells in culture. In *Cell Differentiation and Neoplasia*. ed. G. F. Saunders, pp. 111–41. New York: Raven Press.

Prescott, D. M. (1968). Regulation of cell reproduction. *Cancer Res.* **28**, 1815–20.

Prigogina, E. L., Fleischman, E. W. (1975). Certain pattern of karyotype evolution in chronic myelogenous leukaemia. *Humangenetik* **30**, 113–19.

Prigogina, E. L., Fleischman, E. W., Volkova, M. A. & Frenkel, M. A. (1978). Chromosome abnormalities and clinical and morphologic manifestations of chronic myeloid leukemia. *Hum. Genet.* **41**, 143–56.

Pringle, J. P. (1978). The use of conditional lethal cell cycle mutants for temporal and functional sequence mapping of cell cycle events. *J. Cell. Physiol.* **95**, 393–406.

Prujansky, A., Ravid, A. & Sharon, N. (1978). Co-operativity of lectin binding to lymphocytes, and its relevance to mitogenic stimulation. *Biochim. Biophys. Acta* **508**, 137–46.

Puck, T. T. (1977). Cyclic AMP, the microfilament system, and cancer, *Proc. Natl. Acad. Sci. USA* **74**, 4491–5.

Purchase, I. F. H., Longstaff, E., Ashby, J., Styles, J. A., Anderson, D., Lefevre, P. A. & Westwood, F. R. (1976). Evaluation of six short term tests for detecting organic chemical carcinogens and recommendations for their use. *Nature, Lond.* **264**, 624–7.

Purchio, A. F., Erikson, E., Brugge, J. S. & Erikson, R. L. (1978). Identification of a polypeptide encoded by the avian sarcoma virus *src* gene. *Proc. Natl. Acad. Sci. USA* **75**, 1567–71.

Puri, E. C. & Turner, D. C. (1978). Serum-free medium allows chicken myogenic cells to be cultivated in suspension and separated from attached fibroblasts. *Exp. Cell Res.* **115**, 159–73.

Putman, D. L., Park, D. K., Rhim, J. S., Steuer, A. F. & Ting, R. C. (1977). Correlation of cellular aggregation of transformed cells with their growth in soft agar and tumorigenic potential. *Proc. Soc. Exp. Biol. Med.* **155**, 487–94.

Quigley, J. P. (1976). Association of a protease (plasminogen activator) with a specific membrane fraction isolated from transformed cells. *J. Cell Biol.* **71**, 472–86.

Quinlan, D. C. & Hochstadt, J. (1977). The regulation by fibroblast growth factor of early transport changes in quiescent 3T3 cells. *J. Cell. Physiol.* **93**, 237–46.

Rabinowitz, Z. & Sachs, L. (1968). Reversion of properties in cells transformed by polyoma virus. *Nature, Lond.* **220**, 1203–6.

Raizada, M. K. & Perdue, J. F. (1976). Mitogen receptors in chick embryo fibroblast. Kinetics, specificity, unmasking and synthesis of ^{125}I-insulin binding sites. *J. Biol. Chem.* **251**, 6445–55.

Rajaraman, R., Rounds, D. F., Yen, S. P. S. & Rembaum, A. (1974). A scanning electron microscope study of cell adhesion and spreading *in vitro*. *Exp. Cell Res.* **88**, 327–9.

Rapp, F. & Westmoreland, D. (1976). Cell transformation by DNA-containing viruses. *Biochim. Biophys. Acta* **458**, 167–211.

Ravid, A. & Novogrodsky, A. (1976). Kinetics of the induction of DNA synthesis in lympocytes by periodate. *Exp. Cell Res.* **97**, 1–5.

Ravid, A., Novogrodsky, A. & Wilchek, M. (1978). Grafting of triggering sites into lymphocytes. Requirement of multivalency in the stimulation of dinitrophenyl-modified thymocytes by anti-dinitrophenyl antibody. *Eur. J. Immunol.* **8**, 289–94.

Reaven, E. P. & Axiline, S. G. (1973). Subplasmalemmal micrifilaments and micro-tubules in resting and phagocytizing cultivated macrophages. *J. Cell Biol.* **59**, 12–27.

Rechler, M. M., Bruni, C. B., Podskalny, J. M., Warner, W. & Carchman, R. A. (1977*a*). Modulation of serum-stimulated DNA synthesis in cultured human fibroblasts by cAMP. *Exp. Cell Res.* **104**, 411–22.

Rechler, M. M., Nissley, S. P., Podskalny, J. M., Moses, A. C. & Fryklund, L. (1977*b*). Identification of a receptor for somatomedin-like polypeptides in human fibroblasts. *J. Clin. Endocr. Metab.* **44**, 820–31.

Reddi, A. H. (1976). Collagen and cell differentiation. In *Biochemistry of Collagen*, ed. G. N. Ramachandran & A. H. Reddi, pp. 449–78. New York: Plenum.

Reddy, V. B., Thimmappaya, B., Dhar, R., Subramanian, K. N., Zain, B. S., Pan, J., Ghosh, P. K., Celma, M. L. & Weissman, S. M. (1978). The genome of Simian virus 40. *Science* **200**, 494–502.

Rees, L. H. (1975). The biosynthesis of hormones by non-endocrine tumours – a review. *J. Endocrinol.* **67**, 143–75.

Revel, J. P. (1974). Contacts and junctions between cells. *Symp. Soc. Exp. Biol.* **27**, 447–61.

Revel, J. P. & Wolken, K. (1973). Electron microscope investigations of the underside of cells in culture. *Exp. Cell Res.* **78**, 1–14.

Revesz, L. (1958). Effect of lethally damaged tumor cells upon the development of admixed viable cells. *J. Natl. Cancer Inst.* **20**, 1157–86.

Révész, T. & Greaves, M. (1975). Ligand-induced redistribution of lymphocyte membrane ganglioside GM1. *Nature, Lond.* **257**, 103–7.

Reznikoff, C. A., Bertram, J. S., Brankow, D. W. & Heidelberger, C. (1973b). Quantitative and qualitative studies of chemical transformation of cloned C3H mouse embryo cells sensitive to postconfluence inhibition of cell division. *Cancer Res.* **33**, 3239–49.

Reznikoff, C. A., Brankow, D. W. & Heidelberger, C. (1973a). Establishment and characterization of a cloned line of C3H mouse embryo cells sensitive to postconfluence inhibition of division. *Cancer Res.* **33**, 3231–8.

Rheinwald, J. G. & Green, H. (1975a). Formation of a keratinizing epithelium in culture by a cloned line derived from a teratoma. *Cell* **6**, 317–30.

Rheinwald, J. G. & Green, H. (1975b). Serial cultivation of human epidermal keratinocytes: the formation of keratinizing colonies from single cells. *Cell* **6**, 331–44.

Rheinwald, J. G. & Green, H. (1977). Epidermal growth factor and the multiplication of cultured human epidermal keratinocytes. *Nature, Lond.* **265**, 421–4.

Rhim, J. S. & Huebner, R. J. (1973). Transformation of rat embryo cells *in vitro* by chemical carcinogens. *Cancer Res.* **33**, 695–700.

Rhim, J. S., Kim, C. M., Arnstein, P., Huebner, R. J., Weisburger, E. K. & Nelson-Rees, W. A. (1975). Transformation of human osteosarcoma cells by a chemical carcinogen. *J. Natl. Cancer Inst.* **55**, 1291–4.

Rhim, J. S., Kim, C. M., Okigaki, T. & Huebner, R. J. (1977). Transformation of rat liver epithelial cells by Kirsten murine sarcoma virus. *J. Natl. Cancer Inst.* **59**, 1509–18.

Rhodin, J. A. C. (1968). Ultrastructure of mammalian capillaries, venules, and small collecting veins. *J. Ultrastruc. Res.* **25**, 452–500.

Ribbert, H. (1914). *Geschwulstlehre. 2 Auflage.* Bonn: Cohen.

Riccardi, V. M. (1977). Cellular interaction as a limiting factor in the expression of oncogenic mutations: a hypothesis. In: *Genetics of human cancer* ed. J. J. Mulvihill, R. W. Miller & J. F. Fraumeni, Jr., pp. 383–385. New York: Raven Press.

Rice, R. H. & Green, H. (1978). Relation of protein synthesis and transglutaminase activity to formation of the cross-linked envelope during terminal differentiation of cultured human epidermal keratinocytes. *J. Cell Biol.* **76**, 705–11.

Richart, R. M. (1973). Cervical intra-epithelial neoplasia. In *Pathology annual*, ed. S. C. Sommers, vol. 8, pp. 301–28. New York: Appleton-Century-Crofts.

Rifkin, D. B. & Pollack, R. (1977). Production of plasminogen activator by established cell lines of mouse origin. *J. Cell Biol.* **73**, 47–55.

Rinderknecht, E. & Humbel, R. E. (1976a). Polypeptides with non-suppressible insulin-like and cell-growth promoting activities in human serum: Isolation, chemical characterization, and some biological properties of forms I and II. *Proc. Natl. Acad. Sci. USA* **73**, 2365–9.

Rinderknecht, E. & Humbel, R. E. (1976b). Amino-terminal sequences of two polypeptides from human serum with non-suppressible insulin-like and cell-growth-promoting activities: evidence for structural homology with insulin B chain. *Proc. Natl. Acad. Sci. USA* **73**, 4379–81.

Risser, R. & Pollack, R. (1974). A non-selective analysis of SV-40 transformation of mouse 3T3 cells. *Virology* **59**, 477–89.

Rizzino, A. & Sato, G. (1978). Growth of embryonal carcinoma cells in serum-free medium. *Proc. Natl. Acad. Sci. USA* **75**, 1844–8.

Robbins, J. C. & Nicolson, G. L. (1975). Surfaces of normal and transformed cells. In *Cancer: a Comprehensive Treatise*, ed. F. F. Becker, vol. 4, pp. 3–45. New York: Plenum Press.

Romano, A. H. (1976). Is glucose transport enhanced in virus-transformed mammalian cells? A dissenting view. *J. Cell. Physiol.* **89**, 737–44.

Romeo, D., Zabucchi, G., Miani, N. & Rossi, F. (1975). Ion movement across leucocyte plasma membrane and excitation of their metabolism. *Nature, Lond.* **253**, 542–4.

Rose, S. P., Pruss, R. M. & Herschman, H. R. (1975). Initiation of 3T3 fibroblast cell division by epidermal growth factor. *J. Cell. Physiol.* **86**, 593–8.

Rosen, J. J. & Culp, L. A. (1977). Morphology and cellular origins of substate-attached material from mouse fibroblasts. *Exp. Cell Res.* **107**, 139–49.

Rosenberg, N., Schaeffer, W. J. & Diamond, L. (1975). Isolation of variant cells from SV-40-transformed human diploid fibroblasts. *Cancer Res.* **35**, 1970–4.

Rosenblith, J. Z., Ukena, T. E., Yin, H. H., Berlin, R. D. & Karnovsky, M. J. (1973). A comparative evaluation of the distribution of concanavalin A-binding sites on the surfaces of normal, virally-transformed, and protease-treated fibroblasts. *Proc. Natl. Acad. Sci. USA* **70**, 1625–9.

Rosenthal, K. L., Tompkins, A. F., Frank, G. L., McCulloch, P. & Rawls, W. E. (1977). Variants of a human colon adenocarcinoma cell line which differ in morphology and carcinpembryonic antigen production. *Cancer Res.* **37**, 4024–30.

Rosenthal, S. L., Zucker, D. & Davidson, R. L. (1978). Dissociation of plasminogen activator from the transformed phenotype in a 5-bromodeoxyuridine-dependent mutant of Syrian hamster melanoma cells. *J. Cell. Physiol.* **95**, 275–86.

Ross, R., Glomset, J., Kariya, B. & Harker, L. (1974). A platelet-dependent serum factor that stimulates the proliferation of arterial smooth muscle cells *in vitro*. *Proc. Natl. Acad. Sci. USA* **71**, 1207–10.

Ross, R. & Vogel, A. (1978). The platelet-derived growth factor. Review. *Cell* **14**, 203–10.

Rossini, M. & Baserga, R. (1976). Effects of prolonged quiescence on nuclei and chromatin of WI-38 fibroblasts. *J. Cell. Physiol.* **88**, 1–12.

Rossowski, W., Komitowski, D., Darai, G. & Munk, K. (1977). Scanning electron microscopic studies of herpes simplex virus transformed cells. *Oncology* **34**, 1–5.

Roth, S. (1968). Studies on intercellular adhesive selectivity. *Devl. Biol.* **18**, 602–31.

Rothstein, A., Grinstein, S., Ship, S. & Knauf, P. A. (1978). Asymmetry of functional sites of the erythrocyte anion transport protein. *Trends Biochem. Sci.* **3**, 126–8.

Rous, P. & Kidd, J. G. (1939). A comparison of virus-induced rabbit tumors with tumors of unknown cause elicited by tarring. *J. Exp. Med.* **69**, 399–424.

Rous, P. & Kidd, J. G. (1941). Conditional neoplasms and subthreshold neoplastic states. *J. Exp. Med.* **73**, 365–89.

Rovensky, J. A., Slavnaya, I. L. & Vasiliev, J. M. (1971). Behavior of fibroblast-like cells on grooved surfaces. *Exp. Cell Res.* **65**, 193–201.

Rovera, G. & Baserga, R. (1971). Early changes in the synthesis of acidic nuclear proteins in human diploid fibroblasts stimulated to synthesize DNA by changing the medium. *J. Cell. Physiol.* **77**, 201–12.

Rowley, J. D. (1975). Abnormalities of chromosome 1 in myeloproliferative disorders. *Cancer* **36**, 1748–57.

Rowley, J. (1976). The role of cytogenetics in hematology. *Blood* **48**, 1–6.

Rozengurt, E. (1976). Co-ordination of early membrane changes in growth stimulation. *J. Cell. Physiol.* **89**, 627–32.

Rozengurt, E. & Heppel, L. A. (1975). Serum rapidly stimulates ouabain-sensitive $^{86}Rb^+$ influx in quiescent 3T3 cells. *Proc. Natl. Acad. Sci. USA* **72**, 4492–5.

Rozengurt, E. & Po, C. C. (1976). Selective cytotoxicity for transformed 3T3 cells. *Nature, Lond.* **26**, 701–2.

Rozengurt, E., Stein, W. D. & Wigglesworth, N. M. (1977). Uptake of nucleosides in density-inhibited cultures of 3T3 cells. *Nature, Lond.* **267**, 442–4.

Rubin, H. (1971). pH and population density in the regulation of animal cell multiplication. *J. Cell Biol.* **51**, 686–702.

Rubin, H. (1972). Inhibition of DNA synthesis in animal cells by ethylene diamine tetraacetate and its reversal by zinc. *Proc. Natl. Acad. Sci. USA* **69**, 712–16.

Rubin, H. (1973). pH, serum and Zn^{2+} in the regulation of DNA synthesis in cultures of chick embryo cells. *J. Cell. Physiol.* **82**, 231–8.

Rubin, H. (1975*a*). Non-specific nature of the stimulus to DNA synthesis in cultures of chick embryo cells. *Proc. Natl. Acad. Sci. USA* **72**, 1676–80.

Rubin H. (1975*b*). A central role for magnesium in co-ordinate control of metabolism and growth in animal cells. *Proc. Natl. Acad. Sci. USA* **72**, 3551–5.

Rubin, H. (1976). Magnesium deprivation reproduces the co-ordinate effects of serum removal or cortisol addition on transport and metabolism in chick embryo fibroblasts. *J. Cell. Physiol.* **89**, 613–26.

Rubin, H. (1977*a*). Antagonistic effects of insulin and cortisol on co-ordinate control of metabolism and growth in cultured fibroblasts. *J. Cell Physiol.* **91**, 249–60.

Rubin, H. (1977*b*). Specificity of the requirements for magnesium and calcium in the growth and metabolism of chick embryo fibroblasts. *J. Cell. Physiol.* **91**, 449–58.

Rubin, H. & Chu, B. (1978). Reversible regulation by magnesium of chick embryo fibroblast proliferation. *J. Cell. Physiol.* **94**, 13–20.

Rubin, H. & Koide, T. (1975). Early cellular response to diverse growth stimuli independent of protein and RNA synthesis. *J. Cell. Physiol.* **86**, 47–58.

Rubin, H. & Koide, T. (1976). Mutual potentiation by magnesium and calcium of growth in animal cells. *Proc. Natl. Acad. Sci. USA* **73**, 168–72.

Rubin, H. & Sanui, H. (1977). Complexes of inorganic pyrophosphate, orthophosphate, and calcium as stimulants of 3T3 cell multiplication. *Proc. Natl. Acad. Sci. USA* **74**, 5026–30.

Rubin, H. & Steiner, R. (1975). Reversible alteration in the mitotic cycle of chick embryo cells in various states of growth regulation. *J. Cell. Physiol.* **85**, 261–70.

Rubin, K., Kjellen, L. & Öbrink, B. (1977). Intercellular adhesion between juvenile liver cells. *Exp. Cell Res.* **109**, 413–22.

Rubin, R. W., Warren, R. H., Lukeman, D. S. & Clements, E. (1978). Actin content and organization in normal and transformed cells in culture. *J. Cell Biol.* **78**, 28–35.

Rubinstein, P. A. & Spudich, J. A. (1977). Actin heterogeneity in chick embryo fibroblasts. *Proc. Natl. Acad. Sci. USA* **74**, 120–3.

Rudland, P. S., Eckhart, W., Gospodarowicz, D. & Seifert, W. (1974*a*). Cell transformation mutants are not susceptible to growth activation by fibroblast growth factor at permissive temperatures. *Nature, Lond.* **250**, 337–9.

Rudland, P. S., Seifert, W. & Gospodarowicz, D. (1974*b*). Growth control in cultured mouse fibroblasts: induction of the pleiotypic and mitogenic responses by a purified growth factor. *Proc. Natl. Acad. Sci. USA* **71**, 2600–4.

Rutherford, R. B. & Ross, R. (1976). Platelet factors stimulate fibroblasts and smooth muscle cells quiescent in plasma serum to proliferate. *J. Cell Biol.* **69**, 196–203.

Ryan, G. B., Borysenko, J. Z. & Karnovsky, M. J. (1974). Factors affecting the redistribution of surface-bound concanavalin A on human polymorphonuclear leukocytes. *J. Cell Biol.* **62**, 351–65.

Ryan, W. L. & Heidrick, M. L. (1971). Role of cyclic nucleotides in cancer. *Adv. Cyclic Nucl. Res.* **4**, 81–116.

Sachs, L. (1978). Control of normal cell differentiation and the phenotypic reversion of malignancy in myeloid leukemia. *Nature, Lond.* **274**, 535–9.

Sakakibara, K., Takaoka, T., Katsuta, H., Umeda, M. & Tsukada, Y. (1978). Collagen fiber formation as a common property of epithelial liver cell lines in culture. *Exp. Cell Res.* **111**, 63–71.

Sakakibara, K., Umeda, M., Saito, S. & Nagase, S. (1977). Production of collagen and acidic glycosaminoglycans by an epithelial liver cell clones in culture. *Exp. Cell Res.* **110**, 159–65.

Salyamon, L. S. (1974). *Cancer and Cellular Disfunction.* Leningrad: Nauka (in Russian).

Samoilov, V. I., Rovensky, J. A., Slavnaya, I. L. & Slovatchevsky, M. S. (1978). Alterations of 'shape reactions' in cell nuclei due to neoplastic transformation. *Tsitologija* **20**, 321–8 (in Russian).

Samoilov, V. I., Slovatchevsky, M. S., Rovensky, J. A. & Slavnaja, I. L. (1975). Shape and orientation of nuclei in embryo fibroblast-like cells on substrata with regular relief. *Tsitologija* **17**, 433–57 (in Russian).

Sanders, F. K. & Smith, J. D. (1970). Effect of collagen and acid polysaccharides on the growth of BHK/21 cells in semi-solid media. *Nature, Lond.* **227**, 513–15.

Sanford, K. K. (1958). Clonal studies on normal cells and on their neoplastic transformation *in vitro*. *Cancer Res.* **18**, 747–52.

Sanford, K. K. (1965). Malignant transformation of cells *in vitro*. *Int. Rev. Cytol.* **18**, 249–311.

Sanford, K. K., Barker, B. E., Parshad, R., Westfall, B. B., Woods, M. W., Jackson, J. L., King, D. R. & Peppers, E. V. (1970). Neoplastic conversion *in vitro* of mouse cells: cytologic, chromosomal, enzymatic, glycolytic, and growth properties. *J. Natl. Cancer Inst.* **45**, 1071–96.

Sanford, K. K., Handleman, S. L. & Jones, G. M. (1977a). Morphology and serum dependence of cloned cell lines undergoing spontaneous malignant transformation in culture. *Cancer Res.* **37**, 821–30.

Sanford, K. K., Jones, G. M., Tarone, R. E. & Fox, C. H. (1977b). Direction of locomotion in clones of non-neoplastic fibroblasts and their neoplastic derivatives. *Exp. Cell Res.* **109**, 454–9.

Sanford, K. K., Likely, G. D. & Earle, W. R. (1954). The development of variations in transplantability and morphology within a clone of mouse fibroblasts transformed to sarcoma-producing cells *in vitro*. *J. Natl. Cancer Inst.* **15**, 215–38.

Sanger, J. W. & Holtzer, H. (1972). Cytochalasin B. Effects on cell morphology, cell adhesion and mucopolysaccharide synthesis. *Proc. Natl. Acad. Sci. USA* **69**, 253–7.

Sanui, H. & Rubin, H. (1977). Correlated effects of external magnesium on cation content and DNA synthesis in cultured chicken embryo fibroblasts. *J. Cell. Physiol.* **92**, 23–32.

Sanui, H. & Rubin, H. (1978). Membrane bound and cellular cationic changes associated with insulin stimulation of cultured cells. *J. Cell. Physiol.* **96**, 256–78.

Sattler, C. A., Michalopoulos, G., Sattler, G. L. & Pitot, H. C. (1978). Ultrastructure of adult rat hepatocytes cultured on floating collagen membranes. *Cancer Res.* **38**, 1539–49.

Schaeffer, W. J. & Polifka, M. D. (1975). A diploid rat liver in culture. III. Characterization of the heteroploid morphological variants which develop with time in culture. *Exp. Cell Res.* **95**, 167–75.

Schaffhausen, B., Silver, J. & Benjamin, T. (1978). Tumor antigen(s) in cells productively infected by wild-type poloma virus and mutant NG-18. *Proc. Natl. Acad. Sci. USA* **75**, 79–83.

Scher, C. D., Pledger, W. J., Martin, P., Antoniades, H. & Stiles, C. D. (1978). Transforming viruses directly reduce the cellular growth requirements for a platelet derived growth factor. *J. Cell. Physiol.* **97**, 371–80.

Schertz, G. L. & Marsh, J. C. (1977). Application of cell kinetic techniques to human malignancies. In *Cancer: a Comprehensive Treatise*, ed. F. F. Becker, vol. 5, pp. 29–59. New York: Plenum Press.

Schiaffonati, L. & Baserga, R. (1977). Different survival of normal and transformed cells exposed to nutritional conditions non-permissive for growth. *Cancer Res.* **37**, 541–5.

Schlegel, R. & Benjamin, T. L. (1978). Cellular alterations dependent upon the polyoma virus Hr-t function: separation of mitogenic from transforming capacities. *Cell* **14**, 587–99.

Schlessinger, J., Barak, L. S., Hammes, G. G., Yamada, K. M., Pastan, I., Webb, W. W. & Elson, E. L. (1977*a*). Mobility and distribution of a cell surface glycoprotein and its interaction with other membrane components. *Proc. Natl. Acad. Sci. USA* **74**, 2901–13.

Schlessinger, J., Elson, E. L., Webb., W. W., Yahara, J., Rutishauser, U. & Edelman, G. M. (1977*b*). Receptor diffusion on cell surfaces modulated by locally bound concanavalin A. *Proc. Natl. Acad. Sci. USA* **74**, 1110–14.

Schlessinger, J., Shechter, Y., Willingham, M. C. & Pastan, I. (1978). Direct visualization of binding, aggregation, and internalization of insulin and epidermal growth factor on living fibroblastic cells. *Proc. Natl. Acad. Sci. USA* **75**, 2659–63.

Schlessinger, J., Webb, W. W. & Elson, E. L. (1976). Lateral motion and valence of Fc receptors on rat peritoneal mast cells. *Nature, Lond.* **264**, 550–2.

Schloss, J. A. & Goldman, R. D. (1978). A high molecular weight protein from BHK-21 cells binds to and crosslinks actin. *J. Cell Biol.* **79**, 270a.

Schmidt-Ullrich, R., Wallach, D. F. H. & Davis, F. D. G., II. (1976). Membranes of normal hamster lymphocytes and lymphoid cells neoplastically transformed by Simian virus 40. II. Plasma membrane proteins analyzed by dodecyl sulfate-polyacrylamide gel electrophoresis and two-dimensional immune electrophoresis. *J. Natl. Cancer Inst.* **57**, 1117–26.

Schneider, J. A., Diamond, I. & Rozengurt, E. (1978). Glycolysis in quiescent cultures of 3T3 cells. Addition of serum, epidermal growth factor and insulin increases the activity of phosphofructokinase in a protein synthesis-independent manner. *J. Biol. Chem.* **253**, 872–7.

Schollmeyer, J. E., Furcht, L. T., Goll, D. E., Robson, R. M. & Stromer, M. H. (1976). Localization of contractile proteins in smooth muscle cells and in normal and transformed fibroblasts. In *Cell Motility*, ed. R. Goldman, T. Pollard & J. Rosenbaum. Cold Spring Harbor Conferences on Cell Proliferation, vol. 3, pp. 361–88. New York: Cold Spring Harbor Laboratory.

Schreiner, G. F., Fujiwara, K., Pollard, T. D. & Unanue, E. R. (1977). Redistribution of myosin accompanying capping of surface Ig. *J. Exp. Med.* **145**, 1393–8.

Schröder, C. & Hsie, A. (1973). Morphological transformation of anucleated Chinese hamster cells by dibutyryl cyclic AMP and hormones. *Nature New Biol.* **246**, 58–60.

Schubert, D., Humphreys, S., Vitry, F. & Jacob, F. (1971). Induced differentiation of a neuroblastoma. *Dev. Biol.* **52**, 514–46.

Schultz, A. R. & Culp, L. A. (1973). Contact inhibited revertant cell lines isolated from SV-40-transformed cells. V. Contact inhibition of sugar transport. *Exp. Cell Res.* **81**, 95–103.

Schultz, D. R., Wu, M.-C. & Yunis, A. A. (1975). Immunologic relationship among fibrinolysins secreted by cultured mammalian tumor cells. *Exp. Cell Res.* **96**, 45–57.

Sefton, B. M. & Rubin, H. (1970). Release from density-dependent growth inhibition by proteolytic enzymes. *Nature, Lond.* **227**, 843–5.

Sefton, B. M. & Rubin, H. (1971). Stimulation of glucose transport in cultures of density inhibited chick embryo cells. *Proc. Natl. Acad. Sci. USA* **68**, 3154–7.

Segal, D. M., Taurog, J. D. & Metzger, H. (1977). Dimeric immunoglobulin E serves as a unit signal for mast cell degranulation. *Proc. Natl. Acad. Sci. USA* **74**, 2993–7.

Seglen, P. O. & Fossa, J. (1978). Attachment of rat hepatocytes *in vitro* to substrata of serum protein, collagen, or concanavalin A. *Exp. Cell Res.* **116**, 199–206.

Seif, R. & Cuzin, F. (1977). Temperature-sensitive growth regulation in one type of transformed rat cells induced by the *tsa* mutant of polyoma virus. *J. Virol.* **24**, 721 8.

Seifert, W. E. & Rudland, P. S. (1974a). Possible involvement of cyclic GMP in growth control of cultured mouse cells. *Nature, Lond.* **248**, 138–40.

Seifert, W. E. & Rudland, P. S. (1974b). Cyclic nucleotides and growth control in cultured mouse cells: correlation with a specific phase of the cell cycle. *Proc. Natl. Acad. Sci. USA* **71**, 4920–4.

Sekely, L. I., Malejka-Giganti, D., Gutmann, H. R. & Rydell, R. E. (1973). Malignant transformation of rat embryo fibroblasts by carcinogenic fluorenylhydroxamic acids *in vitro. J. Natl. Cancer Inst.* **50**, 1337–45.

Sell, S. & Becker, F. F. (1978). Alpha-fetoprotein. *J. Natl. Cancer Inst.* **60**, 19–26.

Sell, S., Becker, F. F., Leffert, H. L. & Watabe, H. (1976). Expression of an oncodevelopmental gene product (α-fetoprotein) during fetal development and adult oncogenesis. *Cancer Res.* **36**, 4239–49.

Setlow, R. B. (1978). Repair deficient human disorders and cancer. *Nature, Lond.* **271**, 713–17.

Shabad, L. M. (1967). *Pre-cancer in experimental-morphological aspect.* Moscow: Medizina 1967 (in Russian).

Shapot, V. S. (1972). Some biochemical aspects of the relationship between the tumor and the host. *Adv. Cancer Res.* **15**, 253–86.

Shapot, V. S. (1979). On the multiform relationship between the tumor and the host. *Adv. Cancer Res.* **30** (in press).

Sharon, N. & Lis, H. (1972). Lectins: cell-agglutinating and sugar-specific proteins. *Science* **177**, 949–59.

Shein, H. M., Enders, J. F., Palmer, L. & Grogan, E. (1964). Further studies on SV-40-induced transformation in human renal cell cultures: I. Eventual failure of subcultivation despite a continuing high rate of cell division. *Proc. Soc. Exp. Biol. Med.* **115**, 618–21.

Shields, R. (1978). Growth factors for tumours. *Nature, Lond.* **272**, 670–1.

Shields, R. & Smith, J. A. (1977). Cells regulate their proliferation through alterations in transition probability. *J. Cell. Physiol.* **91**, 345–56.

Shin, S., Freedman, V. H., Risser, R. & Pollack, R. (1975). Tumorogenicity of virus-transformed cells in nude mice is correlated specifically with anchorage independent growth *in vitro. Proc. Natl. Acad. Sci. USA* **72**, 4435–9.

Shingleton, H. M., Richart, R. M., Wiener, J. & Spiro, D. (1968). Human cervical intra-epithelial neoplasia; fine structure of dysplasia and carcinoma *in situ. Cancer Res.* **28**, 695–706.

Shingleton, H. M. & Wilbanks, G. D. (1974). Fine structure of human cervical intra-epithelial neoplasia *in vivo* and *in vitro. Cancer* **33**, 981–9.

Shiu, R. P. C., Pouysségur, J. & Pastan, I. (1977). Glucose depletion accounts for the induction of two transformation-sensitive membrane proteins in Rous sarcoma virus-transformed chick embryo fibroblasts. *Proc. Natl. Acad. Sci. USA* **74**, 3840–4.

Shizuta, Y., Shizuta, H., Gallo, M., Davies, P. & Pastan, I. (1976). Purification and properties of filamin, and actin binding protein from chicken gizzard. *J. Biol. Chem.* **251**, 6562–7.

Shodell, M. (1972). Environmental stimuli in the progression of BHK/21 through the cell cycle. *Proc. Natl. Acad. Sci. USA* **69**, 1455–9.

Siddiqi, M. & Jype, P. T. (1975). Studies on the uptake of 2-deoxy-D-glucose in normal and malignant rat epithelial liver cells in culture. *Int. J. Cancer* **15**, 773–80.

Simmons, J. L., Fishman, P. H., Freese, E., & Brady, R. O. (1975). Morphological alterations and ganglioside sialyltransferase activity induced by small fatty acids in HeLa cells. *J. Cell Biol.* **66**, 414–24.

Simpson, D. L., Thorne, D. R. & Loh, H. H. (1977). Developmentally regulated lectin in neonatal rat brain. *Nature, Lond.* **266**, 367–9.

Simpson, D. L., Thorne, D. R. & Loh, H. H. (1978). Lectins: endogenous carbohydrate-binding proteins from vertebrate tissues: functional role in recognition processes? *Life Sci.* **22**, 727–48.

Sims, P. (1976). The metabolism of polycyclic hydrocarbons to dihydrodiols and diol-epoxides by human and animal tissues. In *Screening Tests in Chemical Carcinogenesis*, ed. R. Montesano, H. Bartsch & L. Tomatis, pp. 211–24. Lyon: International Agency for Research on Cancer.

Sinex, F. M. (1977). Theoretical mechanisms of in-vitro senescence. In *Senescence. Dominant or Recessive in Somatic Cell Crosses?* ed. W. W. Nichols, & D. G. Murphy, pp. 1–11. New York: Plenum Press.

Singer, D., Cooper, M., Maniatis, G., Marks, P. A. & Rifkind, R. A. (1974). Erythropoietic differentiation in colonies of cells transformed by Friend virus. *Proc. Natl. Acad. Sci. USA* **71**, 2668–70.

Sivak, A. (1973). Induction of cell division: role of cell membrane sites. *J. Cell Physiol.* **80**, 167–74.

Sivak, A. (1977). Comparison of the biological activity of the tumor promotor phorbol myristate acetate and a metabolite, phorbol myristate acetate, in the cell culture. *Cancer Lett.* **2**, 285–90.

Skerrow, C. J. (1978). Intercellular adhesion and its role in epidermal differentiation. *Invest. Cell Pathol.* **1**, 23–37.

Skrabanek, P. & Powell, D. (1978). Unifying concept of non-pituitary ACTG-secreting tumors. Evidence of common origin of neural-crest tumors, carcinoids, and oat-cell carcinomas. *Cancer* **42**, 1263–9.

Slaga, T. J., Scribner, J. D., Thompson, S. & Viaje, A. (1976). Epidermal cell proliferation and promoting ability of phorbol esters. *J. Natl. Cancer Inst.* **57**, 1145–9.

Slavnaya, I. L. & Rovensky, J. A. (1975). Quantitative estimation of migration ability of different kinds of fibroblast-like cells grown on the substratum with the ordered relief. *Tsitologiya* **17**, 309–13 (in Russian).

Slavnaya, I. L. & Rovensky, J. A. (1977). The migration ability of transformed fibroblast-like cells grown on the substrate with order relief (quantitative estimation). *Tsitologiya* **19**, 1011–17 (in Russian).

Slavnaya, I. L., Rovensky, J. A., Smurova, E. V. & Novikova, S. P. (1974). Normal and malignant fibroblast-like cells cultivated on different polymer substrates. *Tsitologiya* **16**, 1289–300 (in Russian).

Small, J. V. & Celis, J. E. (1978a). Direct visualization of the 10-nm (100-Å) filament network in whole and anucleated cultured cells. *J. Cell Sci.* **31**, 393–409.

Small, J. V. & Celis, J. E. (1978b). Filament arrangements in negatively stained cultured cells: the organization of actin. *Cytobiol. Eur. J. Cell Biol.* **16**, 308–25.

Small, J. V., Isenberg, G. & Celis, J. E. (1978). Polarity of actin at the leading edge of cultured cells. *Nature, Lond.* **272**, 638–9.

Smets, L. A. (1973). Activation of nuclear chromatin and the release from contact-inhibition of 3T3 cells. *Exp. Cell Res.* **79**, 239–43.

Smets, L. A. & De Ley, L. (1974). Cell cycle dependent modulations of the surface membrane of normal and SV-40 transformed 3T3 cells. *J. Cell. Physiol.* **83**, 343–8.

Smets, L. A., van Beek, W. P., van Rooy, H. & Homburg, Ch. (1978). The relationship between membrane glycoprotein alterations and anchorage-independent growth in neoplastic transformation. *Cancer Biochem. Biophys.* **2**, 203–7.

Smith, G. L. & Temin, H. M. (1974). Purified multiplication stimulating activity from rat liver cell conditioned medium: comparison of biological activities with calf serum, insulin and somatomedin. *J. Cell. Physiol.* **84**, 181–92.

Smith, H. W., Owens, R. B., Hiller, A. J., Nelson-Rees, W. A. & Johnston, J. O. (1976). The biology of human cells in tissue culture. I. Characterization of cells derived from osteogenic sarcomas. *Int. J. Cancer* **17**, 219–34.

Smith, J. A. & Martin, L. (1973). Do cells cycle? *Proc. Natl. Acad. Sci. USA* **70**, 1263–7.

Smith, J. B. & Rozengurt, E. (1978). Lithium transport by fibroblastic mouse cells: characterization and stimulation by serum and growth factors in quiescent cultures. *J. Cell. Physiol.* **97**, 441–50.

Smith, J. R. & Hayflick, L. (1974). Variation in the lifespan of clones derived from human diploid cell strains. *J. Cell Biol.* **62**, 48–53.

Smith, R. T. & Landy, M. (eds.) (1975). *Immunology of the Tumor–Host Relationship.* New York: Academic Press.

Sokal, G., Michaux, J.-L. & van den Berghe, H. (1975). Anemie réfractaire et chromosome 5a: un nouveau syndrome. *Bull. Mém. Acad. Roy. Méd. Belg.* **130**, 368–86.

Somers, K. D., Rachmeler, M. & Christensen, M. (1975). Cyclic AMP-mediated transformation of rat cells transformed by temperature-sensitive mouse sarcoma virus. *Nature, Lond.* **257**, 58–9.

Somers, K. D., Weberg, A. D. & Steiner, Sh. (1977). Cyclic AMP-induced morphological transformation of cells infected by temperature-sensitive mouse sarcoma virus. Expression of transformation-associated markers. *J. Cell Biol.* **74**, 707–16.

Spandidos, D. A. & Siminovitch, L. (1977). Transfer of anchorage independence by isolated metaphase chromosomes in hamster cells. *Cell* **12**, 675–82.

Spandidos, D. A. & Siminovitch, L. (1978a). Transfer of the marker for morphologically transformed phenotype by isolated metaphase chromosomes in hamster cells. *Nature, Lond.* **271**, 259–60.

Spandidos, D. A. & Simonivitch, L. (1978b). The relationship between transformation and somatic mutation in human and chinese hamster cells. *Cell* **13**, 651–62.

Spataro, A. C., Morgan, H. R. & Bosmann, H. B. (1976). Neutral protease activity of Rous sarcoma virus (RSV) transformed chick embryo fibroblasts. *J. Cell Sci.* **21**, 407–13.

Spudich, J. A. & Cooke, R. (1975). Supramolecular forms of actin from amoebae of *Dictyostelium discoideum. J. Biol. Chem.* **250**, 7485–91.

Stackpole, C. W., De Milio, L. T., Hämmerling, U., Jacobson, J. B. & Lardis, M. P. (1974). Hybrid antibody-induced topographical redistribution of surface immunoglobulins, alloantigens, and concanavalin A receptors on mouse lymphoid cells. *Proc. Natl. Acad. Sci. USA* **71**, 932–6.

Staehelin, L. A. (1974). Structure and function of intercellular junctions. *Int. Rev. Cytol.* **39**, 191–283.

Stafl, A. & Mattingly, R. F. (1975). Angiogenesis of cervical neoplasia. *Am. J. Obstet, Gynecol.* **121**, 845–52.

Stanbridge, E. J. & Wilkinson, J. (1978). Analysis of malignancy in human cells: malignant and transformed phenotypes are under separate genetic control *Proc. Natl. Acad. Sci. USA* **75**, 1466–9.

Stanners, C. P. (1978). Characterization of temperature-sensitive mutants of animal cells. *J. Cell. Physiol.* **95**, 407–16.

Starger, J. M., Brown, W. E., Goldman, A. E. & Goldman, R. D. (1978). Biochemical and immunological analysis of rapidly purified 10-nm filaments from baby hamster kidney (BHK-21) cells. *J. Cell Biol.* **78**, 93–109.

Starikova, V. B. & Vasiliev, J. M. (1962). Action of 7,12-dimethylbenz(a)anthracene on the mitotic acitivity of normal and malignant rat fibroblasts *in vitro*. *Nature, Lond.* **195**, 42–3.

Starling, J. J., Capetillo, S. C., Neri, G. & Walborg, E. F., Jr. (1977a). Surface properties of normal and neoplastic rat liver cells. *Exp. Cell Res.* **104**, 177–90.

Starling, J. J., Hixson, D. C., Davis, E. M. & Walborg, E. F., Jr. (1977b). Surface properties of adult rat hepatocytes. Mechanism of increased concanavalin A-induced agglutinability following papain digestion. *Exp. Cell Res.* **104**, 165–75.

Steel, G. G. & Lamerton, L. F. (1966). The growth rate of human tumours. *Br. J. Cancer* **20**, 74–86.

Steeves, R. A., Bubbers, J. E., Plata, F. & Lilly, F. (1978). Origin of spleen colonies generated by Friend virus-infected cells in mice. *Cancer Res.* **38**, 2729–33.

Steinberg, B., Pollack, R., Topp, W. & Botchan, M. (1978). Isolation and characterization of T-antigen-negative revertants from a line of transformed rat cells containing one copy of the SV-40 genome. *Cell* **13**, 19–32.

Steinberg, M. S. (1963). Reconstruction of tissues by dissociated cells. *Science* **141**, 401–8.

Steinberg, M. S. (1964). The problem of adhesive selectivity in cellular interactions. In *Cellular Membranes in Development*, ed. M. Locke, pp. 321–66. New York: Academic Press.

Steinberg, M. S. (1973). Cell movement in confluent monolayers: a re-evaluation of the causes of 'contact inhibition'. In *Locomotion of Tissue Cells*, ed. M. Abercrombie. Ciba Foundation Symposium 14, pp. 333–41. Amsterdam: Associated Scientific Publishers.

Steinberg, M. S. & Garrod, D. R. (1975). Observations on the sorting-out of embryonic cells in monolayer culture. *J. Cell Sci.* **18**, 385–403.

Steiner, M. R., Altenberg, B., Richards, C. S., Dudley, J. P., Medina, D. & Butel, J. S. (1978). Differential response of cultured mouse mammary cells of varying tumorigenicity to cytochalasin B. *Cancer Res.* **38**, 2719–21.

Steinman, R. M., Brodie, S. E. & Cohn, Z. (1976). Membrane flow during pinocytosis. A stereologic analysis. *J. Cell Biol.* **68**, 665–87.

Stenman, S., Wartiovaara, J. & Vaheri, A. (1977). Changes in the distribution of a major fibroblast protein fibronectin, during mitosis and interphase. *J. Cell Biol.* **74**, 453–67.

Stephenson, E. M. & Stephenson, N. G. (1978). Invasive locomotory behaviour between malignant human melanoma cells and normal fibroblasts filmed *in vitro*. *J. Cell Sci.* **32**, 389–418.

Stephenson, J. R., Reynolds, R. K. & Aaronson, S. A. (1973). Characterization of morphologic revertants of murine and avian sarcoma virus transformed cells. *J. Virol.* **11**, 218–22.

Steuer, A. F., Hentosh, P. M., Diamond, L. & Ting, R. C. (1977a). Survival differences exhibited by normal and transformed rat liver epithelial cell lines in the aggregate form. *Cancer Res.* **37**, 1864–7.

Steuer, A. F., Rhim, J. S., Hentosh, P. M. & Ting, R. C. (1977b). Survival of human cells in the aggregate form: potential index of in-vitro cell transformation. *J. Natl. Cancer Inst.* **58**, 917–20.

Steuer, A. F. & Ting, R. C. (1976). Formation of larger cell aggregates by transformed cells: an in-vitro index of cell transformation. *J. Natl. Cancer Inst.* **56**, 1279–80.

Stevens, L. C. (1967). The biology of teratomas. *Adv. Morphog.* **6**, 1–32.

Stewart, H. L. (1975). Comparative aspects of certain cancers. In *Cancer: a Comprehensive Treatise*, ed. F. F. Becker, vol. 4, pp. 303–74. New York: Plenum Press.

Stich, H. F., San, R. H. C., Lam, P. P. S., Koropatnick, D. J., Lo, L. W. & Laishes,

B. A. (1976). DNA fragmentation and DNA repair as an in-vitro and in-vivo assay for chemical procarcinogens, carcinogens and carcinogenic nitrosation products. In *Screening Tests in Chemical Carcinogenesis*, ed. R. Montesano, H. Bartsch & L. Tomatis, pp. 617–36. Lyon: International Agency for Research on Cancer.

Stiles, C. D., Desmond, W., Jr., Sato, G. & Saier, M. J., Jr. (1975). Failure of human cells transformed by Simian virus 40 to form tumors in athymic nude mice. *Proc. Natl. Acad. Sci. USA* **72**, 4971–5.

Stiles, C. D., Pledger, W. J. & Scher, C. D. (1978). Control of the Balb/c-3T3 cell cycle by nutrients and serum factors: analysis using platelet-derived growth factor and platelet-poor plasma. *J. Cell Biol.* **79**, 6a.

Stoker, M. (1964). Regulation of growth and orientation in hamster cells transformed by polyoma virus. *Virology* **24**, 165–74.

Stoker, M. (1968). Abortive transformation by polyoma virus. *Nature, Lond.* **218**, 234–38.

Stoker, M. G. P. (1973). Role of diffusion boundary layer in contact inhibition of growth. *Nature, Lond.* **246**, 200–3.

Stoker, M. & Macpherson, I. (1964). Syrian hamster fibroblast cell line BHK-21 and its derivatives. *Nature, Lond.* **203**, 1355–7.

Stoker, M., O'Neill, C., Berryman, S. & Waxman, V. (1968). Anchorage and growth regulation in normal and virus-transformed cells. *Int. J. Cancer* **3**, 683–93.

Stoker, M. & Piggott, D. (1974). Shaking 3T3 cells: further studies on diffusion boundary effects. *Cell* **3**, 207–15.

Stoker, M. G. P., Piggott, D. & Riddle, P. (1978). Movement of human mammary tumour cells in culture: exclusion of fibroblasts by epithelial territories. *Int. J. Cancer* **21**, 268–73.

Stoker, M. G. P., Piggott, D. & Taylor-Papadimitriou, J. (1976). Response to epidermal growth factors in cultured human mammary epithelial cells from benign tumours. *Nature, Lond.* **264**, 764–7.

Stoker, M. G. P., Shearer, M. & O'Neill, C. (1966). Growth inhibition of polyoma-transformed cells by contact. *J. Cell Sci.* **1**, 297–310.

Storrie, B., Puck, T. T. & Wenger, L. (1978). The role of butyrate in the reverse transformation reaction in mammal cells. *J. Cell. Physiol.* **94**, 69–76.

Storti, R. V. & Rich, A. (1976). Chick cytoplasmic actin and muscle actin have different structural genes. *Proc. Natl. Acad. Sci. USA* **73**, 2346–50.

Stossel, T. P. & Hartwig, I. H. (1975). Interactions between actin, myosin and new actin-binding protein of rabbit alveolar macrophages. Macrophage myosin Mg^{2+}-adenosine triphosphatase requires a cofactor for activation by actin. *J. Biol. Chem.* **250**, 5706–12.

Stossel, T. P. & Hartwig, J. H. (1976). Interactions of actin, myosin, and a new actin-binding protein of rabbit pulmonary macrophages. II. Role in cytoplasmic movement and phagocytosis. *J. Cell Biol.* **68**, 602–19.

Sun, T. T. & Green, H. (1977). Cultured epithelial cells of cornea, conjuncti and skin: absence of marked intrinsic divergence of their differentiated states. *Nature, Lond.* **269**, 489–93.

Sundarraj, N. & Church, R. (1978). Alterations of post-translational modifications of procollagen by SV-40-transformed human fibroblasts. *FEBS Lett.* **85**, 47–51.

Sutherland, R. M., MacDonald, H. R. & Howell, R. (1977). Multicellular spheroids: a new model target for in-vitro studies of immunity to solid tumor allografts: brief communication. *J. Natl. Cancer Inst.* **58**, 1849–53.

Svitkina, T. M. (1977). The effects of colcemid on the morphology of transformed fibroblast-like cells in culture. *Tsitologiya* **19**, 671–5 (in Russian).

Swierenga, S. H. H., MacManus, J. P. & Whitfield, J. F. 1976*a*). Regulation by calcium of the proliferation of heart cells from young adult rats. *In Vitro* **12**, 31–6.

Swierenga, S. H. H., Whitfield, J. F. & Gillan, D. H. (1976*b*). Alteration by malignant transformation of the calcium requirements for cell proliferation *in vitro*. *J. Natl. Cancer Inst.* **57**, 125–30.

Tannock, J. F. (1968). The relation between cell proliferation and the vascular system in transplanted mouse mammary tumour. *Br. J. Cancer* **22**, 258–73.

Tao, T.-W., Burger, M. M. (1977). Non-metastasising variants selected from meta-stasising melanoma cells. *Nature, Lond.* **270**, 437–8.

Tarone, G. & Comoglio, P. M. (1977). Plasma membrane proteins exposed on the outer surface of control and Rous sarcoma virus-transformed hamster fibroblasts. *Exp. Cell Res.* **110**, 143–52.

Taylor, A. C. (1961). Attachment and spreading of cells in culture. *Exp. Cell Res.* Suppl. **8**, 154–73.

Taylor, D. L. (1976). Motile model systems of amoeboid movement. In *Cell Motility*, ed. R. Goldman, T. Pollard & J. Rosenbaum. Cold Spring Harbor Conferences on Cell Proliferation, vol. 3, pp. 797–821. New York: Cold Spring Harbor Laboratory.

Taylor-Papadimitriou, J., Shearer, M. & Stoker, M. G. P. (1977). Further studies on growth requirements of human mammary epithelial cells in culture. *Int. J. Cancer* **20**, 903–8.

Taylor-Papadimitriou, J. Shearer, M. & Walting, D. (1978). Growth requirements of calf lens epithelium in culture. *J. Cell. Physiol.* **95**, 95–104.

Temin, H. M. (1960). The control of cellular morphology in embryonic cells infected with Rous sarcoma virus *in vitro*. *Virology* **10**, 182–97.

Temin, H. M. (1961). Mixed infection with two types of Rous sarcoma virus. *Virology* **13**, 158–63.

Temin, H. M. (1966). Studies on carcinogenesis by avian sarcoma viruses. III. The differential effect of serum and polyanions on multiplication of uninfected and converted cells. *J. Natl. Cancer Inst.* **37**, 167–75.

Temin, H. M. (1967). Studies on carcinogenesis by avian sarcoma viruses. VI. Differential multiplication of uninfected and of converted cells in response to insulin. *J. Cell. Physiol.* **69**, 377–84.

Temin, H. M. (1968). Carcinogenesis by avian sarcoma viruses. X. The decreased requirement for insulin-replaceable activity in serum for cell multiplication. *Int. J. Cancer* **3**, 771–87.

Temin, H. M. (1971). The protovirus hypothesis: speculations on the significance of RNA-directed DNA synthesis for normal development and for carcinogenesis. *J. Natl. Cancer Inst.* **46**, N 2, III–VII.

Temin, H. M. (1976). The DNA provirus hypothesis. *Science* **192**, 422–8.

Temin, H. M. (1977*a*). RNA viruses and cancer. *Cancer* **39**, 422–8.

Temin, H. M. (1977*b*). The relationship of tumor virology to an understanding of non-viral cancers. *BioScience* **27**, 170–6.

Temmink, J. H. M. & Spiele, H. (1978). Preservation of cytoskeletal elements for electron microscopy. *Cell Biol. Int. Rep.* **2**, 51–9.

Tenen, D. G., Martin, R. G., Anderson, J. & Livingston, D. M. (1977). Biological and biochemical studies of cells transformed by Simian virus 40 temperature-sensitive gene A mutants and A mutant revertants. *J. Virol.* **22**, 210–18.

Teng, M., Bartholomew, J. C. & Bissel, M. J. (1976). Insulin effect on the cell cycle: analysis of the kinetics of growth parameters in confluent chick cells. *Proc. Natl. Acad. Sci. USA* **73**, 3173–7.

Teng, M., Bartholomew, J. C. & Bissel, M. J. (1977). Synergism between antimicrotubule agents and growth stimulants in enhancement of cell cycle traverse. *Nature, Lond.* **268**, 739–41.

Teng, N. N. H. & Chen, L. B. (1975). The role of surface proteins in cell proliferation as studied with thrombin and other proteases. *Proc. Natl. Acad. Sci. USA* **72**, 413–17.

Terskikh, V. V. & Malenkov, A. G. (1973). Kinetics of ion composition and of RNA and protein synthesis during proliferation induction in the steady-state culture of Chinese hamster cells. *Tsitologiya* **15**, 868–74 (in Russian).

Terzaghi, M. & Little, J. B. (1976). X-radiation-induction of transformation in a C3H mouse embryo-derived cell line. *Cancer Res.* **36**, 1367–74.

Thrash, C. R. & Cunningham, D. D. (1973). Stimulation of division of density-inhibited fibroblasts by glucocorticoids. *Nature, Lond.* **242**, 399–401.

Thrash, C. A. & Cunningham, D. D. (1974). Dissociation of increased hexose transport from initiation of fibroblast proliferation. *Nature, Lond.* **252**, 45–7.

Thyberg, J., Moskalewski, S. & Friberg, U. (1978). Effects of antimicrotubular agents on the fine structure of the Golgi complex in embryonic chick osteoblasts. *Cell Tissue. Res.* **193**, 247–57.

Thyberg, J., Moskalewski, S. & Friberg, U. (1978). Effects of antimicrotubular agents on the fine structure of the Golgi complex in embryonic chick osteoblasts. *Cell Tissue. Res.* **193**, 247–57.

Tickle, C., Summerbell, D. & Wolpert, L. (1975). Positional signalling and specification of digits in chick limb morphogenesis. *Nature, Lond.* **254**, 199–202.

Tilney, L. G. (1976*a*). Non-filamentous aggregates of actin and their association with membranes. In *Cell Motility*, ed. R. Goldman, T. Pollard & J. Rosenbaum. Cold Spring Harbor Conferences on Cell Proliferation, vol. 3, pp. 513–28. New York: Cold Spring Harbor Laboratory.

Tilney, L. G. (1976*b*). The polymerization of actin. III. Aggregates of non-filamentous actin and its associated proteins: a storage form of actin. *J. Cell Biol.* **69**, 73–89.

Tilney, L. G., Hatano, S., Ishikawa, H. & Mooseker, M. (1973). The polymerization of actin: its role in the generation of the acrosomal process of certain echinoderm sperm. *J. Cell Biol.* **59**, 109–26.

Todaro, G. J. (1975). Evolution and modes of transmission of RNA tumor viruses. *Am. J. Pathol.* **81**, 590–606.

Todaro, G. J. (1978). RNA-tumor-virus genes and transforming genes: patterns of transmission. *Br. J. Cancer* **37**, 139–58.

Todaro, G. J. & De Larco, J. E. (1978). Growth factors produced by sarcoma virus-transformed cells. *Cancer Res.* **38**, 4147–54.

Todaro, G. J., De Larco, J. E. & Cohen, S. (1976). Transformation by murine and feline sarcoma viruses specifically blocks binding of epidermal growth factor (EGF) to cells. *Nature, Lond.* **264**, 26–31.

Todaro, G. J., De Larco, J. E., Nissley, S. P. & Rechler, M. M. (1977). MSA and EGF receptors on sarcoma virus transformed cells and human fibrosarcoma cells in culture. *Nature, Lond.* **267**, 526–8.

Todaro, G. J. & Green, H. (1963). Quantitative studies of the growth of mouse embryo cells in culture and their development into established lines. *J. Cell Biol.* **17**, 299–313.

Todaro, G. J., Lazar, G. K. & Green, H. (1965). The initiation of cell division in a contact-inhibited mammalian cell line. *J. Cell. Comp. Physiol.* **66**, 325–33.

Todaro, G. J., Scher, C. D. & Smith, H. S. (1971). SV-40 transformation and cellular growth control. In *Growth Control in Cell Cultures*, ed. G. E. W. Wolstenholme & J. Knight, pp. 151–62. Edinburgh: Churchill Livingstone.

Toh, B. H. & Hard, G. C. (1977). Actin co-caps with concanavalin A receptors. *Nature, Lond.* **269**, 695–7.

Tomei, L. D. & Bertram, J. S. (1978). Restoration of growth control in malignantly transformed mouse fibroblasts grown in a chemically defined medium. *Cancer Res.* **38**, 444–51.

Tomida, M., Koyama, H. & Ono, T. (1975). Induction of hyaluronic acid synthetase activity in rat fibroblasts by medium change of confluent cultures. *J. Cell. Physiol.* **86**, 121–30.

Topp, W., Hall, J. D., Marsden, M., Teresky, A. K. Rifkin, D., Levine, A. J. & Pollack, R. (1976). In-vitro differentiation of teratomas and the distribution of creatine phosphokinase and plasminogen activator in teratocarcinoma-derived cells. *Cancer Res.* **36**, 4217–23.

Toustanovsky, A. A. & Vasiliev, J. M. (1957). Mammary gland stroma changes in mouse during pregnancy, lactation and involution. *Voprosi Onkologii* **3**, N 2, 139–45 (in Russian).

Trinkaus, J. P. (1976). On the mechanism of metazoan cell movements. In *The Cell Surface in Animal Embryogenesis and Development*, ed. G. Poste & G. L. Nicolson, pp. 225–329. Amsterdam: North-Holland.

Trinkaus, J. P., Betshaku, T. & Krulikowski, L. S. (1971). Local inhibition of ruffling during contact inhibition of cell movement. *Exp. Cell Res.* **64**, 291–300.

Tsan, M. F. & Berlin, R. D. (1971). Effect of phagocytosis on membrane transport of non-electrolytes. *J. Exp. Med.* **134**, 1016–35.

Tubiana, M. (1971). The kinetics of tumour cell proliferation and radiotherapy. *Br. J. Radiol.* **44**, 325–47.

Tucker, R. W., Sanford, K. K. & Frankel, F. R. (1978). Tubulin and actin in paired non-neoplastic and spontaneously transformed neoplastic cell lines *in vitro:* fluorescent antibody studies. *Cell* **13**, 629–42.

Tucker, R. W., Sanford, K. K., Handleman, S. L. & Jones, G. M. (1977). Colony morphology and growth in agarose as tests for spontaneous neoplastic transformation *in vitro*. *Cancer Res.* **37**, 1571–9.

Tupper, J. T., Del Rosso, M., Hazelton, B. & Zorgniotti, F. (1978). Serum-stimulated changes in calcium transport and distribution in mouse 3T3 cells and their modification by dibutyryl cyclic AMP. *J. Cell. Physiol.* **95**, 71–84.

Tupper, J. T., Zorgniotti, F. & Mills, B. (1977). Potassium transport and content during G_1 and S phase following serum stimulation of 3T3 cells. *J. Cell. Physiol.* **91**, 429–40.

Ukena, T. E. & Berlin, R. D. (1972). Effect of colchicine and vinblastine on the topographical separation of membrane functions. *J. Exp. Med.* **131**, 1–7.

Ukena, T. E., Borysenko, J. Z., Karnovsky, M. J. & Berlin, R. D. (1974). Effects of colchicine, cytochalasin B, and 2-deoxyglucose on the topographical organization of surface-bound concanavalin A in normal and transformed fibroblasts. *J. Cell Biol.* **61**, 70–82.

Ukena, T. E. & Karnovsky, M. J. (1977). The role of microvilli in the agglutination of cells by concanavalin A. *Exp. Cell Res.* **106**, 309–25.

Unanue, E. R., Perkins, W. D. & Karnovsky, M. J. (1972). Ligand-induced movements of lymphocyte membrane macromolecules. I. Analysis by immuno-fluorescence and ultrastructural radioautography. *J. Exp. Med.* **136**, 885–906.

Underhill, C. B. & Keller, J. M. (1976). Heparan sulfates of mouse cells. Analysis of parent and transformed 3T3 cell lines. *J. Cell. Physiol.* **90**, 53–60.

Unkeless, J., Danø, K., Kellerman, G. M. & Reich, E. (1974). Fibronolysis associated with oncogenic transformation. Partial purification and characterization of the cell factor, a plasminogen activator. *J. Biol. Chem.* **249**, 4295–305.

Unkeless, J. C., Tobia, A., Ossowski, L., Quigley, J. P., Rifkin, D. B. & Reich, E. (1973). An enzymatic function associated with transformation of fibroblasts by oncogenic viruses. I. Chick embryo fibroblast cultures transformed by avian RNA tumor viruses. *J. Exp. Med.* **137**, 85–111.

Uriel, J. (1976*a*). Cancer, retrodifferentiation, and the myth of Faust. *Cancer Res.* **36**, 4269–75.

Uriel, J. (1976*b*). Fetal characteristics of cancer. In *Cancer: a Comprehensive Treatise*, ed. F. F. Becker, vol. 3, pp. 21–55. New York: Plenum Press.

Vaheri, A. & Mosher, D. F. (1978). High molecular weight cell surface-associated glycoprotein (fibronectin) lost in malignant transformation. *Biochim. Biophys. Acta* **516**, 1–25.

Vaheri, A. & Ruoslahti, E. (1975). Fibroblast surface antigen produced but not retained by virus-transformed human cells. *J. Exp. Med.* **142**, 530–5.

Vaheri, A., Ruoslahti, E. & Hovi, T. (1974). Cell surface and growth control of chick embryo fibroblasts in culture. In *Control of Proliferation in Animal Cells*, ed. B. Clarkson & R. Baserga, pp. 305–12. New York: Cold Spring Harbor Laboratory.

Vaheri, A., Ruoslahti, E., Hovi, T. & Nordling, S. (1973). Stimulation of density-inhibited cell cultures by insulin. *J. Cell. Physiol.* **81**, 355–64.

Vaheri, A., Ruoslahti, E. & Nordling, S. (1972). Neuraminidase stimulates division and sugar uptake in density-inhibited cell structures. *Nature New Biol.* **238**, 211–12.

Vaheri, A., Ruoslahti, E., Westermark, B. & Pontén, J. (1976). A common cell-type specific surface antigen in cultured human glial cells and fibroblasts: loss in malignant cells. *J. Exp. Med.* **143**, 64–72.

Vaitkevicius, V. K., Sugimoto, M. & Brennan, M. J. (1962). The effect of inflammation on tumor establishment and growth. In *Biological Interactions in Normal and Neoplastic Growth*, ed. M. J. Brennan & W. L. Simpson. Henry Ford Hospital International Symposium, pp. 767–76. Boston: Little, Brown & Co.

Vaitukaitis, J. L. (1976). Peptide hormones as tumor markers. *Cancer* **37**, 567–72.

van Beek, W. P., Emmelot, P. & Homburg, C. (1977). Comparison of cell-surface glycoproteins of rat hepatomas and embryonic rat liver. *Br. J. Cancer.* **36**, 157–65.

van den Berg, K. J. & Betel, J. (1973). Increased transport of 2-aminoisobutyric acid in rat lymphocytes stimulated with concanavalin A. *Exp. Cell Res.* **76**, 63–72.

van Diggelen, O. P., Shin, S. & Phillips, D. M. (1977). Reduction in cellular tumorigenecity after mycoplasma infection and elimination of mycoplasma from infected cultures by passage in nude mice. *Cancer Res.* **37**, 2680–7.

Vannucchi, S. & Chiarugi, V. P. (1976). Surface exposure of glycosaminoglycans in resting, growing and virus transformed 3T3 cells. *J. Cell. Physiol.* **90**, 501–10.

van Zaane, D. & Bloemers, H. P. J. (1978). The genome of the mammalian sarcoma viruses. *Biochim. Biophys. Acta* **516**, 249–68.

Vasiliev, J. M. (1958). The role of connective tissue proliferation in invasive growth of normal and malignant tissues: a review. *Brit. J. Cancer* **12**, 524–36.

Vasiliev, J. M. (1961). *Connective tissue and experimental tumour growth.* Moscow: Medizina (in Russian).

Vasiliev, J. M. (1962). The local stimulatory effect of normal tissues upon the growth of tumor cells. In *Biological Interactions in Normal and Neoplastic Growth*, ed. M. J. Brennan & W. L. Simpson. Henry Ford Hospital International Symposium, pp. 229–309. Boston: Little, Brown & Co.

Vasiliev, J. M. (1976). Morphogenetic reactions of transformed and carcinogen-treated cells. In *Screening Test in Chemical Carcinogenesis*, ed. Montesano, H. Bartsch & L. Tomatis, pp. 449–62. Lyon: International Agency for Research on Cancer.

Vasiliev, J. M. & Gelfand, I. M. (1968). Surface changes disturbing intracellular homeostasis as a factor inducing cell growth and division. *Curr. Mod. Biol.* **2**, 43–55.

Vasiliev, J. M. & Gelfand, I. M. (1973). Interactions of normal and neoplastic fibroblasts with the substratum. In *Locomotion of Tissue Cells*, ed. M. Abercrombie.

Ciba Foundation Symposium 14, pp. 311–29. Amsterdam: Associated Scientific Publishers.

Vasiliev, J. M. & Gelfand, I. M. (1976). Effect of colcemid on morphogenetic processes and locomotion of fibroblasts. In *Cell Motility*, ed. R. Goldman, T. Pollard & J. Rosenbaum. Cold Spring Harbor Conferences on Cell Proliferation, vol. 3, pp. 279–304. New York: Cold Spring Harbor Laboratory.

Vasiliev, J. M. & Gelfand, I. M. (1977). Mechanisms of morphogenesis in cell cultures. *Int. Rev. Cytol.* **50**, 159–274.

Vasiliev, J. M., Gelfand, I. M., Domnina, L. V., Dorfman, N. A. & Pletyushkina, O. Y. (1976). Active cell edge and movements of concanavalin A receptors of the surface of epithelial and fibroblastic cells. *Proc. Natl. Acad. Sci. USA* **73**, 4085–9.

Vasiliev, J. M., Gelfand, I. M., Domnina, L. V., Ivanova, O. Y., Komm, S. G. & Olshevskaja, L. V. (1970). Effect of Colcemid on the locomotory behavior of fibroblasts. *J. Embryol. Exp. Morphol.* **24**, 625–40.

Vasiliev, J. M., Gelfand, I. M., Domnina, L. V. & Rappoport, R. I. (1969). Wound healing processes in cell cultures. *Exp. Cell Res.* **43**, 83–93.

Vasiliev, J. M., Gelfand, I. M., Domnina, L. V., Zakharova, O. S. & Lyubimov, A. V. (1975a). Contact inhibition of phagocytosis in epithelial sheets: alterations of cell surface properties induced by cell–cell contacts. *Proc. Natl. Acad. Sci. USA* **72**, 719–22.

Vasiliev, J. M., Gelfand, I. M. & Erofeeva, L. V. (1966a). Behaviour of fibroblasts in cell culture after removal of a part of the monolayer. *Dokl. Acad. Nauk SSSR* **171**, 721–4 (in Russian).

Vasiliev, J. M., Gelfand, I. M. & Guelstein, V. I. (1971). Initiation of DNA synthesis in cell cultures by colcemid. *Proc. Natl. Acad. Sci. USA* **68**, 977–9.

Vasiliev, J. M., Gelfand, I. M., Guelstein, V. I. & Fetisova, E. K. (1970). Stimulation of DNA synthesis in culture of mouse embryo fibroblast-like cells. *J. Cell. Physiol.* **75**, 305–14.

Vasiliev, J. M., Gelfand, I. M., Guelstein, V. I. & Malenkov, A. G. (1966b). Inter-relationships of contacting cells in the cell complexes of mouse ascites hepatoma. *Int. J. Cancer* **1**, 451–62.

Vasiliev, J. M., Gelfand, I. M., Pletyushkina, O. Y. & Fetisova, E. K. (1975b). Insensitivity of stationary cultures of the mouse transformed fibroblasts to action of agents known to stimulate DNA synthesis in cultures of normal cells. *Tsitologiya* **17**, 442–5 (in Russian).

Vasiliev, J. M., Gelfand, I. M. & Tint, I. S. (1975c). Processes causing cell shape alteration after cell detachment from the substratum. *Tsitologiya* **17**, 633–8 (in Russian).

Vasiliev, J. M. & Guelstein, V. I. (1963). Sensitivity of normal and neoplastic cells to the damaging action of carcinogenic substances: a review. *J. Natl. Cancer Inst.* **31**, 1123–51.

Vasiliev, J. M. & Guelstein, V. I. (1966). Local cell interactions in neoplasms and in the foci of carcinogenesis. In *Progress in Experimental Tumor Research*, ed. F. Homburger, vol. 8, pp. 26–65. Basel: S. Karger.

Vasiliev, J. M., Olshevskaja, L. V., Raikhlin, N. T. & Ivanova, O. J. (1962). Comparative study of alterations induced by 7,12-dimethylbenz(a)anthracene and polymer films in the subcutaneous connective tissue of rats. *J. Natl. Cancer Inst.* **28**, 515–59.

Venter, B. R., Venter, J. C. & Kaplan, N. O. (1976). Affinity isolation of cultured tumor cells by means of drugs and hormones covalently bound to glass and Sepharose beads. *Proc. Natl. Acad. Sci. USA* **73**, 2013–17.

Verkleij, A. J. & Ververgaert, P. H. J. Th. (1978). Freeze-fracture morphology of biological membranes. *Biochim. Biophys. Acta* **515**, 303–27.

Vesely, P. (1972). Tumour cell surface specialization in the uptake of nutrients evidenced by cinemicrography as a phenotypic condition for density independent growth. *Folia biologica, Praha* **18**, 395–401.

Vesely, P. & Boyde, A. (1973). The significance of SEM evaluation of the cell surface for tumor cell biology. In *Scanning electron microscopy, 1973*, ed. O. Johari & I. Corvin, pp. 189–196. Chicago: IIT Research Institute.

Vesely, P., Křen, V., Wyke, J. Plaisner, V. & Sladka, M. (1978). Description of cell populations isolated *in vitro* from LEW/CUB rat embryos. *Folia biologica, Praha* **24**, 391.

Vesely, P. & Weiss, R. A. (1973). Cell locomotion and contact inhibition of normal and neoplastic rat cells. *Int. J. Cancer* **11**, 64–76.

Vicker, M. G. (1977). On the origin of the phagocytic membranes. *Exp. Cell Res.* **109**, 127–38.

Virtanen, I., Miettinen, A. & Wartiovaara, J. (1978). Lectin-binding sites are found in rat liver cell plasma membrane only on its extracellular surface. *J. Cell Sci.* **29**, 287–96.

Vlodavsky, I., Brown, K. D. & Gospodarowicz, D. (1978a). A comparison of the binding of epidermal growth factor to cultured granulosa and luteal cells. *J. Biol. Chem.* **253**, 3744–50.

Vlodavsky, I., Fielding, P. E., Fielding, C. J. & Gospodarowicz, D. (1978b). Role of contact inhibition in the regulation of receptor-mediated uptake of low density lipoprotein in cultured vascular endothelial cells. *Proc. Natl. Acad. Sci. USA* **75**, 356–60.

Vogel, A. & Pollack, R. (1973). Isolation and characterization of revertant cell lines. IV. Direct selection of serum revertant sublines of SV-40-transformed 3T3 mouse cells. *J. Cell Physiol.* **82**, 189–98.

Vogel, A. & Pollack, R. (1975). Isolation and characterization of revertant cell lines. VII. DNA synthesis and mitotic rate of serum-sensitive growth conditions. *J. Cell. Physiol* **85**, 151–62.

Vogel, A., Raines, E., Kariya, B., Rivest, M. J. & Ross, R. (1978). Co-ordinate control of 3T3 cell proliferation by platelet-derived growth factor and plasma components. *Proc. Natl. Acad. Sci. USA* **75**, 2810–14.

Vogel, A., Risser, R. & Pollack, R. (1973). Isolation and characterization of revertant cell lines. III. Isolation of density-revertants of SV-40-transformed 3T3 cells using colchicine. *J. Cell. Physiol.* **82**, 181–88.

Vogel, K. G. (1978). Effects of hyaluronidase, trypsin, and EDTA on surface composition and topography during detachment of cells in culture. *Exp. Cell Res.* **113**, 345–57.

Vogt, M. & Dulbecco, R. (1963). Steps in the neoplastic transformation of hamster embryo cells by polyoma virus. *Proc. Natl. Acad. Sci. USA* **49**, 171–9.

Volckaert, G., van der Voorde, A. & Fiers, W. (1978). Nucleotide sequence of the Simian virus-40 small-t gene. *Proc. Natl. Acad. Sci. USA* **75**, 2160–4.

Vollet, J. J., Brugge, J. S., Noonan, C. A. & Butel, J. S. (1977). The role of SV-40 gene A in the alteration of microfilaments in transformed cells. *Exp. Cell Res.* **105**, 119–26.

Walker, P. R., Boynton, A. L. & Whitfield, J. F. (1977). The inhibition by colchicine of the initiation of DNA synthesis by hepatocytes in regenerating rat liver and by cultivated WI-38 and C3HIOT 1/2 cells. *J. Cell. Physiol.* **93**, 89–98.

Wall, R. T., Harker, L. A., Quadrucci, L. J. & Striker, G. E. (1978). Factors influencing endothelial cell proliferation *in vitro*. *J. Cell Physiol.* **96**, 203–14.

Walsh, F. S. & Crumpton, M. J. (1977). Orientation of cell-surface antigens in the lipid bilayer of lymphocyte plasma membrane. *Nature, Lond.* **269**, 307–11.

Wands, J. R., Podolsky, D. K. & Isselbacher, K. J. (1976). Mechanism of human lymphocyte stimulation by concanavalin A: role of valence and surface binding sites. *Proc. Natl. Acad. Sci. USA* **73**, 2118–22.

Wang, E. & Goldberg, A. R. (1976). Changes in microfilament organization and surface topography upon transformation of chick embryo fibroblasts with Rous sarcoma virus. *Proc. Natl. Acad. Sci. USA* **73**, 4065–9.

Wang, K., Ash, J. F. & Singer, S. J. (1975). Filamin, a new high-molecular-weight protein found in smooth muscle and non-muscle cells. *Proc. Natl. Acad. Sci. USA* **72**, 4483–6.

Wang, T., Sheppard, J. R. & Foker, J. E. (1978). Rise and fall of cyclic AMP required for onset of lymphocyte DNA synthesis. *Science* **210**, 155–7.

Wang-Peng, J. (1977). Banding in Leukemia: techniques and implications. *J. Natl. Cancer Inst.* **58**, 3–8.

Wanson, J.-C., Drochmans, P., Mosselmans, R. & Ronveaux, M. F. (1977). Adult rat hepatocytes in primary monolayer culture. Ultrastructural characteristics of intercellular contacts and cell membrane differentiations. *J. Cell Biol.* **74**, 858–77.

Warren, L., Buck, C. A. & Tuszynki, G. P. (1978). Glycopeptide changes and malignant transformation. A possible role of carbohydrate in malignant behaviour. *Biochim. Biophys. Acta* **516**, 97–127.

Webb, T., Harnden, D. G. & Harding, M. (1977). The chromosome analysis and susceptibility to transformation by Simian virus 40 of fibroblasts from ataxia-telangiectasia. *Cancer Res.* **37**, 997–1002.

Weber, K., Rathke, P. C. & Osborn, M. (1978). Cytoplasmic microtubular images in glutaraldehyde-fixed tissue culture cells by electron microscopy and by immuno-fluorescence microscopy. *Proc. Natl. Acad. Sci. USA* **75**, 1820–4.

Weber, K., Rathke, P. C., Osborn, M. & Franke, W. W. (1976). Distribution of actin and tubulin in cells and in glycerinated cell models after treatment with cytochalasin B. *Exp. Cell Res.* **102**, 285–97.

Weber, M. J. (1973). Hexose transport in normal and in Rous sarcoma virus-transformed cells. *J. Biol. Chem.* **248**, 2978–83.

Weber, M. J. (1975). Inhibition of protease activity in cultures of Rous sarcoma virus-transformed cells: effect on the transformed phenotype. *Cell* **5**, 253–61.

Weber, M. J. & Edlin, G. (1971). Phosphate transport, nucleotide pools, and ribonucleic acid synthesis in growing and in density-inhibited 3T3 cells. *J. Biol. Chem.* **246**, 1828–33.

Weinstein, I. B., Jeffrey, A. M., Jennette, K. W., Blobstein, S. H., Harvey, R. G., Harris, C., Autrup, H., Kasai, H. & Nakanishi, K. (1976). Benzo(a)pyrene diol epoxides as intermediates in nucleic acid binding *in vitro* and *in vivo*. *Science* **193**, 592–5.

Weinstein, I. B., Orenstein, J. M., Gebert, R., Kaighn, M. E. & Stadler, U. C. (1975). Growth and structural properties of epithelial cultures established from normal rat liver and chemically induced hepatomas. *Cancer Res.* **35**, 253–63.

Weisburger, J. H. & Williams, G. M. (1975). Metabolism of chemical carcinogens. In *Cancer: a Comprehensive Treasise*, ed. F. F. Becker, vol. 1, pp. 185–233. New York: Plenum Press.

Weiss, L. (1967). *The Cell Periphery, Metastasis and Other Contact Phenomena*. Amsterdam: North-Holland.

Weiss, L. (1976). Biophysical aspects of the metastatic cascade. In *Fundamental Aspects of Metastasis*, ed. L. Weiss, pp. 51–70. Amsterdam: North-Holland.

Weiss, L. (1977*a*). A pathobiologic overview of metastasis. *Semin. Oncol.* **4**, 5–17.

Weiss, L. (1977*b*). Tumor necrosis and cell detachment. *Int. J. Cancer* **20**, 87–92.

Weiss, L., Poste, G., MacKearnin, A. & Willett, K. (1975). Growth of mammalian cells on substrates coated with cellular microexudate. I. Effect on cell growth at low population densities. *J. Cell Biol.* **64**, 135–45.

Weiss, P. (1933). Functional adaptation and the role of ground substances in development. *Am. Nat.* **67**, 322–40.

Weiss, P. (1958). Cell contact. *Int. Rev. Cytol.* **7**, 391–423.

Weiss, P. (1961). Guiding principles in cell locomotion and cell aggregation. *Exp. Cell Res.* Suppl. **8**, 260–81.

Weiss, P & Garber, B. (1952). Shape and movement of mesenchyme cells as a function of the physical structure of the medium: contributions to a quantitative morphology. *Proc. Natl. Acad. Sci. USA* **38**, 264–80.

Weiss, R. A. (1970). The influence of normal cells on the proliferation of tumour cells in culture. *Exp. Cell Res.* **63**, 1–18.

Weiss, R. A. & Njeuma, D. L. (1971). Growth control between dissimilar cells in culture. In *Growth in Cell Cultures*, ed. G. E. W. Wolstenholme & J. Knight, pp. 169–184. Edinburgh: Churchill Livingstone.

Weiss, R. A., Vesely, P. & Sindelarova, J. (1973). Growth regulation and tumor formation of normal and neoplastic rat cells. *Int. J. Cancer* **11**, 77–89.

Weiss, R. L., Goodenough, D. A. & Goodenough, U. W. (1977). Membrane particle arrays associated with basal body and with contractile vacuole secretion in *Chlamydomonas. J. Cell Biol.* **72**, 133–43.

Wessels, N. K., Spooner, B. S., Ash, J. F., Bradley, M. O., Luduena, M. A., Taylor, E. L., Wrenn, J. T. & Yamada, K. M. (1971). Microfilaments in cellular and developmental processes. *Science* **171**, 135–43.

Wessels, N. K., Spooner, B. S. & Luduena, M. A. (1973). Surface movements, microfilaments and cell locomotion. In *Locomotion of Tissue Cells*, ed. M. Abercrombie. Ciba Foundation Symposium 14, pp. 53–77. Amsterdam: Associated Scientific Publishers.

Westermark, B. (1973*a*). Growth regulatory interactions between stationary human glia-like cells and normal and neoplastic cells in culture. I. Normal cells. *Exp. Cell Res.* **81**, 195–206.

Westermark, B. (1973*b*). Induction of a reversible G1 block in human glia-like cells by cytochalasin B. *Exp. Cell Res.* **82**, 341–50.

Westermark, B. (1976). Density dependent proliferation of human glial cells stimulated by epidermal growth factor. *Biochem. Biophys. Res. Commun.* **69**, 304–10.

Westermark, B. & Wasteson, A. (1976). A platelet factor stimulating human normal glial cells. *Exp. Cell Res.* **98**, 170–4.

Weston, J. A. & Hendricks, K. L. (1972). Reversible transformation by urea of contact-inhibited fibroblasts. *Proc. Natl. Acad. Sci. USA* **68**, 3727–31.

Wetzel, B., Jones, G. M. & Sanford, K. K. (1977*a*). Cell cycle and topography in non-synchronized monolayers: the use of autoradiography and time-lapse for more rigorous SEM studies. In *Scanning Electron Microscopy 1977*, Pt 1, ed. O. Johari, pp. 545–52. Chicago: IIT Research Institute.

Wetzel, B., Sanford, K. K., Fox, C. H., Jones, G. M., Westbrook, E. W. & Tarone, R. E. (1977*b*). Topography of non-neoplastic and neoplastic cells of common origin. *Cancer Res.* **37**, 831–42.

Whalen, R. G., Butler-Browne, G. S. & Gros, F. (1976). Protein synthesis and actin heterogeneity in calf muscle cells in culture. *Proc. Natl. Acad. Sci. USA* **73**, 2018–22.

Whaley, W. G., Dauwalder, M. & Kephart, J. E. (1972). Golgi apparatus: influence

344   *References*

on cell surfaces. A role in the assembly of macromolecules makes the organelle a determinant of cell function. *Science* **175**, 596–9.

Whitfield, J. F., MacManus, J. P. & Gillan, D. J. (1973). The ability of calcium to change cyclic AMP from a stimulator to an inhibitor of thymic lymphoblast proliferation. *J. Cell. Physiol.* **81**, 241–50.

Whitfield, J. F., MacManus, J. P., Rixon, R. H., Boynton, A. L., Youdate, T. & Swierenga, S. (1976). The positive control of cell proliferation by the interplay of calcium ions and cyclic nucleotides: a review. *In vitro* **12**, 1–16.

Whittenberger, B. & Glaser, L. (1977). Inhibition of DNA synthesis in cultures of 3T3 cells by isolated surface membranes. *Proc. Natl. Acad. Sci. USA* **74**, 2251–5.

Wickus, G., Gruenstein, E., Robbins, P. W. & Rich, A. (1975). Decrease in membrane-associated actin of fibroblasts after transformation by Rous sarcoma virus. *Proc. Natl. Acad. Sci. USA* **72**, 746–9.

Wigler, M. & Weinstein, I. B. (1976). Tumour promotor induces plasminogen activator. *Nature, Lond.* **259**, 232–3.

Wigley, C. B. (1975). Differentiated cells *in vitro*. *Differentiation* **4**, 25–55.

Wilbanks, G. D. (1976). In-vivo and In-vitro 'markers' of human cervical intra-epithelial neoplasia. *Cancer Res.* **36**, 2485–94.

Williams, G. M. (1976). The use of liver epithelial cultures for study on chemical carcinogenesis. *Am. J. Pathol.* **85**, 739–51.

Williams, G. M. (1977). The deficiency of cylic AMP responsive G_1 controls in cultured malignant liver epithelial cells. *Cancer Lett.* **2**, 239–46.

Williams, J. A. & Lee, M (1976). Microtubules and pancreatic amylase release by mouse pancreas *in vitro*. *J. Cell Biol.* **71**, 795–806.

Willingham, M. C. (1976). Cyclic AMP and cell behavior in cultured cells. *Int. Rev. Cytol.* **44**, 319–63.

Willingham, M. C. & Pastan, I. (1975a). Cyclic AMP modulates microvillus formation and agglutinability in transformed and normal mouse fibroblasts. *Proc. Natl. Acad. Sci. USA* **72**, 1263–7.

Willingham, M. C. & Pastan, I. (1975b). Cyclic AMP and cell morphology in cultured fibroblasts: effect on cell shape, microfilament and microtubule distribution, and orientation to substratum. *J. Cell Biol.* **67**, 146–59.

Willmer, E. N. (1965). Morphological problems of cell type, shape and identification. In *Cell and Tissues in Culture. Methods, Biology and Physiology*, ed. E. N. Willmer, vol. 1, pp. 143–76. London: Academic Press.

Wilson, M. J. & Poirier, L. A. (1978). An increased requirement for methionine by transformed rat liver epithelial cells *in vitro*. *Exp. Cell Res.* **111**, 397–400.

Winkelhake, J. L. & Nicolson, G. L. (1976). Determination of adhesive properties of variant metastatic melanoma cells to BALB/3T3 cells and their virus-transformed derivatives by a monolayer attachment assay. *J. Natl. Cancer Inst.* **56**, 285–91.

Winterbourne, D. J. & Mora, P. T. (1977). Distribution of glycoconjugates in mouse fibroblasts with varying degrees of tumorigenicity. *J. Supramol. Struc.* **7**, 91–100.

Witkowski, J. A. & Brighton, W. D. (1971). Stages of spreading of human diploid cells on glass surface. *Exp. Cell Res.* **68**, 372–80.

Wohlfarth-Bottermann, K. E. & Isenberg, G. (1976). Dynamics and molecular basis of the contractile system of *Physarum*. In *Contractile Systems in Non-muscle Tissues*, ed. S. V. Perry, A. Margreth & R. S. Adelstein, pp. 297–308: Amsterdam: Elsevier/North-Holland Biomedical Press.

Wolf, B. A. & Goldberg, A. R. (1976). Rous sarcoma virus-transformed fibroblasts having low levels of plasminogen activator. *Proc. Natl. Acad. Sci. USA* **73**, 3613–17.

Wolf, D. E., Schlessinger, J., Elson, E. L., Webb, W. W., Blumenthal, R. & Henkart,

P. (1977). Diffusion and patching of macromolecules on planar lipid bilayer membranes. *Biochemistry* **16**, 3476–83.

Wolman, S. R. & Horland, A. R. (1975). Genetics of tumor cells. In *Cancer: a Comprehensive Treatise*, ed. F. F. Becker, vol. 3, pp. 155–98. New York: Plenum Press.

Wolosewick, J. J. & Porter, K. B. (1976). Stereo high-voltage electron microscopy of whole cells of the human diploid line, WI-38. *Am. J. Anat.* **147**, 303–34.

Wolpert, L. (1976). Mechanisms of limb development and malformation. *Br. Med. Bull.* **32**, 65–70.

Wood, S., Jr. (1971). Mechanisms of establishment of tumor metastases. In *Pathobiology Annual*, ed. H. L. Ioachim, pp. 281–307. New York: Appleton-Century-Crofts.

Wood, S., Jr. & Sträuli, P. (1973). Tumor invasion and metastasis. In *Cancer Medicine*, ed. E. Frei & J. Holland, pp. 140–51. Philadelphia: Lea & Febiger.

Wright, E. D., Goldfarb, P. S. G. & Subak-Sharpe, J. H. (1976). Isolation of variant cells with defective metabolic co-operation (mec⁻) from polyoma virus transformed Syrian hamster cells. *Exp. Cell Res.* **103**, 63–77.

Wright, E. D., Slack, C., Goldfarb, P. S. G. & Subak-Sharpe, J. H. (1976). Investigation of the basis of reduced metabolic co-operation in mec⁻ cells. *Exp. Cell Res.* **103**, 79–91.

Wright, J. A. (1973). Morphology and growth rate changes in Chinese hamster cells cultured in presence of sodium butyrate. *Exp. Cell Res.* **78**, 456–60.

Wright, T. C., Ukena, T. E., Campbell, R. & Karnovsky, M. J. (1977). Rates of aggregation, loss of anchorage dependence, and tumorigenicity of cultured cells. *Proc. Natl. Acad. Sci. USA* **74**, 258–62.

Wright, W. E. & Hayflick, L. (1972). Formation of anucleate and multinucleate cells in normal and SV-40 transformed WI-38 by cytochalasin B. *Exp. Cell Res.* **74**, 187–94.

Wu, M.-C., Schultz, D. R., Arimura, G. K., Gross, M. A. & Yunis, A. A. (1975). Characteristics of fibrinolysin secreted by cultured rat breast carcinoma cells. *Exp. Cell Res.* **96**, 37–46.

Wyke, J. A. (1975). Temperature sensitive mutants of avian sarcoma viruses. *Biochim. Biophys. Acta* **417**, 91–121.

Yahara, I. & Edelman, G. M. (1976). Modulation of lymphocyte receptor mobility by locally bound concanavalin A. *Proc. Natl. Acad. Sci. USA* **72**, 1579–83.

Yamada, K. M. (1978). Immunological characterization of a major transformation-sensitive fibroblast cell surface glycoprotein. *J. Cell Biol.* **78**, 520–41.

Yamada, K. M. & Olden, K. (1978). Fibronectin-adhesive glycoproteins of cell surface and blood. *Nature, Lond.* **275**, 179–84.

Yamada, K. M., Yamada, S. S. & Pastan, I. (1976). Cell surface protein partially restores morphology, adhesiveness, and contact inhibition of movement to transformed fibroblasts. *Proc. Natl. Acad. Sci. USA* **73**, 1217–21.

Yamada, K. M., Yamada, S. S. & Pastan, I. (1977). Quantitation of a transformation-sensitive, adhesive cell surface glycoprotein. *J. Cell Biol.* **74**, 649–54.

Yamaguchi, N. & Weinstein, I. B. (1975). Temperature-sensitive mutants of chemically transformed epithelial cells. *Proc. Natl. Acad. Sci. USA* **72**, 214–18.

Yamakagawa, T. & Nagai, Y. (1978). Glycolipids at the cell surface and their biological functions. *Trends Biochem. Sci.* **3**, 128–31.

Yasuda, H., Hanai, N., Kurata, M. & Yamada, M. (1978). Cyclic GMP metabolism in relation to the regulation of cell growth in Balb/c 3T3 cells. *Exp. Cell Res.* **114**, 111–16.

Yen, A. (1978). Requirement for plasma and platelet derived growth and stabilization factors in 3T3 cells. *J. Cell Biol.* **79**, 3a.

Yen, A. & Pardee, A. B. (1978a). Exponential 3T3 cells escape in mid-G$_1$ from their high serum requirement. *Exp. Cell Res.* **116**, 103–13.

Yen, A. & Pardee, A. B. (1978b). Arrested states produced by isoleucine deprivation and their relationship to the low serum produced arrested state in Swiss 3T3 cells. *Exp. Cell Res.* **114**, 389–95.

Yerna, M. J., Aksoy, M. O., Hartshorne, D. J. & Goldman, R. D. (1978). BHK-21 myosin: isolation, biochemical characterization and intracellular localization. *J. Cell Sci.* **31**, 411–29.

Yerna, M. J. & Goldman, R. D. (1978). Calcium-sensitive regulation of actin and myosin interactions in cultured BHK-21 cells. *J. Cell Biol.* **79**, 274a.

Yoshida, M. C., Sasaki, M., Takeichi, N. & Boone, C. W. (1976). Karyotypes of vasoformative sarcomas arising from BALB/3T3 cells attached to polycarbonate plates. *Cancer Res.* **36**, 2235–40.

Yuhas, J. M. & Li, A. P. (1978). Growth fraction as the major determinant of multicellular tumor spheroid growth rates. *Cancer Res.* **38**, 1528–32.

Yuhas, J. M., Li, A. P., Martinez, A. O. & Ladman, A. J. (1977). A simplified method for production and growth of multicellular tumor spheroids. *Cancer Res.* **37**, 3639–43.

Zagyansky, Y. & Edidin, M. (1976). Lateral diffusion of concanavalin A receptors in the plasma membrane of mouse fibroblasts. *Biochim. Biophys. Acta* **433**, 209–14.

Zakharova, O. S. (1976). Features of the upper cell surface of normal and SV-40 transformed epithelium of mouse kidney. A scanning electron microscope study. *Tsitologiya* **18**, 1311–15 (in Russian).

Zankl, H., Zang, K. G. (1972). Cytological and cytogenetical studies of brain tumors. IV. Identification of the missing G chromosome in human meningeomas as No. 22 by fluorescence technique. *Humangenetik* **14**, 167–9.

Zavadina, S. P. & Khesina, A. Y. (1971). Toxic effect and metabolism of carcinogenic hydrocarbons in cultures of normal and neoplastic cells of golden hamster. *Voprosi onkologii* **17**, 32–66 (in Russian).

Zeidman, I. (1957). Metastasis: a review of recent advances. *Cancer Res.* **17**, 157–62.

Zetter, B. R., Sun, T.-T., Chen, L. B. & Buchanan, J. M. (1977). Thrombin potentiates the mitogenic response of cultured fibroblasts to serum and other growth promoting agents. *J. Cell. Physiol.* **92**, 233–40.

Zetterberg, A. & Auer, G. (1970). Proliferative activity and cytochemical properties of nuclear chromatin related to local density of epithelial cells. *Exp. Cell Res.* **62**, 262–70.

Zuna, R. E. & Lehman, J. M. (1977). Heterogeneity of karyotype and growth potential in Simian virus 40-transformed Chinese hamster cell clones. *J. Natl. Cancer Inst.* **58**, 1463–72.

Author Index

The numbers in italics indicate the pages on which names are mentioned in the reference list.

Subject Index

actin, 63–5, 86, 89, 92, 137, 169
Actin-binding protein, 64
α-actinin, 63, 89
activation stage of life cycle, 222, 223, 265
   see also growth activation
active cell edge, 75, 111, 112, 124
adenomatosis of colon and rectum (ACR),
   heriditary, 48, 49, 271
adenylate cyclase, *see* cyclic nucleotides,
adhesion factors, 101
agar, cultivation of cells in, 182
agarose, cultivation of cells in, 182
agglutination of cells by lectins, 150–2, 159
aggregates
   cell survival and proliferation in, 186, 187
   formation by normal cells, 100, 149, 150
   formation by transformed cells, 149, 150, 159
   sorting of cells in, 100
amino acids
   as nutrients, 203, 210
   uptake of, 224, 228, 229, 237
anchorage dependence of growth, 182–7, 215,
   218, 255
   loss of, 23, 35, 37, 41, 43, 54, 179, 184, 259,
   268, 269
   possible mechanisms, 187, 232
anchorage revertants, 47
anchoring of patched receptors, 87, 88
angiogenesis factor, 255
antitubulins, 92, 94, 128
   effects on capping, 95, 96
   effects on phagocytosis, 95
   effects on spreading, 94, 122
   effects on transformed cells, 139
   effects on growth, 205
ascites tumours, 259

basic morphogenetic reactions, 61, 104, 129,
   266
   alterations in transformed cells, 160, 161
blast transformation of lymphocytes, 236, 237
blebs, 75, 106, 115, 136

calcium
   altered requirement by transformed cells,
   209
   content in the cell, 226
   effects on cell growth, 202
   regulation of contraction by, 65
   role in growth activation, 234
   role in spreading, 113, 168
   uptake, 224, 226, 237
capping, 79, 85, 95, 160, 236
   see also clearing of patched receptors,
carcinoma *in situ*, 6, 7, 10, 246, 256
cell–cell communication, *see* intercellular
   communication
cell–cell contacts
   formation of, 99, 123
   in epithelial cultures, 123
   in mixed cultures of normal and transformed
   cells, 157
   morphology of, 98, 99
   in neoplasms growing *in vivo*, 246
   in transformed cultures, 147, 148, 159–61
   specificity of, 100, 101
cell surface, 61–73, 108, 135–7, 146
cellular senescence, 26–8
   alterations of proliferation, 27
   hypotheses on mechanisms, 27, 28, 267
centrioles, 92
centripetal movement of particles, 78, 79, 87,
   88
chemical carcinogens
   metabolism of, 32, 37
   mutagenic effects, 17, 33, 34
   toxic effects, 33, 38, 261
   transforming action in culture, 34–8, 268
chromosomal changes in neoplasms, 15, 17
clearing of patched receptors, 79, 87, 88
   see also capping
cloning efficiency of cells, 182, 183, 196
close contacts, 78, 136
colcemid, *see* antitubulins
colchicine, *see* antitubulins

367

DATE DUE

UCR AUG 1 1 1982